Studies in Mathematical Physics

Princeton Series in Physics

edited by Arthur S. Wightman
and John J. Hopfield

Quantum Mechanics for Hamiltonians Defined as Quadratic Forms *by Barry Simon*

Lectures on Current Algebra and Its Applications *by Sam B. Treiman, Roman Jackiw, and David J. Gross*

Physical Cosmology *by P. J. E. Peebles*

The Many-Worlds Interpretation of Quantum Mechanics *edited by B. S. DeWitt and N. Graham*

The $P(\Phi)_2$ Euclidean (Quantum) Field Theory *by Barry Simon*

Homogeneous Relativistic Cosmologies *by Michael P. Ryan, Jr., and Lawrence C. Shepley*

Studies in Mathematical Physics: Essays in Honor of Valentine Bargmann *edited by Elliott H. Lieb, B. Simon, and A. S. Wightman*

Valentine Bargmann

Studies in Mathematical Physics

Essays in Honor of Valentine Bargmann

edited by
E. H. Lieb, B. Simon,
and A. S. Wightman

Princeton Series in Physics

Princeton University Press
Princeton, New Jersey 1976

Published by Princeton University Press, Princeton, New Jersey
IN THE UNITED KINGDOM:
Princeton University Press, Guildford, Surrey

Library of Congress Cataloging in Publication Data will
be found on the last printed page of this book

Printed in the United States of America
by Princeton University Press
Princeton, New Jersey

CONTENTS

INTRODUCTION

This volume is dedicated to Valentine Bargmann on the occasion of his retirement as Professor of Mathematical Physics at Princeton University.

Valentine Bargmann was born in Berlin, Germany on April 6, 1908. He studied at the University of Berlin from 1926 to 1933. He moved to Zurich on Hitler's rise to power and wrote his doctor's thesis at the University under the guidance of Gregor Wentzel. On the completion of his degree, Bargmann emigrated to the United States. (That flat statement is correct, but does not evoke the temper of the times. Bargmann received a five-year German passport in 1931, before the National Socialists came to power, and used it to go to Switzerland to study. After Hitler took office, administrative regulations were issued withdrawing the citizenship of persons of the wrong "race." For that reason, if the German government had succeeded in finding Bargmann, it would have invalidated his passport. Nevertheless, the passport was accepted by the United States government as a valid basis for an immigration visa. The passport expired two days after he reached the United States in 1937.)

On his arrival in the United States, Bargmann looked up I. I. Rabi, who recommended he attend the symposium on physics at the University of Michigan in the summer of 1937. There he met a number of theoretical physicists, in particular, George Uhlenbeck and Gregory Breit. Breit suggested that he look into the Institute for Advanced Study as a possible place to work. After conversations with John von Neumann, Bargmann was accepted at the Institute. He was soon drawn into the work that Albert Einstein was carrying out on unified field theories of gravitation and electromagnetism. For several years he and Peter Bergmann were Einstein's scientific assistants and coworkers in this enterprise. This work continued until 1943, when he undertook war work in collaboration with John von Neumann. After the war, he joined von Neumann's computer project, working with von Neumann and Deane Montgomery on the inversion of matrices of large dimension.

From 1941 on, Bargmann taught graduate courses at Princeton University, but it was only in 1946 that he received a regular faculty appointment as visiting lecturer in physics. Apart from one term spent at the University of Pittsburgh in 1948, he has been at Princeton ever since. Beginning with ASTP (Army Specialized Training Program) and V12 (Navy) courses, and the aforementioned graduate courses during the war, Bargmann has taught physics and mathematics to generations of graduate and undergraduate students. His courses were noted for their clarity and polish. However, for connoisseurs of the post-war period, it was the set of specialized lectures on his own research that were the gems: the lectures on the Lorentz group and its representations of 1948-1949, the lectures on ray representations of Lie groups in 1953-1954, the lectures on second quantization of 1946-1947. These last lectures were recorded in the elegant calligraphy of Oscar Goldman and deposited on the reference shelf in Fine Hall Library. They served several generations of graduate students until the advent of modern library customs, when they were stolen.

Bargmann's interests in mathematical physics have been broad. As students at the University of Berlin, he and Carl Hempel had a common interest in the philosophical problems at the foundations of physics, an interest they shared with Hans Reichenbach, then Professor at Berlin. The reader of Reichenbach's book *Philosophic Foundations of Quantum Mechanics*, University California Press, Berkeley, 1944, will see some typical results of conversations with Bargmann, the sharpening and clarification of ideas by the construction of examples and counterexamples.

One of the undersigned (A. S. Wightman) can vouch for another characteristic example that occurred a decade later. In the year-long efforts that went into the production of what is sometimes called the Bargmann-Hall-Wightman Theorem, Bargmann's remarks played an important role. Yet Bargmann did not want to put his name on the paper as a co-author.

Bargmann's papers are not numerous by the standards of productivity of our day, but many of them have started industries. It suffices to mention his work on the representations of the group SL(2, R) or on the inverse scattering problem. For that reason the contents of this book are somewhat different from those of the typical Festschrift. The Editors made a list of "Bargmann industries" and sought authors who could give a good account of present knowledge or recent developments. A majority of the articles in this volume fall into this category.

Bargmann's name is synonymous in mathematical physics with depth and lucidity. His work is an inspiration to all those who work in the subject. The authors of the present volume join in wishing Valja and Sonja Bargmann many more happy and fruitful years.

ELLIOTT LIEB

BARRY SIMON

ARTHUR WIGHTMAN

PUBLICATIONS

Über eine Verallgemeinerung des Einsteinschen Raumtyps, Zeitschrift für Physik *65*, 830, 1930.

Bemerkungen zur allgemein-relativistischen Fassung der Quantentheorie, Preussische Akademie der Wissenschaften (Sitzungsberichte) 1932, p. 346.

Über den Zusammenhang zwischen Semivektoren and Spinoren, Helvetica Physica Acta *7*, 57, 1934.

Zur Theorie des Wasserstoffatoms, Zeitschrift für Physik *99*, 576, 1936.

Über die durch Elektronenstrahlen in Kristallen angeregte Lichtemission, Helvetica Physica Acta *10*, 361, 1937.

(with A. Einstein and P. G. Bergmann) Five-dimensional representation of gravitation and electricity, Theodore von Karman Anniversary Volume, p. 212, Pasadena, California Institute of Technology, 1941.

(with A. Einstein) Bivector field I, Ann. of Math. *45*, 1, 1944.

On the Glancing Reflection of Shock Waves, Applied Mathematics Panel Report No. 108, 2R, 1945.

(with D. Montgomery and L. von Neumann) Solution of Linear Systems of High Order, Report to the Bureau of Ordinance, Navy Department, October 1946. Reported in J. von Neumann, Collected Works Vol. V, p. 421, Macmillan Company, New York, 1963.

(with H. F. Ludloff) Elastic Limit for Dynamic Loading, Jour. Applied Physics *17*, 63, 1946.

Irreducible Unitary Representations of the Lorentz Group, Annals of Math. *48*, 568, 1947.

(with E. P. Wigner) Group Theoretical Discussion of Relativistic Wave Equations, Natl. Acad. of Sci. (USA), Proceedings *34*, 211, 1948.

Remarks on the Determination of a Central Field of Force from the Elastic Scattering Phase Shifts, Phys. Rev. *75*, 301, 1949.

On the Connection between Phase Shifts and Scattering Potential, Reviews of Modern Physics *21*, 488, 1949.

On the Number of Bound States in a Central Field of Force, Natl. Academy of Science (USA) Proceedings *38*, 961, 1952.

On Unitary Ray Representations of Continuous Groups, Annals of Math. *59*, 1, 1954.

Relativity, Reviews of Mod. Phys. *29*, 161, 1957.

(with L. Michel and V. Telegdi) Precession of the Polarization of Particles Moving in a Homogeneous Electromagnetic Field, Phys. Rev. Letters *2*, 435, 1959.

Relativity, in "Theoretical Physics in the Twentieth Century" (Memorial Volume to Wolfgang Pauli, edited by M. Fierz and V. F. Weisskopf), p. 187. Interscience Publishers, New York, 1960.

(with M. Moshinsky) Group Theory of Harmonic Oscillators I: The Collective Modes, Nuclear Physics *18*, 697, 1960.

(with M. Moshinsky) Group Theory of Harmonic Oscillators II: The Integrals of Motion for the Quadrupole-quadrupole Interaction, Nuclear Phys. *23*, 177, 1961.

On a Hilbert Space of Analytic Functions and an Associated Integral Transform, Part I, Communications on Pure and Applied Mathematics *14*, 187, 1961.

On the Representations of the Rotation Group, Reviews of Modern Physics *34*, 829, 1962.

Remarks on a Hilbert Space of Analytic Functions, Proc. Natl. Acad. Sci., U.S.A., *48*, 1962, 19-22, 2204.

Note on Wigner's Theorem on Symmetry Operations, Jour. Math. Phys. *5*, 862, 1964.

On a Hilbert Space of Analytic Functions and an Associated Integral Transform, Part II, Communications on Pure and Applied Mathematics *20*, 1, 1967.

Group Representations on Hilbert Spaces of Analytic Functions, in "Analytic Methods of Mathematical Physics," edited by R. P. Gilbert and R. G. Newton, p. 27, Gordon and Breach, New York 1970.

Group Representations in Mathematics and Physics (Battelle Seattle 1969 Rencontres) Edited by V. Bargmann Lecture Notes in Physics, Vol. 6, Springer-Verlag, Berlin - Heidelberg - New York (published Dec. 1970).

(with P. Butera, L. Girandello and J. R. Klauder) On the Completeness of Coherent States. Reports on Math. Physics 2, 221, 1971.

Notes on Some Integral Inequalities, Helvetica Physica Acta *45*, 249, 1972.

(with I. T. Todorov) Spaces of Analytic Functions on a Complex Cone as Carriers for the Symmetric Tensor Representations of SO(n), Preprint 1975.

Studies in Mathematical Physics

THE INVERSE r-SQUARED FORCE: AN INTRODUCTION TO ITS SYMMETRIES

Henry D. I. Abarbanel*

Stanford Linear Accelerator Center†
Stanford University, Stanford, California 94305

An introduction to the symmetries of the Kepler or Coulomb problem with r^{-2} forces is presented. The classical problem is discussed for orientation. Then the quantum mechanical situation is explored. For the r^{-2} force in three dimensions we discuss the hidden symmetry of O(4) of the bound states and the greater symmetry of O(1, 4) which connects them. This greater symmetry is realized explicitly in the two dimensional r^{-2} problem using the method of projective representations developed by Bargmann.

I. CLASSICAL MOTION IN A r^{-2} FORCE

The Coulomb or Kepler problem of solving for the motion of a point particle in an inverse square force has remained fundamental and fascinating from the earliest developments of the modern era of science. In a very underspoken sense one may view the achievements of Copernicus, Kepler, Brahe – the giants on whose shoulders Newton stood – and Newton as being the unfolding and solution of this very problem. Remaining underspoken we may note that the techniques and ideas developed in this endeavor had supremely broad implications.

In this article we wish to give a pedagogical development of the symmetry view of the r^{-2} force beginning with the classical theory and proceeding through to conjectures on the importance of r^{-2} forces in the

* Visitor from the Fermi National Accelerator Laboratory, Box 500, Batavia, Illinois 60510.

† Supported by the U. S. Energy Research and Development Administration.

structure and spectrum of elementary particles. It is a pleasure to contribute this to the celebration of Valentine Bargmann whose inspired pedagogy underlies the understanding of so many generations of students.

We begin by considering how the normal nineteenth century physicist would have approached the question of the motion of a point particle of mass m moving in a central force field

$$F(r) = -\frac{k}{r^2}\,\hat{r} \tag{1}$$

where k is a positive constant representing an attractive force and $\hat{r} = r/|r|$, a unit vector in the radial direction.

First one would note that the angular momentum

$$L = r(t) \times p \ , \tag{2}$$

$$p = m\,\frac{dr(t)}{dt} \tag{3}$$

remained constant in time

$$\frac{dL}{dt} = 0 \ , \tag{4}$$

because

$$\frac{dp}{dt} = -\frac{kr}{|r|^3} \ . \tag{5}$$

This leads directly to

$$\frac{d^2r(t)}{dt^2} - \frac{\ell^2}{m^2r^3} = -\frac{k}{mr^2} \ , \tag{6}$$

setting

$$|L| = \ell = mr(t)^2\,\dot{\theta}(t) \ . \tag{7}$$

So far nothing distinguishes this problem from any other central force problem, which differs from the present discussion only by changing the right hand side of (6). The key element for the moment is that with the r^{-2} force, (6) is soluble in closed form – and thus begins its intriguing character.

Change from variation in time to variation in the angular variable θ by

$$\ell\, dt = mr(t)^2\, d\theta \tag{8}$$

and note that

$$Z(\theta) = \frac{1}{r(\theta)} - \frac{mk}{\ell^2}, \tag{9}$$

satisfies

$$\frac{d^2 Z(\theta)}{d\theta^2} + Z(\theta) = 0, \tag{10}$$

so

$$Z(\theta) = \frac{Am}{\ell^2} \cos(\theta - \theta_0), \tag{11}$$

with θ_0 and A some constants, and

$$r(\theta)^{-1} = \frac{mk}{\ell^2}\left[1 + \frac{A}{k}\cos(\theta - \theta_0)\right], \tag{12}$$

which is the general equation of a conic section with one focus at the origin.

So far this has all been elementary, if not trivial. The deep interest in this problem arises when we ask whether there is an underlying reason for the elementary nature of the solution. Most potentials, even central, do not lead to a closed form for orbits.

With extensive *a posteriori* hindsight we seek the answer in the form of a *symmetry* of the Coulomb problem that is not obvious to the eye. Symmetries are signaled by conserved quantities. We already have three: **L**. There are three more in this r^{-2} force. They are the components of the Runge-Lenz vector[2]

$$\mathbf{M} = \frac{d\mathbf{r}}{dt} \times \mathbf{L} - k\hat{\mathbf{r}}, \tag{13}$$

for which

$$\frac{d\mathbf{M}}{dt} = 0, \tag{14}$$

only in the r^{-2} central force.

Now note that

$$\mathbf{M}\cdot\mathbf{L} = 0 \ , \tag{15}$$

and since

$$\mathbf{r}\cdot\mathbf{L} = 0 \ , \tag{16}$$

the motion takes place in a plane orthogonal to $\mathbf{L}$ in which both $\mathbf{r}$ and $\mathbf{M}$ lie. $\mathbf{M}$ is fixed in time and choosing an appropriate coordinate system in the plane of motion we write

$$\mathbf{M} = M(\cos\theta_0, \sin\theta_0) \ , \tag{17}$$

and

$$\mathbf{r}(t) = r(t)(\cos\theta(t), \sin\theta(t)) \ , \tag{18}$$

so

$$\mathbf{r}\cdot\mathbf{M} = Mr\cos(\theta-\theta_0) = \frac{\ell^2}{m} - kr \tag{19}$$

or

$$\frac{1}{r(\theta)} = \frac{km}{\ell^2}\left[1 + \frac{M}{k}\cos(\theta-\theta_0)\right] \ , \tag{20}$$

which is familiar.

Noting that

$$M^2 = |\mathbf{M}|^2 = k^2 + \frac{2E\ell^2}{m} \ , \tag{21}$$

where the total energy E is

$$E = \frac{mv^2}{2} - \frac{k}{r} \ , \tag{22}$$

the orbit equation reads

$$r(\theta)^{-1} = \frac{km}{\ell^2}\left\{1 + \sqrt{1 + \frac{2E\ell^2}{mk^2}}\cos(\theta-\theta_0)\right\} . \tag{23}$$

The nature of the orbit, of course, depends on E. If $E < 0$, then the orbit is closed and is an ellipse or a circle. If $E > 0$, the orbit is open and is a parabola or hyperbola.

To explore the symmetry that the conservation of $\mathbf{M}$ represents, let us form the Poisson brackets of $\mathbf{M}$ and $\mathbf{L}$.[1]. It is known from mechanics that an infinitesimal rotation of a vector $\mathbf{V}$ by $\delta\theta$ about the axis $\hat{n}$ is generated by $\mathbf{L}\cdot\hat{n}$ in the sense

$$\delta V_i = \delta\theta \, \mathrm{PB}(V_i, \mathbf{L}\cdot\hat{n}) \ , \tag{24}$$

where PB means Poisson bracket. Also we know that

$$\mathrm{PB}(L_i, L_j) = \epsilon_{ijk} L_k \ . \tag{25}$$

Since $\mathbf{M}$ is an ordinary three vector,

$$\mathrm{PB}(L_i, M_j) = \epsilon_{ijk} M_k \ , \tag{26}$$

and the real interest lies in

$$\mathrm{PB}(M_i, M_j) = -\frac{2E}{m} \epsilon_{ijk} L_k \ . \tag{27}$$

This shows that the new conserved quantities M_i generate transformations which close to form a group with rotations.

If $E = -|E| < 0$, then the Poisson bracket relations among the L_i and

$$N_i = M_i \left(\frac{m}{2|E|}\right)^{\frac{1}{2}} \tag{28}$$

satisfy the algebra of generators of four dimensional rotations as

$$\mathrm{PB}(N_i, N_j) = \epsilon_{ijk} L_k \ . \tag{29}$$

So we see that for the closed orbits of the r^{-2} force there is a "hidden" 0(4) symmetry in action.

For the open orbits with $E > 0$, the vector

$$\mathbf{J} = \sqrt{\frac{m}{2E}}\,\mathbf{M} \ , \tag{30}$$

satisfies the Poisson bracket relation

$$\mathrm{PB}(J_i, J_j) = -\epsilon_{ijk} L_k \ , \tag{31}$$

which shows that L and J generate the Lie algebra of 0(1, 3) the homogeneous Lorentz group.

It seems appropriate to ask in which space these generators act.[3,4] For $E < 0$, we form the unit four vector

$$V_4 = \frac{p^2 - 2m|E|}{p^2 + 2m|E|}\ , \tag{32}$$

$$\mathbf{V} = 2\sqrt{2m|E|}\,\mathbf{p}/(p^2 + 2m|E|) \tag{33}$$

where $\mathbf{p} = m d\mathbf{r}/dt$. It is clear that rotations generated by L leave V_4 untouched and rotate V as an ordinary three vector. What is less clear, but correct, is that N_i generates transformations among V_4 and V_i which remain on the unit sphere

$$V_4^2 + \mathbf{V}^2 = 1\ . \tag{34}$$

For $E > 0$ we form the time like Minkowski vector

$$W_4 = \frac{2mE + p^2}{2mE - p^2}\ , \tag{35}$$

$$\mathbf{W} = \frac{2\sqrt{2mE}\,\mathbf{p}}{2mE - p^2}\ , \tag{36}$$

$$W^2 = W_4^2 - \mathbf{W}^2 = 1\ . \tag{37}$$

The L_i again generate rotations among the components of $\mathbf{W}$, while the J_i generate transformations of W_4 and W_i which maintain the Minkowski length (37).

All this is a classical physics description of the observations of Bargmann about the quantum mechanical r^{-2} force problem.[4]

II. THE QUANTUM MECHANICAL r^{-2} FORCE

The r^{-2} force in quantum mechanics asks us to find the eigenvalues and eigenfunctions of the hamiltonian operator

$$H = -\frac{\hbar^2}{2m}\nabla^2 - \frac{k}{r} . \tag{38}$$

Because H is invariant under rotations in three space, we know it commutes with

$$\mathbf{L} = \mathbf{r} \times \mathbf{p} , \tag{39}$$

the generator of such rotations. The classical discussion now encourages us to inquire whether the vector operator[5]

$$\mathbf{M} = \frac{1}{2m}(\mathbf{p}\times\mathbf{L} - \mathbf{L}\times\mathbf{p}) - \frac{k\mathbf{r}}{|\mathbf{r}|} , \tag{40}$$

commutes with H. Indeed it does. This means we can find simultaneous eigenfunctions of H and the set of operators formed from L and M which commute with each other.

The Poisson bracket relations above lead directly to

$$[L_i, L_j] = i\hbar\, \epsilon_{ijk} L_k , \tag{41}$$

$$[L_i, M_j] = i\hbar\, \epsilon_{ijk} M_k , \tag{42}$$

and

$$[M_i, M_j] = -\frac{2H}{m}\epsilon_{ijk} L_k \hbar i . \tag{43}$$

If we restrict our attention to eigenfunctions $\psi_E(\mathbf{r})$ of H with eigenvalue $E < 0$, then L and

$$\mathbf{N} = \sqrt{\frac{m}{-2H}}\,\mathbf{M} \tag{44}$$

operating on such eigenfunctions satisfy the Lie algebra of 0(4). The L_i and N_i do not remove us from this space of $\psi_E(\mathbf{r})$ with $E < 0$ since each commutes with H. The restriction to $E < 0$ is acceptable even though the continuum ψ_E for $E > 0$ are also needed for a complete set of eigenfunctions. The operators $L \cdot N$ and $L^2 + N^2$ commute with all L and N and with H form a complete set of commuting observables. They are not independent, however, as $L \cdot N$ operating on $\psi_E(r)$ gives zero and

$$H = -\frac{mk^2}{2}(N^2 + L^2 + \hbar^2)^{-1} \tag{45}$$

so we need only consider $L^2 + N^2$.

The two vector operators

$$A = \frac{1}{2}(L + N) \tag{46}$$

and

$$B = \frac{1}{2}(L - N) \tag{47}$$

commute and each satisfies the algebra of 0(3). An eigenfunction of $L \cdot N$, $L^2 + N^2$, L_3, and N_3 is also an eigenfunction of A^2, A_3, B^2, and B_3. Because $L \cdot N = A^2 - B^2 = 0$,

$$A^2 \psi_{a,a_3;b,b_3} = a(a+1)\hbar^2 \psi_{a,a_3;b,b_3} \tag{48}$$

$$= b(b+1)\hbar^2 \psi_{a,a_3;b,b_3} \tag{49}$$

and $a = b = 0, \frac{1}{2}, 1, \cdots$. Since $A^2 + B^2 = \frac{L^2 + N^2}{2}$, we have

$$H = -\frac{mk^2}{2\hbar^2}(4a(a+1) + 1)^{-1} \tag{50}$$

$$= -\frac{mk^2}{2\hbar^2}\frac{1}{n^2}, \tag{51}$$

with

$$n = 2a + 1 = 1, 2, 3, \cdots \tag{52}$$

operating on $\psi_{a,a_3;a,b_3}$.

We recognize n as the principal quantum number and see here the usual Rydberg formula. Furthermore the degeneracy of the level n is

$(2a+1)^2 = n^2$ since $-a \leq a_3 \leq a$ and $-a \leq b_3 \leq a$ and the energy is independent of a_3 and b_3. Because

$$\mathbf{L} = \mathbf{A} + \mathbf{B} \ , \tag{53}$$

the ordinary rules of addition for angular momentum tell us that the allowed values of ℓ are $0, 1, \cdots, 2a = n-1$.

Next we would like to deduce the eigenfunctions[3,4,6] for H using the 0(4) invariance. This is best done in momentum space after observing that the unit four vector (32) and (33) is transformed to another unit four vector by action of $\mathbf{L}$ and $\mathbf{N}$. We thus seek functions $X(V_\mu)$, $V_\mu = (\mathbf{V}, V_4)$, on which $\mathbf{L}$ and $\mathbf{N}$ realize a representation of 0(4). In momentum space the wave functions satisfy

$$\left(\frac{p^2}{2m} - E\right)\phi(\mathbf{p}) = \frac{k}{2\pi^2\hbar}\int \frac{d^3q\,\phi(\mathbf{q})}{|\mathbf{p}-\mathbf{q}|^2} \ . \tag{54}$$

To exhibit the 0(4) symmetry we go over to the unit four vector V_μ and a similar vector U_μ for $\mathbf{q}$. Then introducing

$$X(V_\mu) = \left[\frac{2m|E| + p^2}{4m|E|}\right]^2 \frac{1}{\sqrt{2m|E|}}\,\phi(\mathbf{p}) \tag{55}$$

and

$$d^3\Omega_V = 2\delta\,(V_\mu^2 - 1)\; d^4V = \left[\frac{2\sqrt{2m|E|}}{p^2 + 2m|E|}\right]^3 d^3p \ , \tag{56}$$

we have

$$X(V_\mu) = \frac{mk}{2\pi^2\hbar(2m|E|)^{\frac{1}{2}}}\int \frac{d^3\Omega_U\,X(U_\mu)}{|U_\mu - V_\mu|^2} \ , \tag{57}$$

where, of course, $V_\mu^2 = V_4^2 + \mathbf{V}^2$. This exhibits the 0(4) symmetry of the problem directly.

This formula is to be compared to that satisfied by the spherical harmonics $Y_{n;\ell,m}(V_\mu)$ on the unit sphere in four dimensions

$$Y_{n;\ell,m}(V_\mu) = \frac{n}{2\pi^2} \int d^3\Omega_\mu \frac{Y_{n;\ell,m}(U_\mu)}{|U_\mu - V_\mu|^2} , \tag{58}$$

where

$$L^2 Y_{n;\ell,m}(V_\mu) = \ell(\ell+1)\hbar^2 Y_{n;\ell,m}(V_\mu) , \tag{59}$$

$$L_3 Y_{n;\ell,m}(V_\mu) = m\hbar\, Y_{n;\ell,m}(V_\mu) , \tag{60}$$

and

$$\Delta_4 Y_{n;\ell,m}(V_\mu) = (n^2-1)\, Y_{n;\ell,m}(V_\mu) \tag{61}$$

with Δ_4 the angular operator on the unit sphere. With the unit four vector parametrized as

$$\begin{aligned} V_4 &= \cos X \\ V_3 &= \sin X \cos\theta \\ V_2 &= \sin X \sin\theta \sin\phi \\ V_1 &= \sin X \sin\theta \cos\phi \end{aligned} \tag{62}$$

(61) reads

$$\frac{1}{\sin^2 X} \frac{\partial}{\partial X}\left(\sin^2 X \frac{\partial}{\partial X} Y_{n;\ell,m}(V_\mu)\right) + \frac{1}{\sin^2 X} \frac{1}{\sin\theta} \frac{\partial}{\partial\theta}\left(\sin\theta \frac{\partial}{\partial\theta} Y_{n;\ell,m}(V_\mu)\right) + \frac{1}{\sin^2 X \sin^2\theta} \frac{\partial^2}{\partial\phi^2} Y_{n;\ell,m}(V_\mu) = (n^2-1)\, Y_{n;\ell,m}(V_\mu) . \tag{63}$$

Explicitly

$$Y_{n;\ell,m}(X,\theta,\phi) = 2^\ell \ell! \left[\frac{2(n)(n-\ell-1)!}{\pi(n+\ell)!}\right]^{\frac{1}{2}} (\sin X)^\ell C^{\ell+1}_{n-\ell-1}(\cos X) Y_{\ell m}(\theta,\phi) , \tag{64}$$

with $C^\nu_\lambda(x)$ the usual Gegenbauer polynomials and $Y_{\ell m}(\theta,\phi)$ the eigenfunctions of L^2. The $Y_{n;\ell,m}(V_\mu)$ are normalized and orthogonal

$$\int_0^\pi \sin^2 X\, dX \int_0^\pi \sin\theta\, d\theta \int_0^{2\pi} d\phi\; Y^*_{n;\ell,m}(V_\mu) Y_{n';\ell',m'}(V_\mu) = \delta_{nn'}\delta_{\ell\ell'}\delta_{mm'} \,. \tag{65}$$

The integral equation for $Y_{n;\ell,m}(V_\mu)$ now shows us that the eigenfunction we seek are just themselves and the eigenvalue condition (51) falls out directly.

This global method of studying the Coulomb problem was developed by Fock[3] elaborating on the infinitesimal method used by Pauli.[5] The connection between the approaches was made by Bargmann[4] who related it to the separability of the quantum mechanical r^{-2} force problem in parabolic coordinates. This second method leads directly to the eigenfunctions of our operators A^2, A_3 and B_3 above as eigenfunctions $D^a_{a_3 b_3}$ of the rotation group with argument $V_4 + i\boldsymbol{\sigma}\cdot\mathbf{V}$ in SU(2).

The power of the symmetry approach is completely exposed now. We have found the eigenenergies of the bound levels $(E = -|E| < 0)$ and the eigenfunctions by appeal to essentially geometric considerations. For $E > 0$ we must study the harmonics of $0(1,3)$ on the unit hyperboloid $W_4^2 - \mathbf{W}^2 = 1$. This yields the continuum eigenfunctions for the r^{-2} force.[6]

III. A GREATER SYMMETRY

As it happens there is a larger set of symmetry operations in the Coulomb problem that allows one to relate all of the bound energy levels to each other. The key mathematical observation is this: transformations on the unit sphere $V_\mu^2 = 1$ in four dimensions are related directly to transformations on the "light cone" in one-time four space dimensions. Suppose $\underset{\sim}{X} = (X_0, X_1, X_2, X_3, X_4)$ is a light like vector in five dimensions

$$\underset{\sim}{X}^2 = X_0^2 - X_\mu^2 = 0 \,. \tag{66}$$

All transformations Λ of the Lorentz group $0(1,4)$ leave (66) invariant:

$$\Lambda \underset{\sim}{X} = \underset{\sim}{Y} \,, \quad \underset{\sim}{Y}^2 = 0 \,. \tag{67}$$

To each vector $\underset{\sim}{X}$ or $\underset{\sim}{Y}$ we may associate a vector on the unit sphere in four dimensions as

$$V_\mu = X_\mu / X_0 \; . \tag{68}$$

The linear transformation $\Lambda\underset{\sim}{X} = \underset{\sim}{Y}$ takes us nonlinearly from one unit vector V_μ to another

$$V'_\mu = \frac{Y_\mu}{Y_0} = (\Lambda\underset{\sim}{X})_\mu / (\Lambda\underset{\sim}{X})_0 \; . \tag{69}$$

A representation of $0(1,4)$ is thus provided by the operation $T(\Lambda)$ for each Λ in $0(1,4)$ operating on functions of V_μ:

$$T(\Lambda) f(V_\mu) = \left[(\Lambda^{-1})_{00} + \sum_{\mu=1}^{4} (\Lambda^{-1})_{0\mu} V_\mu \right]^{-3/2} f\left\{ \frac{(\Lambda^{-1})_{\mu 0} + \sum_{\nu=1}^{4} (\Lambda^{-1})_{\mu\nu} V_\nu}{(\Lambda^{-1})_{00} + \sum_{\nu=1}^{4} (\Lambda^{-1})_{0\nu} V_\nu} \right\} . \tag{70}$$

Projective representations of this sort were utilized by Bargmann in his study of Lorentz groups.[7]

We are now able to define a representation of $0(1,4)$, actually a unitary representation, by a set of operations on our wave functions $Y_{n;\ell,m}(V_\mu)$. The Lorentz group $0(1,4)$ has ten generators. Six of these will be the $\mathbf{L}$ and $\mathbf{N}$ (or $\mathbf{A}$ and $\mathbf{B}$) operators above. These will commute with H and generate the compact subgroup of $0(4)$. The other four operators will not commute with H and will take us from one energy level to another.

Rather than exhausting ourselves with the algebra of $0(1,4)$ we will exhibit the method by studying the r^{-2} force in two dimensions. Here the hidden symmetry is just $0(3)$ the usual rotation group, and the larger group is $0(1,3)$, the Lorentz group.

We'll restrict ourselves to negative eigenvalues of the hamiltonian

$$H = \frac{p_x^2 + p_y^2}{2m} - k/\sqrt{x^2+y^2} \; , \tag{71}$$

and then the three generators of the 0(3) symmetry are

$$J_3 = L_z = xp_y - yp_x , \tag{72}$$

$$J_2 = \sqrt{\frac{m}{-2H}}\left\{\frac{1}{2m}(p_yL_z + L_zp_y) - kx/\sqrt{x^2+y^2}\right\}, \tag{73}$$

and

$$J_1 = \sqrt{\frac{m}{-2H}}\left\{\frac{-1}{2m}(p_xL_z + L_zp_x) - ky/\sqrt{x^2+y^2}\right\}, \tag{74}$$

as in Equation (44). These operators have eigenfunctions $Y_{JM}(\theta,\phi)$ on the unit sphere in three dimensions and the energy eigenvalues are

$$E = -\frac{mk^2}{2\hbar^2}\frac{1}{\left(J+\frac{1}{2}\right)^2}, \qquad J = 0,1,2,\cdots \tag{75}$$

We now seek three operators K_i that combined with the J_i generate the algebra of $0(1,3)$ operating on the functions $Y_{J,M}(\theta,\phi)$. These are the "boost" operators of pure Lorentz transformations in the i direction. Consider the operator K_3 for example. The element of $0(1,3)$ equal $\exp - iK_3\eta$ makes a boost by η in the 3 direction. It takes the light like vector $(1, \sin\theta\cos\phi, \sin\theta\sin\phi, \cos\theta)$ into $(\cosh\eta + \sinh\eta\cos\theta, \sin\theta\cos\phi, \sin\theta\sin\phi, \cosh\eta\cos\theta + \sinh\eta)$. For the infinitesimal transform $= 1 - iK_3\eta$ we have

$$(1 - iK_3\eta)Y_{JM}(\theta,\phi) = (1-\eta\cos\theta)^{-1}Y_{JM}(\theta',\phi') \tag{76}$$

where

$$\tan\theta' = \frac{\sin\theta}{\cos\theta - \eta} \quad \text{and} \quad \phi' = \phi , \tag{77}$$

as determined by (70). This yields

$$K_3Y_{JM}(\theta,\phi) = i\left[\cos\theta + \sin\theta\frac{\partial}{\partial\theta}\right]Y_{JM}(\theta,\phi) \tag{78}$$

$$= i(J+1)\sqrt{\frac{(J+1)^2-M^2}{4(J+1)^2-1}}\,Y_{J+1,M}(\theta,\phi) - iJ\sqrt{\frac{J^2-M^2}{4J^2-1}}\,Y_{J-1,M}(\theta,\phi). \tag{79}$$

Similarly we find for the action of $K_{\pm} = K_1 \pm iK_2$

$$K_+ Y_{JM}(\theta,\phi) = \frac{iJ}{\sqrt{4J^2-1}} \sqrt{(J-M)(J-M-1)}\, Y_{J-1,M+1}(\theta,\phi)$$

$$+\, i\, \frac{J+1}{\sqrt{4(J+1)^2-1}} \sqrt{(J+M+1)(J+M+2)}\, Y_{J+1,M+1}(\theta,\phi)\,, \qquad (80)$$

and

$$K_- Y_{JM}(\theta,\phi) = -i\, \frac{J}{\sqrt{4J^2-1}} \sqrt{(J+M)(J+M-1)}\, Y_{J-1,M-1}(\theta,\phi)$$

$$-i\, \frac{J+1}{\sqrt{4(J+1)^2-1}} \sqrt{(J-M+1)(J-M+2)}\, Y_{J+1,M-1}(\theta,\phi). \qquad (81)$$

One may verify that these operators and the J_i satisfy the 0(1,3) algebra

$$\begin{aligned} [J_i, J_j] &= i\,\epsilon_{ijk}\, J_k \\ [J_i, K_j] &= i\,\epsilon_{ijk}\, K_k \\ [K_i, K_j] &= -i\,\epsilon_{ijk}\, J_k \end{aligned} \qquad (82)$$

as promised.

There are two important observations about the representation of 0(1,3) which we have induced by operations on functions $Y_{JM}(\theta,\phi)$ on the unit sphere in three dimensions. First, it is irreducible. The J_i operators mix all components of the basis function with index J to those with $J \pm 1$. All basis functions for $J = 0, 1, \cdots, -J \leq M \leq J$, are involved in the representation. Second, it is unitary. This is by immediate appeal to the work of Gelfand *et al.*[8] Clearly it is infinite dimensional.

This means that the eigenfunctions $Y_{JM}(\theta,\phi)$ of the r^{-2} force problem in two dimensions carry a unitary irreducible representation of the Lorentz group in four dimensions, 0(1, 3). This projective representation has been constructed above. The unitary, irreducible nature of the 0(1,D) representation created by operations on 0(D) harmonics carries over into 0(4).

IV. USING THESE SYMMETRIES

Two uses are commonly made on the basis of these symmetry considerations. The first is practical. It consists of utilizing the wave functions of the problem, which are just spherical harmonics in $O(4)$, to evaluate integrals encountered in real life atomic physics situations. This topic is amply covered in the article by Wulfman[9] and the monograph of Englefield.[10] Much of the original work has been done by the first author and by Barut[11] as referred to in these articles.

The second is more interesting, namely to speculate on the basis of the r^{-2} force symmetries to more complicated problems. For example, the greater symmetry group of $O(1,p)$ associated with the r^{-2} force in $p-1$ dimensions provides us with a set of operators (the K_i of the previous section) which generate the spectrum by taking us from one energy level to the next. One may then imagine a search for spectrum generating operators[12] in the richer problem of hadron spectroscopy. If one could find even an approximate symmetry which allowed one to connect the octet of pseudoscalar mesons to the octet of vector mesons by essentially a ladder operator, that would be very exciting. Such speculations are further fueled by the idea that these hadron states are bound states of constituents held together by the exchange of massless vector bosons which, in an instantaneous approximation, give rise to a r^{-2} force between the constituents. The real field theories of fundamental fermions (quarks) bound together by massless gauge bosons (gluons) have a much richer and more complex structure than strictly a r^{-2} force.[13] Yet since massless particle exchange does give rise to precisely this force, the symmetry considerations we have now discussed at length, loom as a tantalizing ground to proceed from.

STANFORD UNIVERSITY

REFERENCES

1. H. Goldstein, *Classical Mechanics* (Addison Wesley Co., Reading, Mass., 1950).

2. C. Runge, *Vektoranalysis*, English translation (Dutton, New York, 1919). W. Lenz, Z. Physik *24*, 197 (1924). See also W. Hamilton, Proc. Roy. Irish Acad. *3*, 441 (1847).

3. V. Fock, Z. Physik *98*, 145 (1935).

4. V. Bargmann, Z. Physik *99*, 576 (1936).

5. W. Pauli, Z. Physik *36*, 336 (1926).

6. M. Bander and C. Itzykson, Rev. Mod. Phys. *38*, 330, 346 (1966).

7. V. Bargmann, Ann. of Math. *48*, 568 (1947).

8. I. M. Gelfand, R. A. Minlos, and Z. Ya. Shapiro, *Representations of the Rotation and Lorentz Groups and their Applications*, trans. G. Cummins and T. Boddington (Pergamon Press, Macmillan, New York, 1963); Part II, Section 2.

9. C. E. Wulfman in *Group Theory and its Applications*, ed. E. M. Loebl (Academic Press, New York, 1971); Vol. II, p. 145.

10. M. J. Englefield, *Group Theory and the Coulomb Problem* (Wiley-Interscience, New York, 1972).

11. A. O. Barut and H. Kleinert, Phys. Rev. *160*, 1149 (1967).

12. Y. Ne'eman, *Algebraic Theory of Particle Physics* (Benjamin, New York, 1967); Chapter 10 and references.

13. E. S. Abers and B. W. Lee, "Gauge Theories, "Physics Reports *9C*, 1 (1973).

CERTAIN HILBERT SPACES OF ANALYTIC FUNCTIONS ASSOCIATED WITH THE HEISENBERG GROUP*

Donald Babbitt

TABLE OF CONTENTS

CHAPTER I

THE HEISENBERG GROUP AND THE ASSOCIATED SCHWARTZ SPACE

§1.1. *Introduction*

This survey is concerned with certain Hilbert spaces of analytic functions which arise with a certain realization of the boson creation and annihilation operator as operators on a Hilbert space. In particular, let $\mathfrak{F}_1$ be the entire functions on C with the following property: if $f = \sum_m \alpha_m z^m$, then $\sum m!|\alpha_m|^2 < \infty$. Define an inner product on $\mathfrak{F}_1$ by:

$$(f, g) = \sum_m m!\, \bar{\alpha}_m \beta_m$$

where $g = \sum_m \beta_m z^m$. Define the creation and annihilation operators A^* and A as follows:

$$A^* f = zf$$

and

$$Af = \frac{d}{dz} f\ .$$

(We will leave the domain unspecified for now.) We see that A^* and A satisfy the boson canonical commutation relations (C.C.R.) i.e.

$$[A, A^*] = 1$$

and at least formally they are each others adjoint. This realization of the boson C.C.R. was first discussed by Fock [11].

In the early 1960's, Bargmann [2] realized that the inner product could be defined using a Gaussian measure on C. In particular he showed that:

$$(f, g) = \pi^{-n} \int f(z)\, g(z)\, e^{-|z|^2}\, dx\, dy$$

and he proceeded to study various beautiful properties of this space (for n degrees of freedom) as well as its relation to the Schrödinger realization of the C.C.R. Segal [21] simultaneously and independently constructed the corresponding "holomorphic" realization when there are infinitely many degrees of freedom and developed some of its properties: Bargmann [4] also treated the case with an infinite number of degrees of freedom. His treatment was more in the spirit of Fock's original paper. Subsequently Bargmann applied these Hilbert spaces of analytic functions to the study of angular momentum [3], the theory of tempered distributions [5], and the mataplectic representation of Sp(n, R) and some of its subgroups [6]. For the remainder of the paper the initials B.F.S. will stand for Bargmann, Fock and Segal.

In this paper we will only treat the case of a finite number of degrees of freedom. To avoid technical problems which arise with unbounded operators we will work with representations of the Heisenberg group rather than directly with the C.C.R. Our approach will be global i.e. coordinate free where possible. We occasionally will choose coordinates so as to match up with the notation in [2]. Most proofs will either just be sketched or omitted completely. The only results which may be new will be our technique for characterizing $\mathcal{S}(\mathcal{F}(W^+))$ (§1.3 and §2.5) and our discussion of holomorphic quantization (§3.4). The table of contents will indicate the organization of the paper.

§1.2. *Generalities on the Heisenberg group*

A. We must first introduce some notation which will be used throughout the paper. $\mathbf{C}$ will denote the complex numbers and $\mathbf{C}^n$ the corresponding space of n-tuples. Typical elements will be denoted by $z = (z_1, \cdots, z_n)$ and if z and w are in $\mathbf{C}^n$, then we define a dot product: $z \cdot w = : \sum_{j=1}^n z_j w_j$. If $z \in \mathbf{C}^n$, then $\bar{z} = : (\bar{z}_1, \cdots, \bar{z}_n)$ where $\bar{z}_j$ denotes the complex conjugate of z_j. The norm on $\mathbf{C}^n$ is defined by: $|z|^2 = z \cdot \bar{z}$. $\mathbf{R}$ will denote the real numbers and $\mathbf{R}^n$, the space of real tuples. $\mathbf{R}^n$ can be viewed as a subspace of $\mathbf{C}^n$ and thus it will have the inherited dot product and norm. $\mathbf{Z}$ will denote the ring of integers.

$\mathbf{N}$ will denote the set of non-negative integers and $\mathbf{N}^n$ the corresponding space of n-tuples. Typical elements will be denoted by $\underset{\sim}{k}$, $\underset{\sim}{\ell}$, etc. Let $|\underset{\sim}{k}| = : k_1 + \cdots + k_n$ and $\underset{\sim}{k}! = : k_1! \cdots k_n!$. We write $\underset{\sim}{k} \geq \underset{\sim}{\ell}$ if $k_j \geq \ell_j$, $j = 1, \cdots, n$.

B.. Let V be a 2n dimensional real vector space and $<,>$ a skew-symmetric non-singular bilinear form on V i.e. 1) $<\cdot, v>$ is a linear mapping from V to $\mathbf{R}$ for each $v \in V$; 2) $<v_1, v_2> = -<v_2, v_1>$ for $v_1, v_2 \in V$ and 3) $<\cdot, v> \equiv 0$ implies $v = 0$. Then the pair $(V, <,>)$ is called a *symplectic vector space* and $<,>$ is called the *symplectic form* on V.

EXAMPLE. Let $V = \mathbf{R}^n \times \mathbf{R}^n$. If $v_j = (q_j, p_j)$, q_j and p_j in $\mathbf{R}^n$, $j = 1,2$, then let $<,>$ be defined by:

$$<v_1, v_2>_s =: q_1 \cdot p_2 - q_2 \cdot p_1 .$$

This example is prototypical in the sense that if $(V, <\ >)$ is a 2^n dimensional symplectic vector space, then there is a linear isomorphism $T : V \to \mathbf{R}^n \times \mathbf{R}^n$ such that:

$$<Tv_1, Tv_2>_s = <v_1, v_2> .$$

However this isomorphism depends on a choice of basis in V and therefore is *not canonical.*

C. We can associate with $(V, <,>) = V$ in a canonical way an interesting simply-connected nilpotent Lie group $N(V)$ which now-a-days is called the *Heisenberg* group associated with the symplectic space $(V, <,>)$. As we shall see below it at least formally is intimately related to the commutation relations of quantum mechanics. As a manifold, $N(V) =: V \times \mathbf{R}$. $N(V)$ is clearly simply connected. Its group law is defined as follows:

$$(v_1, t_1)(v_2, t_2) =: (v_1 + v_2, t_1 + t_2 + \tfrac{1}{2}<v_1, v_2>) . \tag{1.2.1}$$

It is routine to check that this defines an analytic group law on $N(V)$ and thus it is a simply connected Lie group. Its Lie algebra will be denoted by $\mathfrak{N}(V)$ and it will be identified with $V \oplus \mathbf{R}$. If we write v for $v \oplus 0$ and E for $0 \oplus 1$, we see that the only non-trivial Lie bracket in $\mathfrak{N}(V)$ is:

$$[v_1, v_2] = <v_1, v_2> E \tag{1.2.2}$$

and thus $\mathfrak{N}(V)$ and hence $N(V)$ are nilpotent. Note that the center $\mathfrak{Z}$ of $\mathfrak{N}(V)$ is just $\mathbf{R}E$.

Because N(V) and $\mathfrak{N}(V)$ can be canonically identified as manifolds, some care must be taken in the notation. If we wish to view (v, tE) as an element in $\mathfrak{N}(V)$, we shall write it as $v + tE$. If we wish to view it as an element in N(V), we shall use the exponential notation of Cartier and write it as:

$$e^{v+tE} \equiv e^v e^{tE} .$$

In this notation (1.2.1) becomes:

$$(1.2.1') \qquad e^{v_1+t_1E}\, e^{v_2+t_2E} =: e^{v_1+v_2 n(t_1+t_2+\frac{1}{2}<v_1,v_2>E)} .$$

D. In order to discuss the B.F.S. realization of N(V) we will need an "admissible" complex structure on (V,< ,>), as well as the complexification V_C of V. Recall that a *complex structure* on V is a linear mapping $J : V \to V$ such that $J^2 = -1$. V_J will denote the complex vector space where as a set and as a group $V_J = V$ and (complex) scalar multiplication is defined by:

$$(a + ib)v =: av + b(Jv) .$$

V_J is clearly an n dimensional complex vector space.

DEFINITION. The complex structure J on V is said to be *admissible* (*with respect to* < ,>) if a) $<Jv_1, Jv_2> = <v_1, v_2>$ for all $v_1, v_2 \in V$ i.e. J is a symplectic mapping; and b)

$$(v, v)_J =: <v, Jv> \geq 0$$

for all $v \in V$. For $v_1, v_2 \in V$, let $(v_1, v_2)_J =: <v_1, Jv_2>$. $(,)_J$ is thus a *Hermitian structure* on V_J such that $\mathrm{Im}(v_1, v_2)_J = <v_1, v_2>$. We denote the corresponding norm on V_J by $\| \; \|_J$.

EXAMPLE. $(V_J <,>) = (R^n \times R^n, < ,>_s)$. Define J_s as follows:

$$J_s(q, p) = (-p, q)$$

for $(q,p) \in \mathbf{R}^n \times \mathbf{R}^n$. Then $((q_1,p_1),(q_2,p_2))_{J_s} = q_1 \cdot q_2 + p_1 \cdot p_2$. It follows since $(V,<\,,>)$ are symplectically isomorphic to $(\mathbf{R}^n \times \mathbf{R}^n, <\,,>_s)$ that it has at least one admissible complex structure.

The *complexification* V_C of V is a complex vector space defined by:

$$V_C =: V \otimes C\ .$$

Its underlying real space can be identified with $V \oplus iV$.

If $\omega = \omega_R + i\omega_I \in V_C$, its conjugate $\bar{\omega}$ is defined by $\bar{\omega} =: \omega_R - i\omega_I$ is conjugate linear and $\bar{\bar{\omega}} = \omega$. Note, we can and often times will identify V with those elements ω in V_C which satisfy $\bar{\omega} = \omega$. We can extend $<\,,>$ to $V_C \times V_C$ by complex bilinearity. We will denote the extension also by $<\,,>$. It should be remarked that the complexification $\mathfrak{N}_C(V)$ of $N(V)$ can now be identified with $V_C \oplus C$ and the only non-trivial Lie bracket is:

(1.2.3) $$[\omega_1, \omega_2] = <\omega_1, \omega_2> E\ .$$

We shall now bring J into the act. It can be extended by complex linearity to V_C and diagonalized there. It will have two eigenvalues $\pm i$. Let $W_J^{\pm}$ be the corresponding eigenspaces which will both be complex n-dimensional spaces and $V_C = W_J^+ \oplus W_J^-$. These spaces will play a prominent role in the sequel. In fact that B.F.S. realization of $N(V)$ will "live on" W_J^+. Note that $\overline{W_J^+} = W_J^-$ and vice versa. Moreover the mapping $\Theta_J : W_J^+ \to V_J : \omega_+ \rightsquigarrow \omega_+ + \bar{\omega}_+$ is a complex linear isomorphism. Thus elements of $\mathfrak{N}(V)$ can be written either as $v + Et$ or $\omega_+ + \bar{\omega}_+ + Et$ where $\Theta_J(\omega_+) = v$. The corresponding elements in $N(V)$ can be written either as e^{v+Et} or $e^{\omega_+ + \bar{\omega}_+ + Et}$. We will have occasion to use both versions.

E. We will often have to work with "standard" bases in $(V,<\,,>,J)$ and V_C.

DEFINITIONS. *A standard basis* $\{\underset{\sim}{e},\underset{\sim}{f}\}$ for V (w.r. to $<\,,>,J$) is one which is obtained as follows: Let $\underset{\sim}{e} =: \{e_1,\cdots,e_n\}$ be an orthonormal basis for $(V_J,(\,,)_J)$ and let $\underset{\sim}{f} =: \{f_1,\cdots,f_n\}$ where $f_j = Je_j$.

Note that:

(1.2.4) $$\langle e_i, f_j\rangle = (e_i, e_j)_J = \delta_{ij}$$

and therefore $\{\underset{\sim}{e}, \underset{\sim}{f}\}$ is a symplectic basis for $(V, \langle\,,\rangle)$. A basis of the form $\{\underset{\sim}{e}, \underset{\sim}{f}, E\}$ is called a *standard basis* for $\mathfrak{N}(V)$ (w.r.t. $(\langle\,,\rangle, J)$). If $\{\underset{\sim}{e}, \underset{\sim}{f}\}$ are as above and $\underset{\sim}{a}^+ = (a_1^+, \cdots, a_n^+)$ where $a_j^+ = \dfrac{e_j - if_j}{\sqrt{2}}$, $j = 1,2,\cdots,n$ and $\underset{\sim}{a}^- = (a_1^-, \cdots, a_n^-)$ where $a_j^- = \dfrac{e_j + if_j}{\sqrt{2}}$, $j = 1,\cdots,n$, then $\underset{\sim}{a}^+$ is a *standard basis for* W_J^+ (induced by $\{\underset{\sim}{e}, \tilde{f}\}$), $\underset{\sim}{a}^-$ is a *standard basis for* W_J^- (induced by $\{\underset{\sim}{e}, \underset{\sim}{f}\}$) and $\{\underset{\sim}{a}^+, \underset{\sim}{a}^-\}$ is a *standard basis for* V_C (induced by $\{\underset{\sim}{e}, \underset{\sim}{f}\}$). The corresponding *standard basis for* $\mathfrak{N}_C$ (induced by $\{\underset{\sim}{e}, \underset{\sim}{f}, E\}$) is $\{\underset{\sim}{a}^+, \underset{\sim}{a}^-, E\}$.

Observe that from (1.2.4) and (1.2.3) we have:

(1.2.5) $$\langle a_i^+, a_j^-\rangle = \delta_{ij}\, iE$$

and thus viewing a_i^+ and a_j^- as elements in $N_C(V)$ we have from that:

(1.2.6) $$[a_i^+, a_j^-] = i\delta_{ij} E\ .$$

F. Using standard bases, we now discuss the formal connection between an irreducible representation of the Heisenberg group and the commutation relations. According to the classical theorem of Stone [22] and von Neumann [25], there is up to unitary equivalence, exactly one irreducible representation π of $N(V)$ such that $\pi(e^{tE}) = e^{i\lambda t}$, $\lambda > 0$. *For convenience let us consider the case where* $\lambda = 1$. Let $\{\underset{\sim}{e}, \underset{\sim}{f}, E\}$ be a standard basis for $\mathfrak{N}(V)$. Then by Stone's theorem [20], p. 266, there exist self-adjoint operators Q_j and P_j, $j = 1,\cdots,n$ such that

(1.2.7) $$\pi(e^{e_j t}) = e^{-iQ_j t}$$

and

(1.2.8) $$\pi(e^{if_j t}) = e^{-iP_j t}\ .$$

Then using (1.2.1′), we obtain Weyl's form of the commutation relations [27]

(1.2.9) $$e^{-iP_jt_1}e^{-Q_kt_2} = e^{\delta_{jk}it_1t_2}e^{-iQ_kt_2}e^{-iP_jt_1} .$$

Then differentiating (1.2.9) w.r.t. to t_1 and t_2 and setting $t_1 = t_2 = 0$, we obtain:

(1.2.10) $$[Q_j, P_k] = i\delta_{jk} .$$

Similarly

(1.2.11) $$[Q_j, Q_k] = [P_j, P_k] = 0 .$$

If we let

$$A_j = \frac{Q_j + iP_j}{\sqrt{2}}$$

and

$$A_j^* = \frac{Q_j - iP_j}{\sqrt{2}} ,$$

then we have (again formally) that:

(1.2.12) $$[A_j, A_k^*] = \delta_{jk}$$

and

(1.2.13) $$[A_j, A_k] = [A_j^*, A_k^*] = 0$$

i.e. $\{A_j, A_j^*, j = 1, \cdots, n\}$ satisfy the Boson C.C.R.

§1.3. *The Spaces* $\mathcal{S}(\mathcal{H}_\pi)$ *and* $\mathcal{S}'(\mathcal{H}_\pi)$

In this section we will discuss the abstract Schwartz spaces $\mathcal{S}(\mathcal{H}_\pi)$ and its dual space $\mathcal{S}'(\mathcal{H}_\pi)$ which are canonically associated with an irreducible unitary representation $(\pi, \mathcal{H}_\pi)$ of N(V). (*For the remainder of this paper with the exception of* §3.3 *we will assume* $\pi(e^{Et}) = e^{it}$.) $\mathcal{S}(\mathcal{H}_\pi)$ will be a countable Hilbertian space in the sense of Gelfand and Vilenkin [12]. Our main aim in this section will be to describe two equivalent families of Hilbert norms on $\mathcal{S}(\mathcal{H}_\pi)$, one of which will be transparent in the Schrödinger realization while the other will be transparent in the B.F.S. realization.

A. $(V,<,>)$ will be a fixed $2n$ dimensional symplectic space and J will be a fixed admissible complex structure on V which will be held fixed throughout this section. Let $\{\underset{\sim}{e},\underset{\sim}{f}\}$ be a standard basis for V (w.r. to $\{<,>,J\}$). Let Q_j, P_j, $j = 1,\cdots,n$ be the self-adjoint operators defined by 1.2.7 and 1.2.8. Let $\mathcal{G}(\mathcal{H}_\pi)$ be the Garding domain for π i.e. $\mathcal{G}(\mathcal{H}_\pi)$ is the space spanned by vectors of the form

$$\int \psi(n)\pi(n)y\,dn, \quad y \in \mathcal{H}_\pi, \ \psi \in C_0^\infty(N(V))$$

where $C_0^\infty(N(V))$ is the space of infinitely differentiable functions on $N(V)$ with compact support and dn denotes a Haar measure on N. It is known that 1) $\mathcal{G}(\mathcal{H}_\pi)$ is dense in $\mathcal{H}_\pi$; 2) Q_j and P_j are essentially self-adjoint on $\mathcal{G}(\mathcal{H}_\pi)$, $j = 1,\cdots,n$; and 3) Q_j and P_j leave $\mathcal{G}(\mathcal{H}_\pi)$ invariant for $j = 1,\cdots,n$. See [8] or [26], Chapter 4 where this is proved for arbitrary Lie groups. Note that Q_j and P_j, $j = 1,\cdots,n$ satisfy (1.2.10) and (1.2.11) on $\mathcal{G}(\mathcal{H}_\pi)$.

Let $(Q_j)_0$ and $(P_j)_0$ denote the restriction of Q_j and P_j to $\mathcal{G}(\mathcal{H}_\pi)$ for $j = 1,\cdots,n$. For $\underset{\sim}{\ell}, \underset{\sim}{m} \in N^n$, denote $(Q_1)_0^{\ell_1}\cdots(Q_n)_0^{\ell_n}(P_1)_0^{m_1}\cdots(P_k)_0^{m_n}$ by $(\underset{\sim}{Q}^{\ell}\underset{\sim}{P}^{m})_0$ and let $\underset{\sim}{P}^{m}\underset{\sim}{Q}^{\ell}$ be $(\underset{\sim}{Q}^{\ell}\underset{\sim}{P}^{m})_0^*$ where $*$ denotes the adjoint in $\mathcal{H}_\pi$. $\underset{\sim}{Q}^{\ell}\underset{\sim}{P}^{m}$ is defined in a similar way. Let $\mathbf{U}$ denote all the operators on $\mathcal{H}_\pi$ which are finite linear combinations of the $\underset{\sim}{Q}^{\ell}\underset{\sim}{P}^{m}$, $\underset{\sim}{\ell}, \underset{\sim}{m} \in N^n$. The domain of a sum of operators is taken to be the intersection of the domains of the summands.

DEFINITION. $\mathcal{S}(\mathcal{H}_\pi) = \bigcap_{u\in\mathbf{U}} \operatorname{Dom}(u)$ and the topology on $\mathcal{S}(\mathcal{H}_\pi)$ is defined by the family of Hilbertian semi-norms: $\|x\|_u =: \|ux\|$, $u \in \mathbf{U}$, where $\|\ \|$ denotes the norm in $\mathcal{H}_\pi$. (Recall that this means $x_n \to 0$ in $\mathcal{S}(\mathcal{H}_\pi)$ if $\|x_n\|_u \to 0$ for all $u \in \mathbf{U}$.) Note that $\mathcal{S}(\mathcal{H}_\pi)$ is dense in $\mathcal{H}_\pi$

because it contains $\mathcal{G}(\mathcal{H}_\pi)$. We single out an especially simple family of these seminorms. For $\underset{\sim}{\ell}, \underset{\sim}{m} \in N^n$, let

$$\|x\|_{\underset{\sim}{\ell},\underset{\sim}{m}} =: \|Q^{\underset{\sim}{\ell}} P^{\underset{\sim}{m}} x\| .$$

Recall that two families of semi-norms $\{\| \ \|_\alpha : \alpha \in \mathfrak{A}\}$, $\{\| \ \|_\beta : \beta \in \mathfrak{B}\}$ on a linear space X are equivalent if they define the same topology on X i.e. the same notion of convergence.

We now state a well-known theorem ([8], Pt. 3, Sec. 2 or [26], Chapter 4).

THEOREM 1.3.1. a) $\mathcal{S}(\mathcal{H}_\pi) = \{x \in \mathcal{H}_\pi : \pi(n)x$ *is an* $\mathcal{H}_\pi$*-valued infinitely differentiable function on* $N(V)\}$.

b) *The families of norms* $\{\| \ \|_u, u \in \mathbf{U}\}$ *and* $\{\| \ \|_{\underset{\sim}{\ell},\underset{\sim}{m}} : \underset{\sim}{\ell}, \underset{\sim}{m} \in N^n\}$ *are equivalent and* $\mathcal{S}(\mathcal{H}_\pi)$ *is complete;*

c) $\mathcal{S}(\mathcal{H}_\pi)$ *is left invariant by* $\mathbf{U}$.

REMARKS. 1) From a) we conclude that $\mathcal{S}(\mathcal{H}_\pi)$ does not depend on the choice of the basis $\{\underset{\sim}{e}, \underset{\sim}{f}\}$.

2) From a) and c) we conclude that the commutation relations (1.2.10-1.2.11) and (1.2.12) and (1.2.13) hold on $\mathcal{S}(\mathcal{H}_\pi)$. We will use these commutation relations extensively in the remainder of this section.

B. The norms $\| \ \|_{\underset{\sim}{\ell},\underset{\sim}{m}}$ are especially transparent in the Schrödinger realization of N(V) where we readily see that $\mathcal{S}(\mathcal{H}_\pi)$ is the usual Schwartz space on n dimensional space. See Section 2.2A for the definition of the Schrödinger realization. We now wish to introduce another equivalent family of norms $\{\| \ \|_{B,k}, k \in N\}$ on $\mathcal{S}(\mathcal{H}_\pi)$ which will be especially transparent in the B.F.S. realization.

As in Part A we will work with the standard basis $\{\underset{\sim}{e}, \underset{\sim}{f}\}$ and the corresponding operators $Q_j, P_j, A_j^* = (Q_j - iP_j)/\sqrt{2}$, and $A_j = (Q_j + iP_j)/\sqrt{2}$, $j = 1, \cdots, n$. We will use the commutation relations (1.2.10-1.2.13) repeatedly and without specific reference to them.

For $k \in \mathbf{N}$, we define positive integers $a_{\underset{\sim}{\ell}}^{k}$, $|\underset{\sim}{\ell}| \le k$ by the formula:

$$(1+|z|^2)^k = \sum_{|\underset{\sim}{\ell}| \le k} a_{\underset{\sim}{\ell}}^{k} |z_1|^{2\ell_1} \cdots |z_n|^{2\ell_n} . \tag{1.3.1}$$

We define an operator K_k by:

$$K_k =: \sum_{|\underset{\sim}{\ell}| \le k} a_{\underset{\sim}{\ell}}^{k} \underset{\sim}{A}^{\underset{\sim}{\ell}} \underset{\sim}{A}^{*\underset{\sim}{\ell}} \tag{1.3.2}$$

where $\underset{\sim}{A}^{\underset{\sim}{\ell}} \underset{\sim}{A}^{*\underset{\sim}{\ell}} =: A_1^{\ell_1} \cdots A_n^{\ell_n} A_1^{*\ell_1} \cdots A_n^{\ell_n}$ and an inner product on $\mathcal{S}(\mathcal{H}_\pi)$ by

$$(x,y)_{B,k} =: (x, K_k y) . \tag{1.3.3}$$

Note that

$$\|x\| \le \|x\|_{B,k} \cdot \tag{1.3.4}$$

THEOREM 1.3.2. *The family of semi-norms* $\{\| \ \|_{\underset{\sim}{\ell},\underset{\sim}{m}} : \underset{\sim}{\ell}, \underset{\sim}{m} \in \mathbf{N}^n\}$ *and the family of norms* $\{\| \ \|_{B,k} : k \in \mathbf{N}\}$ *are equivalent.*

PROOF. By expressing A_j and A_j^* in terms of Q_j and P_j and using the commutation relations, we can write:

$$K_k = \sum_{\substack{|\underset{\sim}{\ell}| \le L_k \\ |\underset{\sim}{m}| \le M_k}} b_{\underset{\sim}{\ell},\underset{\sim}{m}}^{k} \underset{\sim}{Q}^{\underset{\sim}{\ell}} \underset{\sim}{P}^{\underset{\sim}{m}} . \tag{1.3.5}$$

Then, using (1.3.4), (1.3.5) and Schwarz's inequality, we have:

$$\|x\|_{B,k}^2 \le \|K_k x\|^2 \le \sum_{\substack{|\underset{\sim}{\ell}| \le L_k \\ |\underset{\sim}{m}| \le M_k}} |b_{\underset{\sim}{\ell},\underset{\sim}{m}}^{k}|^2 \, \|x\|_{\underset{\sim}{\ell},\underset{\sim}{m}}^2$$

and thus if $\|x_n\|_{\ell,\underline{m}} \to 0$ as $n \to \infty$ for all $\underline{\ell}, \underline{m} \in N^n$, then $\|x_n\|_{B,k} \to 0$ as $n \to \infty$ for all $k \in N$.

Reversing the process if we are given $\underline{\ell}', \underline{m}' \in N^n$, we can express $Q^{\underline{\ell}'} P^{\underline{m}'}$ as operators on $S(\mathcal{H}_\pi)$ as follows:

$$Q^{\underline{\ell}'} P^{\underline{m}'} = \sum_{\substack{|\underline{\ell}| \leq L \\ |\underline{m}| \leq M}} c_{\underline{\ell},\underline{m}} A^{\underline{\ell}} A^{*\underline{m}}$$

by expressing Q_j and P_j in terms of the A_j and A_j^*, $j = 1, \cdots, n$ and using the commutation relations. Thus

$$(1.3.6) \qquad \|x\|_{\underline{\ell}',\underline{m}'} \leq \sum_{\substack{|\underline{\ell}| \leq L \\ |\underline{m}| \leq M}} |c_{\underline{\ell},\underline{m}}| \, \|A^{\underline{\ell}} A^{*\underline{m}} x\| \; .$$

We will need the following lemma.

LEMMA 1.3.3. *Given* $\underline{\ell}, \underline{m} \in N^n$, *there exists positive constants* $d_{\underline{\ell}_1}, \underline{\ell}_1 \leq \underline{\ell}$ *such that*:

$$(1.3.7) \qquad \|A^{\underline{\ell}} A^{*\underline{m}} x\| \leq \sum_{\underline{\ell}_1 \leq \underline{\ell}} d_{\underline{\ell}_1} \|A^{*\underline{\ell}_1 + \underline{m}} x\| \; .$$

PROOF OF THE LEMMA. From the commutation relation:

$$A_j A_j^* = A_j^* A_j + 1, \qquad j = 1, \cdots, n$$

we have:

$$(1.3.8) \qquad \|A_j^* x\|^2 = \|Ax\|^2 + \|x\|^2, \qquad j = 1, \cdots, n \; .$$

We now proceed to prove (1.3.7) by induction on $|\underline{\ell}|$. If $|\underline{\ell}| = 1$, we are done by (1.3.8). Assume we have established (1.3.7) for $|\underline{\ell}| = L$ and assume $|\underset{\sim}{\ell}| = L+1$. Assuming, say $\ell_1 \neq 0$, we have. using (1.3.8) and the commutation relations, that:

$$\|\underset{\sim}{A}^{\underset{\sim}{\ell}_1} \underset{\sim}{A}^{*\underset{\sim}{m}} x\| \leq \|A^* \underset{\sim}{A}^{\underset{\sim}{\ell}_1 - \varepsilon_1} \underset{\sim}{A}^{*\underset{\sim}{m}} x\|$$

$$\leq \|\underset{\sim}{A}^{\underset{\sim}{\ell}_1 - \varepsilon_1} \underset{\sim}{A}^{*\underset{\sim}{m}+\varepsilon_1} x\| + \ell_1 \|\underset{\sim}{A}^{\underset{\sim}{\ell}_1 - \varepsilon_1} \underset{\sim}{A}^{*\underset{\sim}{m}} x\|$$

where $\varepsilon_1 = (1,0,\cdots,0) \in N^n$ and thus we are done by induction.

Return to the proof of Theorem 1.3.2. Using (1.3.7) and substituting in (1.3.6), we see that there exist positive constants $e_{\underset{\sim}{\ell},\underset{\sim}{m}}$ such that:

(1.3.9)
$$\|x\|_{\underset{\sim}{\ell}',\underset{\sim}{m}'} \leq \sum_{\substack{|\underset{\sim}{\ell}| \leq L \\ |\underset{\sim}{m}| \leq K}} e_{\underset{\sim}{\ell},\underset{\sim}{m}} \|\underset{\sim}{A}^{*\underset{\sim}{\ell}+\underset{\sim}{m}} x\|$$

$$= \sum_{\substack{|\underset{\sim}{\ell}| \leq L \\ |\underset{\sim}{m}| \leq K}} e_{\underset{\sim}{\ell},\underset{\sim}{m}} (x, \underset{\sim}{A}^{\underset{\sim}{\ell}+\underset{\sim}{m}} \underset{\sim}{A}^{*\underset{\sim}{\ell}+\underset{\sim}{m}} x)^{\frac{1}{2}} .$$

But if $k \geq L+K$, we have that

$$(x, \underset{\sim}{A}^{\underset{\sim}{\ell}+\underset{\sim}{m}} \underset{\sim}{A}^{*\underset{\sim}{\ell}+\underset{\sim}{m}} x) \leq (x, K_k x)$$

and therefore there exists C_k such that:

(1.3.10)
$$\|x\|_{\underset{\sim}{\ell}',\underset{\sim}{m}'} \leq C_k \|x\|_{B,k}$$

and thus if $\|x_n\|_{B,k} \to 0$ for all $k \in N$, $\|x_n\|_{\underset{\sim}{\ell},\underset{\sim}{m}} \to 0$ for $\underset{\sim}{\ell},\underset{\sim}{m} \in N^n$ and the two families of semi-norms are equivalent.

C. $\mathcal{S}'(\mathcal{H}_\pi)$ is defined as the space of continuous anti-linear functionals on $\mathcal{S}(\mathcal{H}_\pi)$ and is called the abstract tempered distributions. It will be very easy to describe in the B.F.S. realization. It is just the usual space of tempered distributions when $(\pi, \mathcal{H}_\pi)$ is the Schrödinger realization.

CHAPTER II
THE B.F.S. REALIZATION OF THE HEISENBERG GROUP

§2.1. *Generalities on polarizations and holomorphic induction*

In this section we shall discuss the general procedure for constructing irreducible (unitary) realizations of $N(V)$ from positive "polarizations" on $\mathfrak{N}(V)$ via the technique of holomorphic induction. Our reasons for doing so are two-fold. The first reason is that this procedure will show that the Schrödinger and B.F.S. realizations are special cases of this general procedure and thus they are equally as "natural." The second reason is to give a very brief introduction to some important techniques used by Auslander and Kostant [1] in classifying the irreducible (unitary) representations of type I simply connected solvable groups. In particular totally complex positive polarizations and the corresponding holomorphically induced representations are especially important in their theory. The B.F.S. realization of $N(V)$ is an example and a model for such a representation. One has to look no further than the oscillator group [23] to find a solvable group which requires totally complex polarizations to construct its irreducible representations. No proofs will be given in this section.

A. For the remainder of this chapter we shall consider a fixed $2n$ dimensional symplectic space $(V, \langle\ ,\ \rangle)$ and an admissible complex structure J on V. Thus we shall write N rather than $N(V)$, etc. $\mathfrak{N}^*$ will denote the dual of $\mathfrak{N}$ i.e. the space of real linear functionals on $\mathfrak{N}$. If $0 \neq f \in \mathfrak{N}^*$, define $B_f(n_1, n_2) =: f([n_1, n_2])$ for $n_1, n_2 \in \mathfrak{N}$. Note that B_f is antisymmetric. Extend B_f to $\mathfrak{N}_C$ by complex bilinearity and denote the extension by B_f^C. Define $\ker B_f^C =: \{n \in \mathfrak{N}_C : B_f(n', n) = 0 \text{ for all } n' \in \mathfrak{N}_C\}$. It can easily be shown that $\ker B_f = CE$.

DEFINITION. *A positive polarization of* $\mathfrak{N}$ *at* f *is a complex subalgebra* $\mathfrak{H}$ *of* $\mathfrak{N}^C$ which satisfies:

1) $\mathfrak{H}$, viewed as a complex subspace of $\mathfrak{N}^C$ is a maximal totally isotropic subspace for B_f^C, i.e. if $h_1, h_2 \in \mathfrak{H}$, then $B_f^C(h_1, h_2) = 0$ and if $\mathfrak{H}_1 \supset \mathfrak{H}$ and $\mathfrak{H}_1$ has this property, then $\mathfrak{H} = \mathfrak{H}_1$.

2) $\mathfrak{H} + \overline{\mathfrak{H}}$ is a subalgebra of $\mathfrak{N}^C$.

3) $iB_f^C(h, \overline{h}) \geq 0$ for all $h \in \mathfrak{H}$.

H is said to be *real* if $\mathfrak{H} = \overline{\mathfrak{H}}$ and to be *totally complex* if $\mathfrak{H} + \overline{\mathfrak{H}} = \mathfrak{N}^C$. Note that if $\mathfrak{H}$ is real and satisfies 1) and 2), 3) is automatically satisfied and $\mathfrak{H}$ is a positive polarization for f. For the remainder of this chapter we take $f = E^*$ where E^* is defined by:

$$E^*(v \oplus tE) = : t \ .$$

B. We will now discuss the two examples of polarization which will be relevant in the sequel.

EXAMPLE 1. Let Q be a real n-dimensional subspace of V_J such that $(\,,\,)_J$ is real on Q. Let $P = JQ$ and thus $V = Q \oplus P$. Let $\mathfrak{H}_S = CP + CE$. It is easy to check that $\mathfrak{H}_S$ is a real positive polarization at E^*. It is called a *Schrödinger polarization* (w.r. to J). We shall see that it will lead to the usual Schrödinger realization of N.

EXAMPLE 2. Let $W_J^{\pm}$ be as in §1.2D and let $\mathfrak{H}_B = : W_J^- + CE$. It is easy to check that $\mathfrak{H}_B$ is a positive polarization at E^*. It is *totally complex* because $V^C = W_J^+ \oplus \overline{W_J^-}$, $\overline{W_J^-} = W_J^+$ and $\mathfrak{N}_C = V^C \oplus CE$. $\mathfrak{H}_B$ is called the B.F.S. polarization at E^* w.r. to J.

C. We shall now describe how one combines a positive polarization $\mathfrak{H}$ at E^* and the method of holomorphically induced representations to obtain a unitary representation $\pi_{\mathfrak{H}} = \pi$ of N which is irreducible and which satisfied $\pi(e^{Et}) = e^{it}$.

If $n \in \mathfrak{N}$, define $\rho(n)\phi(n) = : \frac{d}{dt}\phi(ne^{tn})|_{t=0}$ for $\phi \in C^\infty(N)$. If $n \in \mathfrak{N}_C$, then write $n = n_1 + in_2$, $n_1, n_2 \in \mathfrak{N}$ and define: $\rho(n) = : \rho(n_1) + i\rho(n_2)$. Let $C^\infty(\mathfrak{H})$ denote the complex vector space of C^∞ functions on N which satisfy:

(2.1.1) $$\rho(h)\phi = -iE^*(h)\phi$$

for $h \in \mathfrak{H}$. Let $\mathfrak{D} = \mathfrak{H} \cap \mathfrak{N}$ and $D = \exp \mathfrak{D}$. Define a character on D by:

$$\chi(e^{\delta}) = e^{iE^*(\delta)} .$$

Note that if $h \in \mathfrak{D}$, then (2.1.1) is equivalent to:

$$\phi(n \exp h) = \chi^{-1}(\exp h)\phi(n) . \tag{2.1.2}$$

N and D are unimodular Lie groups and therefore N/D has a N-left invariant measure $d\dot{n}$. It is characterized by the following property:

$$\int_{N/D} \int_D \phi(n\alpha)\, d\alpha\, d\dot{n} = \int \phi(n)\, dn$$

where $\phi \in C_0^\infty(N)$, $d\alpha$ is Haar measure on D, and we identify D-right invariant functions on N with functions on N/D. Let $L^2(N/D, d\dot{n})$ denote the space of dn measurable functions ϕ on N such that $|\phi(n)|^2$ is right D-invariant and $\int_{N/D} |\phi(\dot{n})|^2 dn < \infty$ where $|\phi(\dot{n})|^2$ denotes the function on N/D uniquely determined by $|\phi(n)|^2$. Note by (2.1.2) that if $\phi \in C^\infty(\mathfrak{H})$, then $|\phi|^2$ is right D invariant.

We now have all of the machinery to define the representation $(\pi_{\mathfrak{H}}, \mathcal{H}(\mathfrak{H}))$ corresponding to $\mathfrak{H}$. Let $\mathcal{H}(\mathfrak{H})$ be the completion of $C^\infty(\mathfrak{H}) \cap L^2(N/D, d\dot{n})$ with respect to the Hilbertian norm.

$$\|\phi\|_{\mathfrak{H}}^2 = : \int_{N/D} |\phi(\dot{n})|^2\, d\dot{n} .$$

Define $\pi_{\mathfrak{H}}$ by:

$$\pi_{\mathfrak{H}}(n)\phi(n') = : \phi(n^{-1}n') .$$

Property 3 of positive polarizations assures us that $\mathcal{H}(\mathfrak{H}) \neq \{0\}$. Property (2.1.1) of $C^\infty(\mathfrak{H})$ assures us that:

$$\pi_{\mathfrak{H}}(e^{Et}) = e^{it} .$$

$\pi_{\mathfrak{H}}$ is clearly unitary. $(\pi_{\mathfrak{H}}, \mathcal{H}(\mathfrak{H}))$ is called the *representation of* N *holomorphically induced from* $(\mathfrak{H}, E^*)$. It can be shown to be irreducible.

REMARK. If $\mathfrak{H}$ is real, the above procedure is equivalent to the usual induction of (D,χ) to N and there really is nothing holomorphic about it. However, if $\mathfrak{H}$ is totally complex, then $\mathcal{H}(\mathfrak{H})$ will be essentially a Hilbert space of holomorphic functions as we shall see in the case of $\mathcal{H}(\mathfrak{H}_B)$.

§2.2. *The Schrödinger and* B.F.S. *realizations*

In this section we shall define the Schrödinger and B.F.S. realization of N and indicate how they are derived from the holomorphic induction process associated with $\mathfrak{H}_S$ and $\mathfrak{H}_B$ respectively. We shall define a unitary intertwining operator between the two realizations.

A. Let Q and P be as in the definition of $\mathfrak{H}_S$. Note that the restriction of $(\,,)_J$ to $Q\times Q$ and $P\times P$ defines a real inner product. Let $d\underset{\sim}{q}$ be the Lebesgue measure on Q which is normalized so that a hypercube in Q with unit sides has volume equal to 1. Let $\mathcal{H}_S = L^2(Q, d\underset{\sim}{q})$. Then the *Schrodinger realization* $T_{Q,P} = T$ of N is defined as follows:

$$(2.2.1)\qquad T(\exp[\underset{\sim}{q}+\underset{\sim}{p}+tE])\psi(\underset{\sim}{q}')$$
$$=:\exp[i(t+\frac{1}{2}<\underset{\sim}{q},\underset{\sim}{p}> - <\underset{\sim}{q}',\underset{\sim}{p}>)]\psi(\underset{\sim}{q}'-\underset{\sim}{q}),$$

where $\underset{\sim}{q}\in Q$, $\underset{\sim}{p}\in P$. A straightforward calculation shows that T is a unitary representation of N such that $T(e^{Et}) = e^{it}$.

Let $\underset{\sim}{e}$ be an orthonormal basis for Q and $\underset{\sim}{f} = J\underset{\sim}{e}$ be the corresponding basis for P. Then we can identify $\underset{\sim}{q} = \sum_{j=1}^n \alpha_j e_j$ with $\alpha = (\alpha_1,\cdots,\alpha_n)$ $\in R^n$ and $\underset{\sim}{p} = \sum_{j=1}^n \beta_j f_j$ with $\beta = (\beta_1,\cdots,\beta_n)$ and thus $L^2(Q, d\underset{\sim}{q})$ with $L^2(R^n)$. Then if we write $T(\exp[\sum_{j=1}^n \alpha_j e_j + \sum_{j=1}^n \beta_j f_j])$ as $T_{\alpha,\beta}$, then (2.2.1) can be written:

$$(2.2.2)\qquad T_{\alpha,\beta}\psi(q) = \exp[i(\frac{1}{2}\alpha\cdot\beta - q\cdot\beta)]\psi(q-\alpha).$$

Note this is just 3.11 as on page 207 in [2].

B. Let W_J^+ and W_J^- be as in the definition of $\mathfrak{H}_B$. We define a Hermitian structure on W_J^+ by:

$$(\omega'_+, \omega_+)_+ =: -i\langle\omega'_+, \bar{\omega}_+\rangle .$$

We denote $(\omega_+, \omega_+)_+$ by $\|\omega_+\|_+^2$. Let $d\omega'_+$ be the corresponding normalized Lebesgue measure on W_J^+. Let $d\mu =: \frac{1}{\pi^n} e^{-\|\omega_+\|^2} d\omega_+$ and let $\mathcal{H}_{B.F.S.} = \mathfrak{F}(W_J^+, d\mu) =$ the space of holomorphic functions on W_J^+ which are square integrable with respect to $d\mu$. The inner product is defined by:

$$(f_1, f_2)_B =: \int \bar{f}_1 f_2 \, d\mu .$$

We define the B.F.S. *realization* $V_J = V$ of N as follows:

$$V(e^{\omega_+ + \bar{\omega}_+ + Et}) f(\omega'_+) \tag{2.2.3}$$

$$=: \exp[it + (\omega'_+, \omega_+)_+ - \tfrac{1}{2}(\omega_+, \omega_+)_+] f(\omega'_+ - \omega_+) .$$

A straightforward calculation shows that V is a unitary representation of N and $V(e^{Et}) = e^{it}$. Let $\{\underset{\sim}{a}^+, \underset{\sim}{a}^-\}$ be a standard basis for W_J^+ and W_J^- respectively. Then we can identify $\omega'_+ = \sum_{j=1}^n z_j e_j$ with $z = (z_1, \cdots, z_n) \in C^n$ and $\mathfrak{F}(W_J^+)$ with $\mathfrak{F}_n$, the space of holomorphic functions on C^n which are square integrable with respect to $(2\pi)^{-n} e^{-|z|^2} dz\, d\bar{z}$. If $\omega_+ = \sum_{j=1}^n c_j a_j^+$, then $(\omega'_+, \omega_+)_+ = z \cdot \bar{c}$ and $(\omega_+, \omega_+)_+ = c \cdot \bar{c}$. If we write V_c for $V(e^{\omega_+ + \omega_-})$, then we can write (2.2.3) as follows:

$$V_c f(z) = \exp[\bar{c} \cdot (z - \tfrac{1}{2}c)] f(z - c) . \tag{2.2.4}$$

Note this is just (3.5) on page 206 in [2].

C. It is a fact that both T and V are irreducible, so by the Stone-von Neumann theorem there must exist a unitary mapping $A : L^2(Q) \to \mathfrak{F}(W_J^+)$ such that

$$V(e^{\omega_+ + \bar{\omega}_+ + Et}) A = A T(e^{q+p+Et}) \tag{2.2.5}$$

where $\omega_+ + \bar{\omega}_+ = \underset{\sim}{q} + \underset{\sim}{p}$. In our concrete situation one can express such an intertwining using an explicit and rather simple kernel. Let Q, P and W_J^+ be as above. Define $K: V_C \to V_C$ by

$$K(\underset{\sim}{q}+\underset{\sim}{p}) = \underset{\sim}{q} - \underset{\sim}{p}\ , \qquad \underset{\sim}{q} \in Q,\ \underset{\sim}{p} \in P$$

and lift K to V_C by complex linearity. Let

(2.2.6) $A(\omega_+, q)$

$$=: \pi^{-n/4} \exp\left(i\left[\frac{1}{2} < \underset{\sim}{q}, iJ\underset{\sim}{q} > + < \underset{\sim}{q}, (I-iJ)\omega_+ > + \frac{1}{2} < K\omega_+, \omega_+ >\right]\right)$$

and define A by:

(2.2.7) $$A\psi(\omega_+) =: \int_Q A(\omega_+, \underset{\sim}{q})\psi(\underset{\sim}{q})d\underset{\sim}{q}\ .$$

A straightforward calculation which is best made by choosing coordinates will show that A maps $\mathcal{K}(Q)$, the continuous functions on Q with compact support, isometrically into $\mathcal{F}(W_J^+)$, and satisfies (2.2.5) on $\mathcal{K}(Q)$. With somewhat more difficulty it can be shown that (2.2.7) maps $L^2(Q)$ into $\mathcal{F}(W_J^+)$ and in fact is unitary. Since we will not make use of A in this paper with the exception of Section 3.2, we will not go into any more detail here. See [2].

If $\underset{\sim}{e}$, $\underset{\sim}{f}$, $\underset{\sim}{a}^+$, $\underset{\sim}{a}^-$ are as above and if we accordingly identify $L^2(Q)$ with $L^2(R^n)$, $\mathcal{F}(W_J^+)$ with $\mathcal{F}_n$, and write $A(\sum_{j=1}^n a_j a_j^+, \sum_{j=1}^n q_j e_j)$ as $A_n(z, q)$, then

$$A_n(z, q) = \pi^{-n/4} \exp\left(-\frac{1}{2}\, q \cdot q + \sqrt{2}\, q \cdot z - \frac{1}{2}\, z \cdot z\right)\ ,$$

which is the kernel on page 189 in [2].

D. We shall now show how the Schrödinger realization (2.2.1) *is derived* from $(\pi_S, \mathcal{H}_S)$. We first must normalize some measures appropriately. First let dv be the Lebesgue measure on V normalized w.r. to $(\ ,\ )_J$ and let $dn = dv \cdot dt$. Let $d\underset{\sim}{p}$ be the Lebesgue measure on P which is normalized w.r.t. to the restriction of $(\,,\,)_J$ to P. Then $d\alpha$ will denote Haar measure on $D = e^{P+RE}$ defined by:

$$\int_D g(\alpha)\,d\alpha =: \int g(e^{\underset{\sim}{p}+Et})\,d\underset{\sim}{p}\,dt\ .$$

One can check that if g is a non-negative dn measurable function which is right D-invariant, then

$$(2.2.8) \qquad \int_{N/D} g(\dot{n})\,d\dot{n} = \int_Q g(e^{\underset{\sim}{q}})\,d\underset{\sim}{q}\ .$$

We now wish to define a unitary mapping $U_S : \mathcal{H}(\mathfrak{H}_S) \to L^2(Q)$ such that

$$(2.2.9) \qquad T(n)U_S = U_S\pi_S(n)$$

for $n \in N$. In fact, define:

$$\underset{\sim}{\psi}(\underset{\sim}{q}) \equiv U_S\psi(\underset{\sim}{q}) =: \underset{\sim}{\psi}(e^{\underset{\sim}{q}})$$

for $\underset{\sim}{\psi}$ a continuous function in $\mathcal{H}(\mathfrak{H}_S)$. From (2.1.2), applied to $\mathfrak{H}_S$, we have that $\underset{\sim}{\psi}$ is right D invariant and therefore by 2.2.8, we have:

$$\int_{N/D} |\psi(\dot{n})|^2\,d\dot{n} = \int |\psi(\underset{\sim}{q})|^2\,d\underset{\sim}{q}\ .$$

Thus U_S is isometric and can be extended to an isometry on $\mathcal{H}(\mathfrak{H}_S)$. If ψ is a continuous function in $L^2(Q)$, define:

$$V_S\psi(e^{\underset{\sim}{q}+\underset{\sim}{p}+Et}) =: \chi^{-1}(e^{\underset{\sim}{p}+E(t-\frac{1}{2}\langle \underset{\sim}{q},\underset{\sim}{p}\rangle)})\psi(\underset{\sim}{q})\ .$$

Then $U_SV_S = I$ on such functions which form a dense set in $L^2(Q)$. Therefore U_S is onto and therefore unitary.

We must now check that (2.2.9) holds. Using (1.2.1′) several times we have:

$$\begin{aligned} U_S\pi_S(e^{\underset{\sim}{q}+\underset{\sim}{p}+Et})\psi(q') &= \pi_S(e^{\underset{\sim}{q}+\underset{\sim}{p}+Et})\underset{\sim}{\psi}(e^{\underset{\sim}{q}'}) = \psi(e^{-\underset{\sim}{q}-\underset{\sim}{p}-Et}e^{\underset{\sim}{q}'}) \\ &= \underset{\sim}{\psi}(e^{[(\underset{\sim}{q}'-\underset{\sim}{q})-\underset{\sim}{p}-E(t+\frac{1}{2}<\underset{\sim}{p},\underset{\sim}{q}'>)]}) \\ &= \psi(e^{-\underset{\sim}{p}-E(t-<a',p>+\frac{1}{2}<q,p>} \cdot e^{q'-q}) \\ &= \exp(i[t+\tfrac{1}{2}<\underset{\sim}{q},\underset{\sim}{p}> - <\underset{\sim}{q}',\underset{\sim}{p}>])\psi(e^{\underset{\sim}{q}'-\underset{\sim}{q}})] \\ &= T(e^{\underset{\sim}{q}+\underset{\sim}{p}+tE})U_S\underset{\sim}{\psi}(\underset{\sim}{q}')\ , \end{aligned}$$

and (2.2.9) thus holds and we have completed the derivation of the Schrödinger realization from $(\pi_S, \mathcal{H}(\mathfrak{H}_S))$.

E. We shall now derive the B.F.S. realization from $(\pi_B, \mathcal{H}(\mathfrak{H}_B))$. First observe that $\mathfrak{D} = \mathfrak{Z}$ and $D \equiv \exp \mathfrak{D} = Z = :\{e^{tE} : t \in \mathbf{R}\}$. Thus since $N = V \times D$, $N/D \cong V$. Moreover if we use dt as the Haar measure on D, we have $d\dot{n} \approx dv$ where dv is the Lebesgue measure on V normalized with respect to $(\,,\,)_J$. Moreover, if Θ_J is the mapping discussed in 1.2D, then

$$(\Theta_J\omega_+, \Theta_J\omega_+)_J = 2(\omega_+, \omega_+)_+ \ . \tag{2.2.10}$$

Using Θ_J^{-1} we identify V_J and W_J^+ and thus by 2.2.10 we have: $dv \approx 2^n d\omega^+$ where $d\omega^+$ is the Lebesgue measure on W_J^+ normalized w.r. to $(\,,\,)_+$. If $\underset{\sim}{g}$ is a non-negative dn measurable function on N which is right D invariant, then:

$$\int_{N/D} \underset{\sim}{g}(\dot{n})\, d\dot{n} = \int_{W_J^+} g(e^{\omega_+ + \bar{\omega}_+})\, d\omega_+ \ . \tag{2.2.11}$$

We will need to introduce an intermediate space $\tilde{\mathfrak{F}}(W_J^+)$ and a corresponding unitary map $\tilde{U}_B : \mathcal{H}(\mathfrak{H}_B) \to \tilde{\mathfrak{F}}(W_J^+)$ before we can construct the unitary map $U_B : \mathcal{H}(\mathfrak{H}_B) \to \mathfrak{F}(W_J^+)$ with the property that

$$V(n)U_B = U_B\pi_B(n) \tag{2.2.12}$$

holds for $n \in N$. For $\underset{\sim}{f} \in C^{\infty}(\mathfrak{H}_B) \cap L^2(N, dn)$; define:

$$\tilde{U}_B \underset{\sim}{f}(\omega_+) \equiv \tilde{f}(\omega_+) = : \underset{\sim}{f}(e^{\omega_+ + \bar{\omega}_+}) .$$

It follows from (2.1.2) that

$$\int_{N/D} |\underset{\sim}{f}(\dot{n})|^2 \, d\dot{n} = \int |\underset{\sim}{f}(\omega_+)|^2 \, d\omega^+$$

and thus $\tilde{U}_B$ can be extended to an isometry of $\mathcal{H}(\mathfrak{H}_B)$ into $L^2(W_J^+, d\omega^+)$.

We must identify the image of $\tilde{U}_B$. Thus we must compute $\tilde{\rho}(\omega^-) = : \tilde{U}_B \rho(\omega^-) \tilde{U}_B^*$ where $\rho(\omega^-)$ is defined in 2.1 C. First observe that if $\omega^- = v_1 + iv_2$, $v_1, v_2 \in V$, then $v_1 = \frac{1}{2}(\omega^- + \bar{\omega}^-)$ and $v_2 = -\frac{1}{2}(i\omega^- + i\omega^-)$. Then for $\underset{\sim}{f} \in C^{\infty}(\mathfrak{H}_B) \cap L^2(N, dn)$, $\tilde{f} = \tilde{U}_B \underset{\sim}{f}$, we have:

$$\begin{aligned}
\tilde{\rho}(\omega^-)\tilde{f}(\omega_+') &= \rho(\omega^-)\underset{\sim}{f}(e^{\omega_+' + \bar{\omega}_+}) \\
&= \frac{d}{dt}(\underset{\sim}{f}(e^{\omega_+' + \bar{\omega}_+'} e^{t/2(\bar{\omega}_- + \omega_-)})|_{t=0} \\
&\quad + i\frac{d}{dt}(\underset{\sim}{f}(e^{\omega_+' + \bar{\omega}_+'} e^{t/2(\bar{\omega}_- + \omega_-)})|_{t=0} \\
&= \frac{1}{2}(\partial(\bar{\omega}^-) + i\partial(i\bar{\omega}^-) - \langle \omega_+', \omega_- \rangle)\tilde{f}(\omega_+')
\end{aligned}$$

where $\partial(\omega_+)\tilde{f}(\omega_+') = : \frac{d}{dt}(\tilde{f}(\omega_+' + \omega^+ t))_{t=0}$. Thus the image $C^{\infty}(\mathfrak{H}_B) \cap L^2(N, dn)$ under $\tilde{U}_B$ is $\tilde{\mathcal{F}}(W_J^+) = \{\tilde{f} \in C^{\infty}(\mathfrak{H}_B) \cap L^2(N, dn) : \frac{1}{2}(\partial(\bar{\omega}^-) + i\partial(i\bar{\omega}^-) - \langle \omega_+', \omega_- \rangle)\tilde{f}(\omega_+') = 0$ for $\omega^- \in W_J^-\}$. It can be shown that $\tilde{\mathcal{F}}(W_J^+)$ is already a Hilbert space. Thus U_B is a unitary map from $\mathcal{H}(\mathfrak{H}_B)$ onto $\tilde{\mathcal{F}}(W_J^+)$.

If we define $\tilde{V}(e^{\omega_+ + \bar{\omega}_+ + Et})$ by:

$$\tilde{V}(e^{\omega_+ + \bar{\omega}_+ + Et})\tilde{f}(\omega_+') = : \exp[it + \frac{1}{2}\{(\omega_+, \omega_+')_+ - (\omega_+', \omega_+)_+\}\tilde{f}(\omega_+' - \omega_+)],$$

then a calculation similar to the one we carried out in the preceding section shows that:

(2.2.13) $$\tilde{V}(n)\tilde{U}_B = \tilde{U}_B \pi_B(n)$$

for $n \in N$.

We now define a mapping $\tilde{\tilde{U}}_B : \tilde{\mathcal{F}}(W_J^+) \to L^2(W_J^+, d\mu)$. In fact, define

$$\tilde{\tilde{U}}_B \tilde{f}(\omega_+) \equiv f(\omega_+) = : (\pi)^{-n/2} \exp[-\tfrac{1}{2} \|\omega_+\|^2] .$$

$\tilde{\tilde{U}}_B$ is clearly an isometry into $L^2(W_J^+, d\mu)$. From the definition of $\tilde{\mathcal{F}}(W_J^+)$ it follows that the image of $\tilde{\tilde{U}}_B$ is $\{f \in C^\infty(W_J^+) \cap L^2(W_J^+, d\mu)\}$ such that:

$$\partial(i\omega_+)f = i\partial(\omega_+)f$$

for $\omega_+ \in W_J^+$. These are just the Cauchy-Riemann equations on W_J^+. Thus $\tilde{\tilde{U}}_B$ is a unitary mapping from $\tilde{\mathcal{F}}(W_J^+)$ onto $\mathcal{F}(W_J^+)$. A direct calculation shows that:

(2.2.14) $$V(n)\tilde{\tilde{U}}_B = \tilde{\tilde{U}}_B \tilde{V}(n)$$

for $n \in N$. Let $U_B = \tilde{\tilde{U}}_B \tilde{U}_B$. Then combining (2.2.13) and (2.2.14), we

$$V(n)U_B = U_B \pi_B(n) ,$$

$n \in N$ which establishes the unitary equivalence of π_B and the B.F.S. realization V.

§2.3. *Properties of* $\mathcal{F}(W_J^+)$ *and the* B.F.S. *realization* V

In this section we will discuss various special properties of the Hilbert space of analytic functions $\mathcal{F}(W_J^+)$ and the associated properties of V. Proofs by and large will only be sketched.

A. We first will show that $\mathcal{F}(W_J^+)$ is complete and therefore a Hilbert space.

THEOREM 2.3.1. *$\mathcal{F}(W_J^+)$ is complete* w.r. *to the norm* $\|f\|_B = [(f,f)_B]^{\frac{1}{2}}$.

PROOF. It is sufficient to show that $\mathcal{F}(W_J^+)$ is isometrically isomorphic to $\ell_2(N^n)$, the Hilbert space of square summable sequences indexed by N^n. Let $\underset{\sim}{a}^+$ be a fixed standard basis for W_J^+ and define:

$$u_{\underset{\sim}{m}}\left(\sum_{j=1}^{n} z_j a_j^+\right) =: \frac{z^{\underset{\sim}{m}}}{(\underset{\sim}{m}!)^{\frac{1}{2}}}$$

for $\underset{\sim}{m} \in N^n$. Let $R > 0$, then, using polar coordinates it is easy to see that:

$$\int_{|z_1|<R} \cdots \int_{|z_n|<R} \overline{u_{\underset{\sim}{m}_1}}\, u_{\underset{\sim}{m}_2}\, d\mu = 0 \qquad (2.3.1)$$

for $\underset{\sim}{m}_1 \neq \underset{\sim}{m}_2$. A standaid Gaussian moment calculation shows that $\int_{|z_1|<R}, \ldots, \int_{|z_n|<R} |u_{\underset{\sim}{m}}|^2\, d\mu =: \mu_{\underset{\sim}{m}}(R)$ converges monotonically to 1 as $R \to \infty$ for $\underset{\sim}{m} \in N^n$. Let $f \in \mathcal{B}(W_J^+)$ the space of holomorphic functions on W_J^+. Then we can write

$$f = \sum_{\underset{\sim}{m}} \alpha_{\underset{\sim}{m}} u_{\underset{\sim}{m}} \qquad (2.3.2)$$

which converges uniformly on compact subsets in W_J^+. Thus, using (2.3.1) we see that

$$\int_{|z_1|<R} \cdots \int_{|z_n|<R} |f|^2\, d\mu = \sum_{\underset{\sim}{m}} |\alpha_{\underset{\sim}{m}}|^2 \mu_{\underset{\sim}{m}}(R)$$

and by the monotone convergence theorem we have:

$$\int_{W_J^+} |f|^2\, d\mu = \sum_{\underset{\sim}{m}} |\alpha_{\underset{\sim}{m}}|^2 \,. \qquad (2.3.3)$$

Thus $f \in \mathcal{F}(W_J^+)$ iff $\sum_{\underset{\sim}{m}} |\alpha_{\underset{\sim}{m}}|^2 < \infty$ and the mapping $f \to \{\alpha_{\underset{\sim}{m}} : \underset{\sim}{m} \in N^n\}$ where the $\{\alpha_{\underset{\sim}{m}}\}$ are defined by (2.3.2) is an isometric isomorphism from $\mathcal{F}(W_J^+)$ onto $\ell_2(N^n)$.

B. One of the most interesting and useful features of $\mathcal{F}(W_J^+)$ is the existence of a family of *principal vectors* $\{e_{\omega_+}, \omega_+ \in W_J^+\}$ in $\mathcal{F}(W_J^+)$ with the property that:

$$f(\omega_+) = (e_{\omega_+}, f)_B \tag{2.3.4}$$

for $f \in \mathcal{F}(W_J^+)$. In fact, let:

$$e_{\omega_+}(\omega_+') =: \exp[(\omega_+', \omega_+)_+] \ . \tag{2.3.5}$$

Note that if $\underset{\sim}{a}^+$ is a standard basis for W_J^+ and if $\omega_+ = \sum_{j=1}^n w_j a_j^+$ and $\omega_+' = \sum_{j=1}^n z_j a_j^+$, then

$$e_{\omega_+}(\omega_+') = \exp[z \cdot \overline{w}] \ .$$

THEOREM 2.3.2. *The family* $\{e_{\omega_+}\}$ *defined by* (2.3.5) *satisfies* (2.3.4).

PROOF. Let $\underset{\sim}{a}^+$ be a standard basis for W_J^+ and let $\{u_{\underset{\sim}{m}} : \underset{\sim}{m} \in N^n\}$ be defined as in the proof of Theorem 2.3.1. Note that in the process of the proof of Theorem 2.3.1 we showed that the $\{u_{\underset{\sim}{m}}\}$ are a complete orthonormal system. By expanding e_{ω_+} in terms of the $u_{\underset{\sim}{m}}$ we see that

$$u_{\underset{\sim}{m}}(\omega_+) = (e_{\omega_+}, u_{\underset{\sim}{m}}) \ . \tag{2.3.6}$$

For $f \in \mathcal{F}(W_J^+)$, let $f = \sum_{\underset{\sim}{m}} \alpha_{\underset{\sim}{m}} u_{\underset{\sim}{m}}$. This series converges both pointwise and in norm and thus by (2.3.6) and the continuity of the inner product we have

$$\alpha_{\underset{\sim}{m}} u_{\underset{\sim}{m}}(\omega_+) = (e_{\omega_+}, \underset{\underset{\sim}{m}}{} \alpha_{\underset{\sim}{m}} u_{\underset{\sim}{m}})$$

and (2.3.4) is established.

REMARK. Note that by (2.3.4) that we have:

$$(f_1, f_2)_B = \int (f_1, e_{\omega_+})_B (e_{\omega_+}, f_2)_B \, d\mu \tag{2.3.7}$$

i.e. the principal vectors have many features enjoyed by a complete orthonormal system.

The following norm computations are useful.

LEMMA 2.3.3. *The following equalities hold:*

$$\|e_{\omega_+}\|^2 = e^{\|\omega_+\|^2} \tag{2.3.8}$$

$$\|(\cdot,\omega_+)_+ e_{\omega'_+}\|^2 = [\,|(\omega'_+,\omega_+)_+|^2 + (\omega_+,\omega_+)_+]\, e^{\|\omega'_+\|^2} \tag{2.3.9}$$

where $(\cdot,\omega_+)_+$ *denotes the function* $\omega''_+ \rightsquigarrow (\omega''_+,\omega_+)_+$.

SKETCH OF THE PROOF. (2.3.8) follows directly from (2.3.4). (2.3.9) follows directly from the identity:

$$((\cdot,\omega_+)e_{\omega'_+},(\cdot,\omega_+)e_{\omega'_+}) = (e_{\omega'_+},\partial(\omega_+)\{(\cdot,\omega_+)e_{\omega'_+}\})$$

and (2.3.4).

C. We will now discuss some applications of the principal vectors. The following is a typical property of Hilbert spaces of analytic functions.

THEOREM 2.3.4. *There exists a reproducing kernel for* $\mathcal{F}(W_J^+)$ *i.e. there exists a holomorphic function* $\mathcal{K}$ *on* $W_J^+ \times W_J^-$ *such that*

$$f(\omega_+) = \int \mathcal{K}(\omega_+,\bar{\omega}'_+)\, f(\omega'_+)\, d\mu(\omega'_+)$$

for $f \in \mathcal{F}(W_J^+)$.

PROOF. Just let

$$\mathcal{K}(\omega_+,\omega_-) =: (e_{\omega_+},e_{\bar{\omega}_-})_B = \exp[-i\langle\omega_+,\omega_-\rangle]\ .$$

Another application of principal vectors is to obtain point estimates for $f \in \mathcal{F}(W_J^+)$) and its derivatives in terms of $\|f\|_B$.

THEOREM 2.3.5. *Let* $f \in \mathfrak{F}(W_J^+)$ *and* $\omega_+ \in W_J^+$, *then*

$$|f(\omega_+')|^2 \leq e^{\|\omega_+'\|_+^2} \|f\|_B^2 \tag{2.3.10}$$

and

$$|\partial(\omega_+) f(\omega_+')|^2 \leq [|(\omega_+', \omega_+)_+|^2 + \|\omega_+\|_+^2] e^{\|\omega_+'\|^2} \|f\|_B^2 \,. \tag{2.3.11}$$

SKETCH OF THE PROOF. (2.3.10) follows from (2.3.4), Schwarz's inequality and (2.3.8).

To prove (2.3.11), assume first of all that $\partial(\omega_+) f \in \mathfrak{F}(W_J^+)$ and observe that

$$\partial(\omega_+) f(\omega_+') = (e_{\omega_+'}, \partial(\omega_+) f)_B = ((\cdot, \omega_+)_+ e_{\omega_+'}, f)_B \,.$$

Then apply Schwarz's inequality and (2.3.9) to obtain (2.3.11). For general f, let $f_\lambda(\omega_+') =: f(\lambda\omega_+')$, $\lambda \in (0,1)$. Then one can show $\partial(\omega_+) f_\lambda \in \mathfrak{F}(W_J^+)$ and (2.3.11) then holds for f_λ. Then let $\lambda \to 1$.

We are able to characterize weak convergence in $\mathfrak{F}(W_J^+)$ in a very simple way.

THEOREM 2.3.6. *A sequence* $\{f_n\}$ *in* $\mathfrak{F}(W_J^+)$ *converges weakly to* $f \in \mathfrak{F}(W_J^+)$ *iff* $\{f_n\}$ *converges to* f *pointwise and* $\{\|f_n\|_B\}$ *are bounded.*

SKETCH OF PROOF. If $\{f_n\}$ converges to f weakly, then automatically there exists $K > 0$ such that $\|f_n\|_B \leq K$ (by the uniform boundedness principle). But since $\{f_n\}$ converges weakly to f, we have by (2.3.8) that:

$$\lim_{n\to\infty} (e_{\omega_+}, f_n)_B = \lim f_n(\omega_+) = f(\omega_+)$$

for $\omega_+ \in W_J^+$ which proves the assertion in one direction.

Now suppose $f_n(\omega_+)$ converges to $f(\omega_+)$ for $\omega_+ \in W_J^+$. This means $(e_{\omega_+}, f_n)_B \to (e_{\omega_+}, f)_B$ for e_{ω_+}, $\omega_+ \in W_J^+$. But the $\{e_{\omega_+}\}$ are clearly a total subset in $\mathfrak{F}(W_J^+)$ and therefore since $\{\|f_n\|_B\}$ is bounded, we have $(g, f_n)_B \to (g, f)_B$ for all $g \in \mathfrak{F}(W_J^+)$ and the theorem is proved.

REMARK. By using Fatou's lemma, we see that if $f_n \to f$ pointwise and $\{\|f_n\|\}$ is bounded, then $f \in L^2(W_J^+, d\mu)$. By (2.3.10) we see that $f_n \to f$ normally and thus f is holomorphic and therefore it is in $F(W_J^+)$.

Finally if B is a bounded operator on $F(W_J^+)$, it will have a holomorphic kernel $B(\omega_+, \omega_-)$. In fact, let

$$B(\omega_+, \omega_-) =: (e_{\omega_+}, Be_{\overline{\omega}_-})_B .$$

Then

$$Bf(\omega_+) = \int B(\omega_+, \overline{\omega}'_+) f(\omega'_+) d\mu .$$

In particular the kernel of the identity I is just the reproducing kernel $\mathcal{K}(\omega_+, \omega_-)$.

D. We now use the principal vectors to prove some basic properties of the B.F.S. realization V.

THEOREM 2.3.7. *V is a continuous unitary representation.*

PROOF. We already know that

$$V(n_1 n_2) = V(n_1)V(n_2) ,$$

$n_1, n_2 \in N$ and that $V(n)$ is unitary for $n \in N$. It remains to show that V is continuous. Since $V(n)$ is unitary for $n \in N$, it is sufficient to show that $(e_{\omega_+}, V(n)e_{\omega_+})$ is continuous for each $\omega_+ \in W_J^+$. But

$$(e_{\omega_+}, V(e^{\omega'_+ + \overline{\omega}'_+ + Et})e_{\omega_+})$$

$$= e^{it} \exp[(\omega_+, \omega'_+)_+ - \tfrac{1}{2}(\omega'_+, \omega'_+) + i(\omega_+ - \omega'_+, \omega_+)_+]$$

which is clearly continuous in ω'_+.

THEOREM 2.3.8. *V is irreducible.*

SKETCH OF THE PROOF. To show that V is irreducible, it is sufficient to show that if B is a bounded operator on $\mathfrak{F}$ and $BV(n) = V(n)B$ for all $n \in N$, then $B = cI$. To see that this is the case, just observe that

$$e_{\omega_+} = \exp[\tfrac{1}{2}\|\omega_+\|^2]\, V(e^{\omega_+ + \bar{\omega}_+})\, e_0$$

Let $B(\omega_+,\omega_-)$ be the kernel of B, then

$$\begin{aligned} B(\omega_+,\bar{\omega}_+) &= e^{\|\omega_+\|^2}(V(e^{\omega_+ + \bar{\omega}_+})e_0, B\,V(e^{\omega_+ + \bar{\omega}_+})e_0)_B \\ &= e^{\|\omega_+\|^2}(V(e^{\omega_+ + \bar{\omega}_+})e_0, V(e^{\omega_+ + \bar{\omega}_+})B\,e_0)_B = e^{-i<\omega_+,\bar{\omega}_+>}<e_0, B\,e_0>. \end{aligned}$$

But B is uniquely determined on the real $2n$ dimensional subspace V of $V_C = W_J^+ \times W_J^-$ and thus $B(\omega_+,\omega_-) = c\,e^{-i<\omega_+,\omega_->} = c\mathcal{K}(\omega_+,\omega_-)$ which of course is the kernel of cI.

REMARK. There are of course other ways of proving irreducibility. However, Bargmann [6] uses this *technique* to elegantly compute the metaplectic representation of $Sp(n, \mathbf{R})$.

§2.4. *Characterization of the infinitesimal generator of* $V(e^{t(\omega_+ + \bar{\omega}_+)})$

A. The infinitesimal generator $L(\omega_+)$ of $V(e^{t(\omega_+ + \bar{\omega}_+)})$ is defined to be the operator with domain

$$\mathfrak{D}(L(\omega_+)) =: \left\{ f \in \mathfrak{F}(W_J^+) : \lim_{t \to 0} \left(\frac{V(e^{t(\omega_+ + \bar{\omega}_+)})f - f}{t} \right) \text{exists} \right\}$$

and for $f \in \mathfrak{D}(L(\omega_+))$,

$$L(\omega_+)f =: i\left(\lim_{t \to 0} \left[\frac{V(e^{t(\omega_+ + \bar{\omega}_+)})f - f}{t} \right] \right). \tag{2.4.1}$$

For $f \in \mathfrak{F}(W_J^+)$, define $\Lambda(\omega_+)f$ by:

$$\Lambda(\omega_+)f(\omega_+') =: (\omega_+', \omega_+)_+ f - \partial(\omega_+)f .$$

The following theorem will show that $L(\omega_+)$ has a very simple characterization.

THEOREM 2.4.1. $\mathfrak{D}(L(\omega_+)) = \{f \in \mathcal{F}(W_J^+) : \Lambda(\omega_+) f \in \mathcal{F}(W_J^+)\}$ *and*

$$L(\omega_+)f = i(\Lambda(\omega_+))f \; . \tag{2.4.2}$$

PROOF. If $f \in \mathcal{F}(W_J^+)$, write:

$$\begin{aligned} f(\omega_+', t) &=: V(e^{t(\omega_+ + \bar{\omega}_+)}) f(\omega_+') \\ &= \exp[t(\omega_+', \omega_+)_+ - \tfrac{1}{2} t^2 \|\omega_+\|_+^2] f(\omega_+' - t\omega_+) \; . \end{aligned}$$

Thus if the limit (2.4.1) exists (in $\mathcal{F}(W_J^+)$), it also exists pointwise and we see that

$$L(\omega_+) f(\omega_+') = i \frac{\partial}{\partial t} f(\omega_+', t)\big|_{t=0} = i\Lambda(\omega_+) f(\omega_+')$$

and thus $\mathfrak{D}(L(\omega_+)) \subset \{f \in \mathcal{F}(W_J^+) : \Lambda(\omega_+) f \in \mathcal{F}(W_J^+)\}$ and

$$L(\omega_+)f = i\Lambda(\omega_+)f \; .$$

Now suppose $f \in \mathcal{F}(W_J^+)$ and $\Lambda(\omega_+) f \in \mathcal{F}(W_J^+)$. Then:

$$\frac{\partial f}{\partial t}(\omega_+', t) = -i(V(e^{t(\omega_+ + \bar{\omega}_+)}) \Lambda(\omega_+) f$$

both pointwise and as an equation in $F(W_J^+)$. This can be integrated to read:

$$i\left(\frac{V(e^{t(\omega_+ + \bar{\omega}_+)}) f - f}{t}\right) = t^{-1} \int_0^t V(e^{\tau(\omega_+ + \omega_+)}) \Lambda(\omega_+) f \, d\tau \; .$$

But because of the strong continuity of V, the limit (in $\mathcal{F}(W_J^+)$) on the right hand side exists as $t \to 0$ and equals $\Lambda(\omega_+) f$ and thus the same statement applies to the left hand side which is what we wanted to prove.

B. Let us now choose a standard basis $\{\underset{\sim}{e},\underset{\sim}{f}\}$ for $\{V,J\}$ and $\{\underset{\sim}{a}^+,\underset{\sim}{a}^-\}$ the corresponding bases for W_J^+ and W_J^- respectively. We shall now compute P_j, Q_j, A_j^*, and A_j for $j=1,\cdots,n$ which were formally defined in §1.2E.

Applying Theorem 2.4.1, we have: P_j, *the infinitesimal generator of* $V(e^{te_j}) = V(e^{(t(a_j^+ + a_j^-))/\sqrt{2}})$ is:

$$P_j = \frac{1}{\sqrt{2}}\{-i<\cdot,a_j^-> - \partial(a_j^+)\} \; ; \tag{2.4.3}$$

Q_j, the infinitesimal generator of $V(e^{tf_j}) = V(e^{t\{(ia_j^+ + i\overline{a_j^-})/\sqrt{2}\}})$ is:

$$Q_j = \frac{i}{\sqrt{2}}\{-i<\cdot,a_j^-> - i\partial(a_j^+)\} \; ; \tag{2.4.4}$$

$$A_j^* =: \frac{Q_j - iP_j}{\sqrt{2}} = -i<\cdot,a_j> \tag{2.4.5}$$

$$A_j =: \frac{Q_j + iP_j}{\sqrt{2}} = \partial(a_j^+) \tag{2.4.6}$$

with $\mathfrak{D}(A_j^*) = \mathfrak{D}(A_j) = \mathfrak{D}(P_j) \cap \mathfrak{D}(Q_j)$, $j=1,\cdots,n$.

If we identify $\mathfrak{F}(W_J^+)$ with $\mathfrak{F}_n$ via $\underset{\sim}{a}^+$, then the above can be written in the familiar form:

$$P_j f = \frac{i}{\sqrt{2}}\left\{z_j f - \frac{\partial f}{\partial z_j}\right\}, \tag{2.4.3$'$}$$

$$Q_j f = \frac{1}{\sqrt{2}}\left\{z_j f + \frac{\partial f}{\partial z_j}\right\} \tag{2.4.4$'$}$$

$$A_j^* f = z_j f \tag{2.4.5$'$}$$

$$A_j f = \frac{\partial f}{\partial z_j}, \tag{2.4.6$'$}$$

$j=1,\cdots,n$.

REMARK. Note the extremely simple nature of A_j and A_j^*. In particular the creation operator A_j^* is diagonalized. *It is this feature of the* B.F.S. realization which makes it so appealing at least when one wants to work with the Boson creation and annihilation operators.

We now wish to give a more convenient characterization of $\mathfrak{D}(A_j)$ and $\mathfrak{D}(A_j^*)$.

THEOREM 2.4.2. $\mathfrak{D}(A_j^*) = \{f \in \mathfrak{F}(W_J^+) : -i<\cdot, a_j^-> f \in \mathfrak{F}(W_J^+)\}$ *and*

$$\mathfrak{D}(A_j) = \{f \in \mathfrak{F}(W_J^+) : \partial(a_j^+) f \in \mathfrak{F}(W_J^+)\} ,$$

$j = 1, 2, \cdots, n.$

SKETCH OF THE PROOF. For $f \in \mathfrak{F}(W_J^+)$, let $\hat{A}_j^* f = : -i<\cdot, a_j^-> f$ and $\hat{A}_j f = : \partial(a_j^+) f$. It is clear that the theorem will follow if we can show:

$$\|\hat{A}_j^* f\|_B^2 = \|\hat{A}_j f\|_B^2 + \|f\|_B^2 . \tag{2.4.7}$$

In fact, it follows directly from the definition of Q_j and P_j that $\mathfrak{D}(P_j) \cap \mathfrak{D}(Q_j) = \{f \in \mathfrak{F}(W_J^+) : \hat{A}_j f \in \mathfrak{F}(W_J^+)\} \cap \{f \in \mathfrak{F}(W_J^+) : \hat{A}_j^* f \in \mathfrak{F}(W_J^+)\}$. But (2.4.7) implies that $\{f \in \mathfrak{F}(W_J^+) : \hat{A}_j f \in \mathfrak{F}(W_J^+)\} \cap \{f \in \mathfrak{F}(W_J^+) : \hat{A}_j^* f \in \mathfrak{F}(W_J^+)\} = \{f \in \mathfrak{F}(W_J^+) : \hat{A}_j f \in \mathfrak{F}(W_J^+)\}$ $= \{f \in \mathfrak{F}(W_J^+) : \hat{A}_j^* f \in \mathfrak{F}(W_J^+)\}$ which of course proves the theorem.

To see (2.4.7), expand f in terms of the $u_{\underset{\sim}{m}}$ (see §2.3 A)) i.e. $f = \sum_{\underset{\sim}{m}} \alpha_{\underset{\sim}{m}} u_{\underset{\sim}{m}}$. Then $A_j f = \sum_{\underset{\sim}{m}} m_j \alpha_{\underset{\sim}{m}} z^{\underset{\sim}{m} - \varepsilon_j}$ and $A_j^* f = \sum_{\underset{\sim}{m}} \alpha_{\underset{\sim}{m}} z^{\underset{\sim}{m} + \varepsilon_j}$ where $\varepsilon_j = \varepsilon_j = (\delta_{1j}, \cdots, \delta_{nj}) \in N^n$. Let $f_N = \sum_{|\underset{\sim}{m}| < N} \alpha_{\underset{\sim}{m}} u_{\underset{\sim}{m}}$, then a direct calculation shows:

$$\|A_j^* f_N\|_B^2 = \|A_j f_N\|_B^2 + \|f_N\|_B^2 .$$

Then let $N \to \infty$ to obtain (2.4.7). Note in (2.4.7) we allow infinity as a possible value.

REMARK. It follows directly from the definition of A_j and A_j^* that they are adjoints to each other on $\mathcal{S}(\mathfrak{F}(W_J^+))$. It is a fact that they are adjoints in the Hilbert space sense. See [2], p. 210.

§2.5. *The characterization of* $\mathcal{S}(\mathcal{F}(W_J^+))$ *and* $\mathcal{S}'(\mathcal{F}(W_J^+))$

A. Let $k \in Z$ and let $\mathcal{F}^k(W_J^+) = \{f \in \mathcal{B}(W_J^+) : \int |f|^2(1+\|\omega_+\|_+^2)^k d\mu < \infty\}$ and define an inner product on $\mathcal{F}^k(W_J^+)$ by:

$$(f_1, f_2)_{B,k} =: \int \overline{f_1}\, f_2 (1+\|\omega_+\|_+^2)^k d\mu \ .$$

THEOREM 2.5.1. $\mathcal{F}^k(W_J^+)$ *is a Hilbert space of analytic functions.*

SKETCH OF THE PROOF. We just need to show that $\mathcal{F}^k(W_J^+)$ is complete. Let V_R be the volume of the ball in W_J^+ with radius R and let

$$\mu_R = \pi^{-n/2} \min_{\|\omega_+\|_+ \leq R} (1+\|\omega_+\|_+^2)^{k/2} e^{-\frac{1}{2}\|\omega_+\|_+^2} \ .$$

Then a direct calculation shows:

$$|f(\omega_+)| \leq \frac{\|f\|_{B,k}}{\mu_{2R}} \sqrt{V_R} \tag{2.5.1}$$

for $\|\omega_+\| \leq R$. Then if $\{f_n\}$ is a Cauchy sequence in $\mathcal{F}^k(W_J^+)$, it converges to an L^2 function f since $L^2(W_J^+, (1+\|\omega_+\|_+^2)^k d\mu) (\supset \mathcal{F}^k(W_J^+))$ is complete. But (2.5.1) implies $\{f_n\}$ converges to f uniformly on compact subsets of W_J^+ and therefore f is holomorphic and hence in $\mathcal{F}^k(W_J^+)$.

THEOREM 2.5.2. $\mathcal{S}(\mathcal{F}(W_J^+)) \subset \mathcal{F}^k(W_J^+)$, $k \in Z$.

PROOF. It is sufficient to show this for $k \in N$. Let $\{\underset{\sim}{a}^+, \underset{\sim}{a}^-\}$ be standard bases for W_J^+ and W_J^- and let A_j^* and A_j, $j = 1, \cdots, n$ be the corresponding creation and annihilation operator. Then identifying W_J^+ with C^n, we see that $\mathcal{F}^k(W_J^+)$ is identified with $\mathcal{F}_n^k = \{f \in \mathcal{B}(C^n) : |f|^2(1+|z|^2)^k d\mu < \infty\}$. If $f \in \mathcal{S}(\mathcal{F}_n)$ and $\underset{\sim}{\ell} \in N^n$, then

$$(f, A^{\underset{\sim}{\ell}} A^{*\underset{\sim}{\ell}} f) = \int |f|^2 |z_1|^{2\ell_1} \cdots |z_n|^{2\ell_n} d\mu \ . \tag{2.5.2}$$

In fact this follows from (2.4.5)′ and the fact that A_j^* and A_j are adjoints on $\mathcal{S}(\mathcal{F}_n)$. But from (2.5.2) we see that for $f \in \mathcal{S}(\mathcal{F}_n)$,

$$(2.5.3) \qquad (f, K_k f) = \int |f|^2 (1+|z|^2)^k d\mu = \|f\|_{B,k}^2$$

where K_k is defined in §1.3B and thus the right hand side is finite i.e. $f \in \mathcal{F}_n^k$.

REMARK. We thus see that the norms $\| \ \|_{B,k}$ introduced in §1.3B and the norms $\| \ \|_{B,k}$ introduced above are the same.

THEOREM 2.5.3. *$\mathcal{S}(\mathcal{F}(W_J^+))$ is dense in $\mathcal{F}^k(W_J^+)$ for $k \in N$.*

SKETCH OF THE PROOF. We again choose a standard bases $\{\underset{\sim}{a}^+, \underset{\sim}{a}^-\}$ and let A_j^* and A_j, $j = 1, \cdots, n$ be the corresponding creation and annihilation operators. Let $\{u_{\underset{\sim}{m}} : \underset{\sim}{m} \in N^n\}$ be as in §2.3A and K_k as in §1.3B (see in particular (1.3.1, 1.3.2). $u_{\underset{\sim}{m}'} \in \mathcal{S}(\mathcal{F}(W_J^+))$, for $\underset{\sim}{m}' \in N^n$, because if $g \in \mathcal{G}(\mathcal{F}(W_J^+))$, we have: $|(\underset{\sim}{P}^{\underset{\sim}{m}} \underset{\sim}{Q}^{\underset{\sim}{\ell}} g, u_{\underset{\sim}{m}'})| \le \|q\|_B \|\underset{\sim}{Q}^{\underset{\sim}{\ell}} \underset{\sim}{P}^{\underset{\sim}{m}} u_{\underset{\sim}{m}'}\|_B$ and then $u_{\underset{\sim}{m}'} \in$ Domain $([\underset{\sim}{P}^{\underset{\sim}{m}} \underset{\sim}{Q}^{\underset{\sim}{\ell}}]^*)$, for all $\underset{\sim}{m}, \underset{\sim}{\ell} \in N^n$.

Note that:

$$K_k u_{\underset{\sim}{m}} = \sum_{|\underset{\sim}{\ell}| \le k} a_{\underset{\sim}{\ell}}^k (m_1 + \ell_1) \cdots (m_1 + 1) \cdots (m_n + 1) u_{\underset{\sim}{m}} .$$

Let $\eta_{\underset{\sim}{m}}^k = : \sum_{|\underset{\sim}{\ell}| \le k} a_{\underset{\sim}{\ell}}^k (m_1 + \ell_1) \cdots (m_n + 1)$ and note that $\eta_{\underset{\sim}{m}}^k > 0$. Let $u_{\underset{\sim}{m}}^k = : (\eta_{\underset{\sim}{m}}^k)^{-\frac{1}{2}} u_{\underset{\sim}{m}}$. We will show that $\{u_{\underset{\sim}{m}}^k\}$ is a complete orthonormal system for $\mathcal{F}^k(W_J^+)$ and since $u_{\underset{\sim}{m}}^k \in \mathcal{S}(\mathcal{F}(W_J^+))$ for $\underset{\sim}{m} \in N^n$, this will imply that $\mathcal{S}(\mathcal{F}(W_J^+))$ is dense in $\mathcal{F}^k(W_J^+)$.

To see that the $\{u_{\underset{\sim}{m}}^k\}$ are orthonormal, just observe that:

$$\begin{aligned}(u_{\underset{\sim}{m}}^k, u_{\underset{\sim}{m}'}^k)_{B,k} &= (u_{\underset{\sim}{m}}^k, K_k u_{\underset{\sim}{m}'}^k)_B \\ &= \eta_{\underset{\sim}{m}'}^k (\eta_{\underset{\sim}{m}}^k)^{-\frac{1}{2}} (\eta_{\underset{\sim}{m}'}^k)^{-\frac{1}{2}} (u_{\underset{\sim}{m}}, u_{\underset{\sim}{m}'})_B = \delta_{\underset{\sim}{m}, \underset{\sim}{m}'} .\end{aligned}$$

To see that $\{u_{\underset{\sim}{m}}^k\}$ is total we need to show that $(f, u_{\underset{\sim}{m}}^k)_{B,k} = 0$, for $\underset{\sim}{m} \in N^n$ implies that $f \equiv 0$. But since $\mathcal{F}^k(W_J^+) \subset \mathcal{F}(W_J^+)$, we can write $f = \sum_{\underset{\sim}{m}} \alpha_{\underset{\sim}{m}} u_{\underset{\sim}{m}}$ where convergence is both in the L^2 and uniformly on compact subsets of W_J^+. Therefore identifying $\mathcal{F}(W_J^+)$ with $\mathcal{F}_n$, we have:

$$\int_{|z_1|\le R} \cdots \int_{|z_n|\le R} u_{\underset{\sim}{m}}^k(z)(1+|z|^2)^k d\mu$$

$$= \alpha_{\underset{\sim}{m}}(\eta_{\underset{\sim}{m}}^k)^{\frac{1}{2}} \int_{|z_1|\le R} \cdots \int_{|z_n|\le R} |u_{\underset{\sim}{m}}^k|^2 (1+|z|^2)^k d\mu$$

where we have used the fact that

$$\int_{|z_1|\le R} \cdots \int_{|z_n|\le R} u_{\underset{\sim}{m}'}(1+|z|^2)\, d\mu = 0$$

if $\underset{\sim}{m} \neq \underset{\sim}{m}'$. Letting $R \to \infty$, we obtain

$$(f, u_{\underset{\sim}{m}}^k)_{B,k} = \alpha_{\underset{\sim}{m}}(\eta_{\underset{\sim}{m}}^k)^{\frac{1}{2}} = 0$$

and therefore $\alpha_{\underset{\sim}{m}} = 0$ for $\underset{\sim}{m} \in N^n$ i.e. $f \equiv 0$.

REMARK. $\mathcal{F}^k(W_J^+)$ has principal vectors and a reproducing kernel just as $\mathcal{F}(W_J^+)$ does. In fact, if we let $\tilde{K}_k$ be the Friedrichs' extension of K_k, then $e_{\omega_+}^k = \tilde{K}_k^{-1} e_{\omega_+}$, $\omega_+ \in W_J^+$ can be shown to be the family of principal vectors for $\mathcal{F}^k(W_J^+)$. The corresponding reproducing kernel is just

$$\mathcal{K}^k(\omega_+, \omega_-) = : e_{\overline{\omega}_-}^k(\omega_+) \,.$$

B. We can now characterize $\mathcal{S}(\mathcal{F}(W_J^+))$.

THEOREM 2.5.4. *$\mathcal{S}(\mathcal{F}(W_J^+)) = \bigcap_{k=0}^{\infty} \mathcal{F}^k(W_J^+)$ and $\{f_n\}$ in $\mathcal{S}(\mathcal{F}(W_J^+))$ converges to 0 iff $\{\|f_n\|_{B,k}\}$ converges to 0 for $k \in N$.*

PROOF. It follows from Theorem 2.5.2 that $\mathcal{S}(\mathcal{F}(W_J^+)) \subset \bigcap_{k=0}^{\infty} \mathcal{F}^k(W_J^+)$. Thus suppose $f_0 \in \bigcap_{k=0}^{\infty} \mathcal{F}^k(W_J^+)$. Let $\underset{\sim}{\ell}', \underset{\sim}{m}' \in N^n$. We must show that $f_0 \in \mathrm{Dom}(\underset{\sim}{Q}^{\ell'} \underset{\sim}{P}^{m'})$. But from (1.3.10) we know there exist $k \in N$, $C_k > 0$ such that

$$\|\underset{\sim}{Q}^{\ell'} \underset{\sim}{P}^{m'} f\|_B \leq C_k \|f\|_{B,k} . \tag{2.5.4}$$

Since $\mathcal{S}(\mathcal{F}(W_J^+))$ is dense in $\mathcal{F}^k(W_J^+)$, there exists a sequence $\{f_n\}$ in $\mathcal{S}(\mathcal{F}(W_J^+))$ such that $\|f_n - f\|_{B,k} \to 0$ as $n \to \infty$. It follows automatically that $\|f_n - f\|_B \to 0$.

If $g \in \mathcal{G}(\mathcal{F}(W_J^+))$, we have; using (2.5.4) that:

$$|(\underset{\sim}{P}^{m'} \underset{\sim}{Q}^{\ell'} g, f_n)_B| = |(g, \underset{\sim}{Q}^{\ell'} \underset{\sim}{P}^{m'} f_n)_B| \leq C_k \|g\|_B \|f_n\|_{B,k} .$$

Thus letting $n \to \infty$, we have:

$$|(\underset{\sim}{P}^{m'} \underset{\sim}{Q}^{\ell'} q, f)| \leq C_k \|g\|_B \|f\|_{B,k}$$

and $f \in \mathrm{Dom}(\underset{\sim}{Q}^{\ell'} \underset{\sim}{P}^{m'})$. Since $\underset{\sim}{\ell}', \underset{\sim}{m}'$ were arbitrary, $f \in \mathcal{S}(\mathcal{F}(W_J^+))$.

It follows from Theorem 1.3.2 and (2.5.3) that convergence is defined by the norms $\{\| \ \|_{B,k}, k \in N\}$.

C. We define a duality between $\mathcal{F}^k(W_J^+)$ and $\mathcal{F}^k(W_J^+)$, $k \in N$, as follows:

$$(f, \chi)_B =: \int \bar{f} \chi \, d\mu .$$

Thus if L is continuous anti-linear functional on $\mathcal{F}^k(W_J^+)$, there exists a unique $\chi \in \mathcal{F}^{-k}(W_J^+)$ such that $L(f) = (f, \chi)_B$.

THEOREM 2.5.5. $\mathcal{S}'(\mathcal{F}(W_J^+)) \cong \bigcup_{k=1}^{\infty} \mathcal{F}^{-k}(W_J^+)$.

PROOF. This follows directly from the fact that if L is a continuous anti-linear functional on $\mathcal{S}(\mathcal{F}(W_J^+))$, then there exists $k \in N$, $C_k > 0$ such that

$$|L(f)| \leq C_k \|f\|_{B,k}$$

for $f \in \mathcal{S}(\mathcal{F}(W_J^+))$, a *dense* subspace of $\mathcal{F}^k(W_J^+)$. Thus there exists $\chi \in \mathcal{F}^{-k}(W_J^+)$ such that

$$L(f) = (f, \chi)_B \ .$$

CHAPTER III
APPLICATIONS

§3.1. *The symmetries of the 3-j symbols (after Bargmann)*

The purpose of this section is to show how $\mathcal{F}_n \equiv \mathcal{F}(C^n)$ can be used to give an elegant derivation of the full symmetry group (including the Regge symmetries) for Wigner 3-j symbols. It is important in atomic and nuclear physics to be able to explicitly compute the 3j symbols and thus the more symmetries one has the fewer the computations one has to make. We will follow the treatment, including notation, given in [3].

SU(n) will denote the Lie group of $n \times n$ unitary matrices U such that $\det U = 1$. It has a natural action on $\mathcal{F}_n$. In fact, if $U \in SU(n)$ and $f \in \mathcal{F}_n$, we define:

$$T_U f(z) = : f(U^{-1}z) \ .$$

The measure $d\mu = \pi^{-n} \exp[-|z|^2]\,dx\,dy$ is obviously left invariant by U and thus T_U, $U \in SU(n)$, is a unitary representation of SU(n).

We now assume $n = 2$. We denote $z \in C^2$ by $z = (\xi, \eta)$. $\mathcal{F}_2$ decomposes into the Hilbert space direct sum $\sum_{2j=0}^{\infty} \oplus O_j$ where O_j is the space of homogeneous polynomials of degree 2j, $2j = 0, 1, 2, 3, \cdots$. It is well known that $T_U|_{O_j} = : \mathfrak{D}^j$ is an irreducible representation of SU(2), $j = 0, \frac{1}{2}, 1, \cdots$. An orthonormal basis for O_j is given by the $2j+1$ polynomials $V_m^j(\xi, \eta) = : (\xi^{j+m}\eta^{j-m})/[(j+m)!(j-m)!]^{\frac{1}{2}}$, $m = -j, \cdots, j$. Note that $V_m^j = u_{(j+m, j-m)}$ in the notation of §2.3 A.

Let $O_{j_1, j_2, j_3} = : O_{j_1} \otimes O_{j_2} \otimes O_{j_3}$ be the $(2j_1+1)(2j_2+1)(2j_3+1)$ dimensional subspace of $\mathcal{F}_6$ spanned by $\{V_{m_1}^{j_1} V_{m_2}^{j_2} V_{m_3}^{j_3}\}$. We then have a natural representation $\mathfrak{D}^{j_1} \otimes \mathfrak{D}^{j_2} \otimes \mathfrak{D}^{j_3}$ of SU(2) on O_{j_1, j_2, j_3} defined by:

$$\mathfrak{D}^{j_1} \otimes \mathfrak{D}^{j_2} \otimes \mathfrak{D}^{j_3}(U|f(z_1,z_2,z_3) = : f(U^{-1}z_1, U^{-1}z_2, U^{-1}z_3) \ .$$

One is interested in the case where O_{j_1,j_2,j_3} has a fixed vector under $\mathfrak{D}^{j_1} \otimes \mathfrak{D}^{j_2} \otimes \mathfrak{D}^{j_3}$.

FACT. There exists a unique one-dimensional subspace R of O_{j_1,j_2,j_3} such that $\mathfrak{D}^{j_1} \otimes \mathfrak{D}^{j_2} \otimes \mathfrak{D}^{j_3}|_R = I$ if $j_3 = j_1 + j_2 - k_3$, $k_3 \in N$ and $j_3 \geq |j_1 - j_2|$. From now on we assume that this is the case. Note that R is generated by:

$$F_{\underset{\sim}{k}} = \frac{\underset{\sim}{\delta}^{\underset{\sim}{k}}}{\underset{\sim}{k}!} \tag{3.1.1}$$

where $\underset{\sim}{\delta} = (\delta_1, \delta_2, \delta_3)$, $\delta_1 = \xi_2\eta_3 - \xi_3\eta_2$, $\delta_2 = \xi_3\eta_1 - \xi_1\eta_3$, $\delta_3 = \xi_1\eta_2 - \xi_2\eta_1$, $\underset{\sim}{k} = (k_1, k_2, k_3)$, $k_\alpha = J - 2j_\alpha$, $\alpha = 1,2,3$ and $J = j_1 + j_2 + j_3$. It can be shown ([3], pp. 839-840) that:

$$\|F_{\underset{\sim}{k}}\|_B^2 = \frac{(J+1)!}{\underset{\sim}{k}!}$$

and thus $H_{\underset{\sim}{k}} = :\Delta(j_1, j_2, j_3)F_{\underset{\sim}{k}}$ with $\Delta(j_1, j_2, j_3) = :[(\underset{\sim}{k}!)/(J+1)!]^{\frac{1}{2}}$, is a normalized generator of R.

The 3-j symbols are now defined as follows: expand $H_{\underset{\sim}{k}}$ in terms of the orthonormal basis $\{V_{m_1}^{j_1} V_{m_2}^{j_2} V_{m_3}^{j_3}\}$

$$H_{\underset{\sim}{k}} = \sum_{m_1,m_2,m_3} \begin{pmatrix} m_1 & m_2 & m_3 \\ j_1 & j_2 & j_3 \end{pmatrix} V_{m_1}^{j_1} V_{m_2}^{j_2} V_{m_3}^{j_3} \ .$$

The numbers $\left\{\begin{pmatrix} m_1 & m_2 & m_3 \\ j_1 & j_2 & j_3 \end{pmatrix}\right\}$ are the 3-j symbols of Wigner. (In many other references they are written $\begin{pmatrix} j_1 & j_2 & j_3 \\ m_1 & m_2 & m_3 \end{pmatrix}$.) Observe that $\begin{pmatrix} m_1 & m_2 & m_j \\ j_1 & j_2 & j_3 \end{pmatrix}$ $= 0$ if $m_1 + m_2 + m_3 \neq 0$. Thus we only consider 3j symbols such that $m_1 + m_2 + m_3 = 0$.

We now wish to find explicit expressions for the 3j symbols. Write:

$$F_{\underset{\sim}{k}} = \sum_{\underset{\sim}{\kappa},\underset{\sim}{\lambda}} f_{\underset{\sim}{k}\underset{\sim}{\kappa}\underset{\sim}{\lambda}} \, \underset{\sim}{\xi}^{\underset{\sim}{\kappa}} \underset{\sim}{\eta}^{\underset{\sim}{\lambda}} \tag{3.1.2}$$

and

$$H_{\underset{\sim}{k}} = \sum_{\underset{\sim}{\kappa},\underset{\sim}{\lambda}} h_{\underset{\sim}{k}\underset{\sim}{\kappa}\underset{\sim}{\lambda}} \frac{\underset{\sim}{\xi}^{\underset{\sim}{\kappa}} \underset{\sim}{\eta}^{\underset{\sim}{\lambda}}}{[\underset{\sim}{\kappa}!\,\underset{\sim}{\lambda}!]^{\frac{1}{2}}}$$

where $\underset{\sim}{\xi} = (\xi_1, \xi_2, \xi_3)$, $\underset{\sim}{\eta} = (\eta_1, \eta_2, \eta_3)$, and

$$\kappa_\alpha =: j_\alpha + m_\alpha\,, \qquad \lambda_\alpha =: j_\alpha - m_\alpha\,, \qquad \alpha = 1, 2, 3\ .$$

Note that:

$$h_{\underset{\sim}{k}\underset{\sim}{\kappa}\underset{\sim}{\lambda}} = \left[\frac{\underset{\sim}{k}!\ \underset{\sim}{\kappa}!\ \underset{\sim}{\lambda}!}{(J+1)!}\right]^{\frac{1}{2}} f_{\underset{\sim}{k}\underset{\sim}{\kappa}\underset{\sim}{\lambda}} \tag{3.1.3}$$

and

$$h_{\underset{\sim}{k}\underset{\sim}{\kappa}\underset{\sim}{\lambda}} = \begin{pmatrix} m_1 & m_2 & m_3 \\ j_1 & j_2 & j_3 \end{pmatrix} . \tag{3.1.4}$$

Let $L = (\ell_{i\alpha})$ be the 3×3 matrix with $\underset{\sim}{k}$, $\underset{\sim}{\kappa}$, and $\underset{\sim}{\lambda}$ as its rows and define $h_L =: h_{\underset{\sim}{k}\underset{\sim}{\kappa}\underset{\sim}{\lambda}}$ and $f_L =: f_{\underset{\sim}{k}\underset{\sim}{\kappa}\underset{\sim}{\lambda}}$.

The important point is that there is a 1:1 correspondence between the 3j symbols $\begin{pmatrix} m_1 & m_2 & m_3 \\ j_1 & j_2 & j_3 \end{pmatrix}$ and the non-negative integral matrices $L = \begin{pmatrix} \underset{\sim}{k} \\ \underset{\sim}{\kappa} \\ \underset{\sim}{\lambda} \end{pmatrix}$ such that the rows and columns add up to a fixed integer J. The correspondence is defined by:

$$k_\alpha = J - 2j_\alpha$$

$$\kappa_\alpha = j_\alpha + m_\alpha$$

$$\lambda_\alpha = j_\alpha - m_\alpha\ .$$

Permutations of rows and columns of L will correspond to permutations of the corresponding 3j symbols. For example if you interchange two columns of L, you permute the corresponding (j_α, m_α)'s. If you interchange the second and third rows of L, you change the sign of all of the m_α's.

We can now compute the f_L and hence the 3j symbols via (3.1.3) and (3.1.4).

THEOREM 3.1.1. *The following identity holds*:

$$f_L = \sum_{\{(\underset{\sim}{q},\underset{\sim}{p}) : Q(\underset{\sim}{q},\underset{\sim}{p}) = L\}} (-1)^{|\underset{\sim}{q}|} (\underset{\sim}{p}!\,\underset{\sim}{q}!)^{-1} \tag{3.1.5}$$

where $\underset{\sim}{q}$ *and* $\underset{\sim}{p}$ *are triples of non-negative integers and*

$$Q(\underset{\sim}{q},\underset{\sim}{p}) =: \begin{pmatrix} q_1 + p_1 & q_2 + p_2 & q_3 + p_3 \\ q_2 + p_3 & q_3 + p_1 & q_1 + p_2 \\ q_3 + p_2 & q_1 + p_3 & q_2 + p_1 \end{pmatrix}.$$

SKETCH OF THE PROOF. Using the binomial theorem we have:

$$\frac{\delta_1^{k_1}}{k_1!} = \sum_{p_1+q_1=k_1} \frac{(\xi_2\eta_3)^{p_1}(-\xi_3\eta_2)^{q_1}}{p_1!\,q_1!}$$

$$\frac{\delta_2^{k_2}}{k_2!} = \sum_{p_2+q_2=k_2} \frac{(\xi_3\eta_1)^{p_2}(-\xi_1\eta_3)^{q_2}}{p_2!\,q_2!}$$

$$\frac{\delta_3^{k_3}}{k_3!} = \sum_{p_3+q_3=k_3} \frac{(\xi_1\eta_2)^{p_3}(-\xi_2\eta_1)^{q_3}}{p_3!\,q_3!}.$$

Using the definition of $F_{\underset{\sim}{k}}$ (3.1.1), the expression (3.1.2) for f_L and matching coefficients, we obtain (3.1.5).

To compute the symmetries of the 3-j symbols, we introduce the generating function for the f_L which is defined as follows:

$$\Phi(\tau,\xi,\eta) =: \sum_{L} f_L\, \tau^{\underset{\sim}{k}} \xi^{\underset{\sim}{\kappa}} \eta^{\underset{\sim}{\lambda}} .$$

It follows from the definition of δ(s, the $F_{\underset{\sim}{k}}$'s and the f_L's that:

$$\Phi(\tau,\xi,\eta) = \exp(\underset{\sim}{\tau}\cdot\delta) = \exp(\det\Theta)$$

where

$$\Theta =: \begin{pmatrix} \tau_1 & \tau_2 & \tau_3 \\ \xi_1 & \xi_2 & \xi_3 \\ \eta_1 & \eta_2 & \eta_3 \end{pmatrix}.$$

We can now state the main theorem of this section.

THEOREM 3.1.2. (Regge's symmetry theorem).

1. *If* P *is an even permutation of rows or of columns or the interchange of rows and columns (of a* 3×3 *matrix), then*

$$f_{P(L)} = f_L \quad \textit{and} \quad h_{P(L)} = h_L .$$

2. *If* P *is an odd permutation of rows or columns, then*

$$f_{P(L)} = (-1)^J f_L \quad \textit{and} \quad h_{P(L)} = (-1)^J h_L .$$

PROOF. If P is a permutation operation on 3×3 matrices, then:

$$\Phi(P(\Theta)) = \sum f_{P(L)} \tau^{\underset{\sim}{k}} \xi^{\underset{\sim}{\kappa}} \eta^{\underset{\sim}{\lambda}} = \exp(P(\Theta)) .$$

Thus if $\det P(\Theta) = \det(\Theta)$ i.e. if 1) holds, then $\Phi(P(\Theta)) = \Phi(\Theta)$ and

$$f_{P(L)} = f_L$$

and thus by (3.1.3),

$$h_{P(L)} = h_L \ .$$

Let $\Phi'(\Theta) = : \exp(-\det(\Theta))$. Then $\Phi'((\Theta)) = \Phi(-\tau, \xi, \eta) = \sum (-1)^J f_L \tau^{\underset{\sim}{k}} \xi^{\underset{\sim}{\kappa}} \eta^{\underset{\sim}{\lambda}}$ since $k_1 + k_2 + k_3 = J$. If $\det P(\Theta) = -\det \Theta$ i.e. if 2.1 holds, we have:

$$f_{P(L)} = (-1)^J f_L$$

and thus

$$h_{P(L)} = (-1)^J h_L \ .$$

REMARK. Similar, but necessarily more complicated, considerations lead to the full symmetry group for the 6-j symbols. See [3] for this treatment.

§3.2. *Applications of the* B.F.S. *realization of* N(V) *to the Fourier transforms (after Bargmann)*

A. Let $(V, \langle\, , \rangle, J)$ be given and let Q be as in Section 2.1 B. Denote $\mathcal{S}(L^2(Q))$ by $\mathcal{S}(Q)$. This is justified because it is easy to show that $\mathcal{S}(L^2(Q))$ is indeed Schwartz space over Q. In fact, just use the semi-norms $\| \ \|_{\underset{\sim}{\ell}, \underset{\sim}{m}}$ of Section 1.3 A. Let A be the unitary intertwining operator between the Schrödinger realization $(V, \mathcal{F}(W_J^+))$. (See Section 2.2A for its definition.) Because A is an intertwining operator between $(T, L^2(Q))$ and $(V, \mathcal{F}(W_J^+))$, it *necessarily* follows from the abstract definition of $\mathcal{S}(\mathcal{H}_\pi)$ that A maps $\mathcal{S}(Q)$ isomorphically in the sense of topological vector spaces onto $\mathcal{S}(\mathcal{F}(W_J^+))$. A powerful technique for studying many operations on or properties of $\mathcal{S}(Q)$, $L^2(Q)$ or $\mathcal{S}'(Q)$ is to transform them by A to $\mathcal{S}(\mathcal{F}(W_J^+))$, study them there where they are more transparent, obtain the desired results, and then transform the results back to $\mathcal{S}(Q)$, $L^2(Q)$ or $\mathcal{S}'(Q)$ by A^{-1}. We shall illustrate this technique by applying it to the Fourier transform on $\mathcal{S}(Q)$ and $L^2(Q)$.

REMARK. One can also realize A_j and A_j^*, $j = 1, \cdots, n$ (and hence N) in $\ell_2(N^n)$. It turns out that $\mathcal{S}(N^n) = : \mathcal{S}(\ell_2(N^n))$ is the space of "rapidly decreasing sequences." Using the Hermite expansion of functions in $L^2(R^n)$ one can explicitly write down the intertwining operator between the Schrödinger realization and the above realization. One can obtain simple proofs of many properties of $\mathcal{S}(R^n)$ and $\mathcal{S}'(R^n)$ by first proving them for $\mathcal{S}(N^n)$ and $\mathcal{S}'(N^n)$ and then "transporting" these results back to $\mathcal{S}(R^n)$ and $\mathcal{S}'(R^n)$ by the intertwining operator. For example one obtains easy proofs of the nuclearity of $\mathcal{S}(R^n)$ and Schwartz's nuclear theorem. See [20], pp. 141-145.

B. If $\psi \in \mathcal{S}(Q)$, we define its Fourier transform $F\psi$ by:

$$F\psi(\underset{\sim}{q}) =: (2\pi)^{-n/2} \int e^{-i(\underset{\sim}{q},\underset{\sim}{q}')_J} \psi(\underset{\sim}{q}')\,d\underset{\sim}{q}' .$$

We define a one-parameter unitary group $\{\hat{F}\psi\}$ on $\mathcal{F}(W_J^+)$ by:

$$\hat{F}_t f(\omega_+) =: f(e^{-it}\omega_+) .$$

$\hat{F}_t$ clearly maps $\mathcal{S}(\mathcal{F}(W_J^+)$ isomorphically onto itself. The key fact is that F is transformed into $\hat{F}_{\pi/2}$ via A. Note that $\hat{F}_{\pi/2}f(\omega_+) = f(-i\omega_+)$, a remarkably simple operation. We state this in a formal theorem.

THEOREM 3.2.1. *If* $\psi \in C_0^\infty(Q)$, *then*

$$F\psi = A^{-1}\hat{F}_{\pi/2}A\psi . \tag{3.2.1}$$

VERY SKETCHY PROOF.

$$A^{-1}\hat{F}_{\pi/2}A\psi(\underset{\sim}{q}) = \lim_{\lambda\uparrow 1} \iint \overline{A(\omega_+,\underset{\sim}{q})}\,A(-i\lambda\omega_+,\underset{\sim}{q}')\psi(\underset{\sim}{q}')\,d\underset{\sim}{q}\,d\mu$$

where $A(\omega_+,\underset{\sim}{q})$ is the kernel of A (Section 2.2 A) and where the double integral is absolutely convergent. Thus

$$\iint \overline{A(\omega_+,\underset{\sim}{q})}\,A(-i\lambda\omega_+,\underset{\sim}{q}')\psi(\underset{\sim}{q}')\,d\underset{\sim}{q}'\,d\mu = \iint \overline{A(\omega_+,\underset{\sim}{q})}\,A(-i\lambda\omega_+,\underset{\sim}{q}')\,d\mu\,\psi(\underset{\sim}{q}')\,d\underset{\sim}{q}'.$$

But a Gaussian integral calculation ([2], Section 1h and page 206) shows that $\int \overline{A(\omega_+, q)}\, A(-i\lambda\omega_+, q')\, d\mu$ converges uniformly as $\lambda \uparrow 1$, to $(2\pi)^{-n/2} e^{-i(q,q')}$ for fixed q and $q' \in \text{support}(\psi)$. This establishes (3.2.1).

COROLLARY 3.2.2. 1. F *maps* $\mathcal{S}(Q)$ *isomorphically onto itself.*
2. $F^{-1}\psi(q) = (2\pi)^{-n/2} \int e^{i(q,q')_J} \psi(q')\, dq'$, *for* $\psi \in \mathcal{S}(Q)$.
3. $\|F\psi\|^2 = \|\psi\|^2$ *for* $\psi \in \mathcal{S}(Q)$.

PROOF. 1) follows from the fact that A is an isomorphism of $\mathcal{S}(Q)$ onto $\mathcal{S}(\mathcal{F}(W_J^+))$ and $\hat{F}_{\pi/2}$ clearly maps $\mathcal{S}(\mathcal{F}(W_J^+))$ isomorphically onto itself. 3) follows from the unitarity of A and the obvious fact that $\|\hat{F}_{\pi/2} f\|_B^2 = \|f\|_B^2$. To prove 2) observe that $\hat{F}_\pi f(\omega_+) = f(-\omega_+)$ and thus $\hat{F}^2\psi(q) = \psi(-q)$ since $A(-\omega_+, q) = A(\omega_+, -q)$. But $(\hat{F}_{\pi/2})^3 \hat{F}_{\pi/2} = I$ and thus $F^3 = F^{-1}$. But

$$F^3\psi(q) = F^2(F\psi)(q) = (2\pi)^{-n/2} \int e^{i(q,q')_J} \psi(q')\, dq'$$

which proves 2).

From property 3) we see that we can extend F to a unitary operator on $L^2(Q)$.

COROLLARY 3.2.3. *There exists a one-parameter unitary group* $\{F_t\}$ *on* $L^2(Q)$ *such that* $F = F_{\pi/2}$.

PROOF. Just let $F_t = A^{-1}\hat{F}_t A$.

The details for the above proof and many other applications of this technique are given in [5].

§3.3. *θ-functions (after Cartier)*

A. It turns out that θ-functions can be described in a very natural way in the B.F.S. setting. We will however have to make two modifications in our treatment of the B.F.S. realization. As before, we assume a fixed $(V,<,>,J)$ and thus V_J has the Hermitian structure $(\,,\,)_J$.

The first modification is that we will work with the "unique" irreducible representation π of $N(V)$ such that

$$\pi(e^{tE}) = e^{2\pi it} .$$

If we carry through the construction of the B.F.S. in this situation, the representation space will be

$$\mathcal{F}_{2\pi}(W_J^+) =: \left\{ f \in \mathcal{Z}(W_J^+) : \int |f|^2 e^{-2\pi \|\omega_+\|_+^2} d\omega_+ d\bar{\omega}_+ \right\}$$

and the B.F.S. realization $V_{2\pi}$ of $N(V)$ is defined by:

$$V_{2\pi}(e^{\omega_+ + \bar{\omega}_+ + Et}) f(\omega_+') =: e^{2\pi it} \exp[2\pi(\omega_+', \omega_+)_+ - \pi(\omega_+, \omega_+)_+] f(\omega_+' - \omega_+) .$$

The whole theory goes through as before. For example,

$$\mathcal{S}'(\mathcal{F}(W_J^+)) = \left\{ f \in \mathcal{Z}(W_J^+) : \exists\, k \in N \text{ such that } \int |f|^2 (1 + \|\omega_+\|_+^2)^{-k} e^{-2\pi \|\omega_+\|_+^2} d\omega_+ d\bar{\omega}_+ < \infty \right\} .$$

The second modification is that we must transport everything over to $(V_J, (\,,\,)_J)$ via θ_J which was defined in §1.2D. Let $\mathcal{F}_\pi(V_J) =: \{\phi \in \mathcal{Z}(V_J) : \int |\phi(v)|^2 e^{-\pi \|v\|_J^2} dv < \infty\}$ with the obvious inner product. Define $U_\Theta : \mathcal{F}_\pi(V_J) \to \mathcal{F}_{2\pi}(W_J^+) : \phi \rightsquigarrow \phi \circ \Theta(\omega_+)$. U_θ is a unitary mapping. Since $(\theta(\omega_+), \theta(\omega_+))_J = 2(\omega_+, \omega_+)_+$. A direct calculation shows that $\omega_J =: U_\Theta^{-1} V_{2\pi} U_\Theta$ is defined by:

$$\text{(3.3.1)} \quad \omega_J(e^{v+tE})\phi(v') =: e^{2\pi it} \exp\left[-\frac{\pi}{2}(v,v)_J + \pi(v',v)_J\right] \phi(v' - v) .$$

Note that ω_J can be extended to $\mathcal{S}'(\mathcal{F}_\pi(V_J))$ just by using the same formula. It is this modified B.F.S. realization which is used when discussing θ functions.

B. Let L be an integral lattice in $(V, <\,,>)$ i.e. L is a discrete subgroup of V with 2n independent generators and such that $<\,,>$ restricted to $L \times L$ is $\mathbb{Z}$-valued. Note V/L is a compact Lie group. A *quasi-character* τ *on* L is a homomorphism from L to the unit circle in $\mathbb{C}$ such that:

$$\tau(\gamma+\gamma') = \tau(\gamma)\tau(\gamma')e^{-i\pi<\gamma,\gamma'>} .$$

DEFINITION. A holomorphic function ϕ on V_J is said to be a theta function of type (J,τ) if it satisfies:

$$\phi(v-\gamma) = \phi(v)\tau(\gamma)\exp\left[\frac{\pi}{2}(\gamma,\gamma)_J - \gamma(v,\gamma)\right] \tag{3.3.2}$$

for $v \in V$, $\gamma \in L$.

Theta functions of type (J,τ) now have the following natural interpretation in $\mathcal{S}'(\mathcal{F}_\pi(V_J))$.

THEOREM 3.3.1. ϕ is a theta function of type (J,τ) iff

1. $\phi \in \mathcal{S}'(\mathcal{F}_\pi(V_J))$;
2. $\omega_J(e^\gamma)\phi = \tau(\gamma)\phi$.

PROOF. Writing out 2) and using (3.3.1), we have:

$$\exp\left[-\frac{\pi}{2}(\gamma,\gamma)_J + \pi(\gamma,v)_J\right]\phi(v-\gamma) = \tau(\gamma)\phi(v)$$

which is just the functional equation (3.3.2). Thus it remains to show that if ϕ is a θ-function of type (J,τ), there exists $k \in \mathbb{N}$ such that:

$$\int |\phi(v)|^2(1+\|v\|_J^2)^{-k} e^{-\pi\|v\|_J^2}\, dv < \infty . \tag{3.3.3}$$

If we let $\psi(v) =: |\phi(v)|^2 \exp(-\pi\|v\|_J^2)$, we see that:

$$\begin{aligned}\psi(v-\gamma) &= |\phi(v)|^2\left|\exp\left[\frac{\pi}{2}(\gamma,\gamma)_J - \pi(v,\gamma)_J\right]\right|^2 \exp[-\pi(v-\gamma,v-\gamma)_J] \\ &= |\phi(v)|^2\exp[-\pi\|v\|_J^2] = \psi(v) .\end{aligned}$$

Thus since ψ is continuous and invariant under L and V/L is compact, $\psi(\cdot)$ is bounded. Therefore if we let $k \in N$ be such that $\int (1+\|v\|_J^2)^{-k} dv < \infty$, then ϕ satisfies (3.3.3) and is in $\mathcal{S}'(\mathcal{F}_\pi(V_J))$.

Using the essential uniqueness of the irreducible representation π of N(V) such that $\pi(e^{tE}) = e^{2\pi i t}$ and the induced representation theory of Mackey, Cartier [9] is able to give a very elegant proof of the following classical theorem of Frobenius.

THEOREM 3.3.2. *The dimension of the θ-functions of type* (J,τ) *is the square root of the discriminant of* $\langle\,,\rangle$ *with respect to* L. *(Recall the definition of the discriminant of* $\langle\,,\rangle$ *with respect to* L *is* $\det(\langle v_i, v_j\rangle)$ *where* $\{v_1, \cdots, v_{2n}\}$ *is a basis for* L.)

PROOF. (Omitted.)

§3.4. *Holomorphic quantization*

A. In this section we will be concerned with the quantization of a certain space $\operatorname{Exp}(V_C)$ of holomorphic functions on complex "phase" space $V_C = W_J^+ \oplus W_J^-$. As usual we will fix $V, \langle\,,\rangle$, and J throughout this section. It will turn out that the B.F.S. space $\mathcal{F}(W_J^+)$ is the natural space in which to carry out this quantization procedure. In particular we want to define a mapping Q from $\operatorname{Exp}(V_C)$ into the (unbounded) operators on $\mathcal{F}(W_J^+)$ such that Q preserves certain classical mechanical algebraic structures on V_C. If we choose a standard basis $\{\underset{\sim}{a}^+, \underset{\sim}{a}^-\}$ for $W_J^+ \oplus W_J^-$ and thus identify it with $C^n \times C^n$, then QF, $F \in \operatorname{Exp}(V_C)$, is usually written as $F(A_1^*, \cdots, A_n^*, A_1, \cdots, A_n)$.

Our definition of QF will exactly parallel the usual Weyl Quantization procedure [27]. However, we will use the Fourier-Borel transform and its inverse in the same way the Fourier transform and its inverse is used in the Weyl procedure. We will first discuss the holomorphic quantization procedure in a purely formal manner hoping to convince the reader that it is the correct one. We will then give a rigorous definition of Q. To do this, we will first need to introduce a dense subspace $\mathrm{Exp}_\infty(W_J^+)$ of $\mathcal{F}(W_J^+)$ which will serve as a natural domain for all of the quantized operators Q(F). In a subsequent paper we will rigorously show that Q has all of the properties which we discuss here only in a formal way.

B. We define a complex symplectic structure $V_C = W_J^+ \oplus W_J^-$ by:

$$[\omega, \omega'] =: -i[\langle \omega_+, \omega'_- \rangle - \langle \omega'_+, \omega_- \rangle]$$

where $\omega = \omega_+ + \omega_-$, $\omega' = \omega'_+ + \omega'_-$. Thus if $\{\underset{\sim}{a}^+, \underset{\sim}{a}^-\}$ is a standard basis for V_C and we identify V_C with $C^n \times C^n$, then:

$$[(z, w), (z', w')] = z \cdot w' - z' \cdot w$$

where $z, w \in C^n$. An *analytic functional* on $\mathcal{Z}(V_C)$ is a continuous linear functional ν on $\mathcal{Z}(V_C)$ where a sequence $\{f_n\}$ in $\mathcal{Z}(V_C)$ converges to 0 if it converges to 0 uniformly on compact subsets of V_C. Thus continuity of ν means that if $\{f_n\}$ converges to 0 in $\mathcal{Z}(V_C)$, then $\nu(f_n) \to 0$.

The *Fourier-Borel transformation* $\hat{\nu}$ of ν is defined by:

$$\hat{\nu}(\omega) =: \nu(\exp_\omega)$$

where $\exp_\omega(\omega') =: \exp([\omega, \omega'])$. It is sometimes suggestive to write:

$$\hat{\nu}(\omega) = \int e^{[\omega, \omega']} d\nu(\omega') .$$

We define a norm on V_C by:

$$\|\omega\|^2 = \|\omega_+\|_+^2 + \|\bar{\omega}_-\|_+^2 .$$

Let $\mathrm{Exp}(V_C) =: \{F \in \mathcal{Z}(V_C) : \exists M_F, K_F > 0 \ni (F(\omega)) \leq M_F \exp(K_F \|\omega\|)\}$. Then it's known [24], p. 266, that the Fourier-Borel transform is an (algebraic) isomorphism between $A(V_C)$, the space of analytic functionals, and $\mathrm{Exp}(V_C)$.

We need to discuss the inverse of the Fourier Borel transform in more detail. Let $\{\underset{\sim}{a}^+, \underset{\sim}{a}^-\}$ be a standard basis for V_C so that we identify V_C with $C^n \oplus C^n$. $\delta^{(\underset{\sim}{n},\underset{\sim}{m})}$ will denote the analytic functional defined by:

$$\delta^{(\underset{\sim}{n},\underset{\sim}{m})}(F) =: \left(-\frac{\partial}{\partial z}\right)^{\underset{\sim}{m}} \left(\frac{\partial}{\partial w}\right)^{\underset{\sim}{n}} F(z,w)\Big|_{z=w=0} .$$

If $F \in \operatorname{Exp}(V_C)$, expand it in its power series:

$$F(z,w) = \sum_{\underset{\sim}{n},\underset{\sim}{m}} \frac{F^{(\underset{\sim}{n},\underset{\sim}{m})}}{\underset{\sim}{n}!\,\underset{\sim}{m}!} z^{\underset{\sim}{n}} w^{\underset{\sim}{n}}$$

where $F^{(\underset{\sim}{n},\underset{\sim}{m})} =: \left[\left(\frac{\partial}{\partial z}\right)^{\underset{\sim}{n}} \left(\frac{\partial}{\partial w}\right)^{\underset{\sim}{m}} F\right](0,0)$. Then the inverse Fourier-Borel transform ν_F of F is:

$$\nu_F = \sum_{\underset{\sim}{n},\underset{\sim}{m}} \frac{F^{(\underset{\sim}{n},\underset{\sim}{m})}}{\underset{\sim}{n}!\,\underset{\sim}{m}!} \delta^{\underset{\sim}{n},\underset{\sim}{m}} .$$

The analogue of the Fourier inversion formula reads:

$$F = \hat{\nu}_F$$

or

$$F(\omega) = \int e^{[\omega,\omega']} d\nu_F(\omega')$$

or

$$F(z,w) = \int e^{z\cdot w' - z'\cdot w}\, d\nu_F(z',w') . \tag{3.4.1}$$

We now proceed formally. Let us assume that we can extend the definition of $V(e^{\omega_+ + \overline{\omega}_+})$ to V_C and thus we can write $V(e^{\omega_+ + \omega_-})$. In Part C we shall see that there is a natural (dense) subspace $\operatorname{Exp}_\infty(W_J^+)$ of $\mathfrak{F}(W_J^+)$ in which $V(e^{\omega_+ + \omega_-})$ does make sense and in fact is "holomorphic on V_C." If we choose a standard basis $\{a^+, a^-\}$ for V_C, we see that formally by "analytic continuation" that:

$$V(e^{z\cdot a^+ + w\cdot a^-}) = \exp[w\cdot A^* - z\cdot A]$$

since:

$$V(e^{z\cdot a^+ + \bar{z} a^-}) = \exp[\bar{z}A^* - z\cdot A] .$$

Note that $z\cdot a^+ =: \sum_{j=1}^{n} z_j a_j^+$, etc.

Following Weyl but using the inversion formula (3.4.1), it is natural to (formally) define QF by:

$$QF =: F(A^*, A) =: \int \exp[w'\cdot A^* - z'\cdot A]\, d\nu_F =: \nu_F(V) .$$

Note that QF does not depend on the choice of basis $\{a^+, a^-\}$.

We will now discuss the algebraic structure and "physical" interpretation of $\mathrm{Exp}(V_C)$. As we have already observed V, the "phase space for a system with n degrees of freedom," is a real 2n dimensional subspace of V_C. Recall it consists of elements of the form $\omega_+ + \bar{\omega}_+$. Elements of $\mathcal{S}(V_C)$ and hence $\mathrm{Exp}(V_C)$ are uniquely determined by their restriction to V. Thus we may identify elements of $\mathrm{Exp}(V_C)$ with certain complex valued functions on V and so we call elements of $\mathrm{Exp}(V_C)$ *(classical) dynamical variables.*

We must determine when the restriction of F to V is real i.e. when is F an observable. Choose a standard basis $\{a^+, a^-\}$ for V_C and expand F in its power series:

$$F = \sum_{n, m} \frac{F^{(n, m)}}{n!\, m!} z^n w^m .$$

One can check that the restriction of F to $V \cong \{(z, \bar{z}) : z \in C^n\}$ is real iff $\overline{F^{(n,m)}} = F^{(m,n)}$. This suggests that we define a $*$ operation on $\mathrm{Exp}(V_C)$ by:

$$F^*(z, w) =: \sum_{n, m} \frac{\overline{F^{(n, m)}}}{n!\, m!} z^m w^n .$$

This $*$-operation is conjugate linear, involutory, and does not depend on the choice of the standard basis $\{a^+, a^-\}$. From the above discussion, it is reasonable to call F a (*classical*) *observable* $F = F^*$ and we shall do so.

Q formally preserves the $*$ operation i.e. $Q(F^*) = Q(F)^*$. In fact, if $\{a^+, a^-\}$ is a standard basis for V_C and if we identify V_C with $C^n \times C^n$, then $\mu_{P_{n,m}} = \delta^{(n,m)}$ where $P_{n,m} = z^n w^m$ and

$$(3.4.2) \quad QP_{n,m} = \delta^{n,m}(e^{w \cdot A^* - z \cdot A})$$

$$= \left(\frac{\partial}{\partial w}\right)^n \left(-\frac{\partial}{\partial z}\right)^m (e^{w \cdot A^*} e^{-z \cdot A} e^{-\frac{1}{2} z \cdot w})|_{z=w=0}$$

$$= \sum_{\substack{0 \le m_1 \le m \\ 0 \le n_1 \le n \\ m - m_1 = n - n_1}} \binom{n}{n_1}\binom{m}{m_1}\left(\frac{1}{2}\right)^{(m-m_1)/2}\left(\frac{1}{2}\right)^{(n-n_1)/2} \frac{A^{*n_1} A^{m_1}}{\sqrt{(n-n_1)!}\sqrt{(m-m_1)!}} .$$

Thus $Q(P^*_{n,m}) = Q(P_{m,n}) = Q(P_{n,m})^*$ since $((A^*)^{n_1} A^{m_1})^* = (A^*)^{m_1} A^{n_1}$ and the symmetry in m_1 and n_1 in the last expression in (3.4.2). Since Q is linear and using power series expansions, we see that $Q(F^*) = Q(F)^*$.

There is a natural twisted convolution $\nu_1 \boxed{X} \nu_2$ of two analytic functionals ν_1, ν_2 which depends on the symplectic structure $[\,,]$ on V_C. This is the holomorphic analogue of the twisted convolution on the phase space V treated by many authors (e.g. see [17], [19]). It is defined as follows:

$$\nu_1 \boxed{X} \nu_2(g) = : \nu_{2,\omega'}[\nu_{1,\omega}(g(\omega + \omega')e^{\frac{1}{2}[\omega', \omega]})]$$

where $\nu_{1,\omega}$ means ν_1 is acting on the ω variable while treating ω' as a fixed parameter while $\nu_{2,\omega'}$ means ν_2 is acting on the ω' variable. Using the isomorphism between $A(V_C)$ and $\mathrm{Exp}(V_C)$, we can define a "twisted product" of dynamical variables as follows:

$$F =: F_1 \text{ (M) } F_2$$

where $\nu_F = \nu_{F_1} \boxed{X} \nu_{F_2}$. $\operatorname{Exp}(V_C)$ now becomes a $*$ algebra. A direct formal calculation shows that:

$$Q(F_1 \text{ (M) } F_2) = Q(F_1)Q(F_2) .$$

The quantization mapping Q thus formally maps the $*$-algebra of classical dynamical variables homomorphically into the $*$-algebra of quantum dynamical variables and in particular Q maps the classical observables into the quantum observables.

REMARK 1. The bilinear mapping $F_1, F_2 \to F_1 \text{ (M) } F_2 - F_2 \text{ (M) } F_1$ is called the *holomorphic Moyal bracket* of F_1, F_2 and "to first order" is the holomorphic Poisson bracket of F_1 and F_2.

REMARK 2. Two other possible quantization maps are:

$$Q_1 F =: \int \exp[w \cdot \underset{\sim}{A}^*] \exp[-z \cdot \underset{\sim}{A}] \, d\nu_F$$

and

$$Q_2 F =: \int \exp[-z\underset{\sim}{A}] \exp[w\underset{\sim}{A}^*] \, d\nu_F .$$

These would lead to the normal ordering and "anti-normal" ordering quantizations respectively.

C. We now turn to the rigorous definition of Q. We first must construct a convenient subspace of $\mathcal{F}(W_J^+)$ on which the quantized operators will act. Let $k \in R$ and let $\operatorname{Exp}_k(W_J^+) = \{f \in \mathcal{Z}(W_J^+) : \int |f|^2 e^{k\|\omega_+\|_+} d\mu < \infty\}$. $\operatorname{Exp}_k(W_J^+)$ can be shown to be a Hilbert space of analytic functions in the same way as it was for $\mathcal{F}^k(W_J^+)$. $\| \; \|_k$ will denote the norm on $\operatorname{Exp}_k(W_J^+)$ defined by:

$$\|f\|_k^2 =: \int |f|^2 e^{k\|\omega_+\|^2} d\mu .$$

Note that the polynomials are in $\mathrm{Exp}_k(W_J^+)$ for all k. Since $\mathrm{Exp}_k(W_J^+) \subset \mathcal{F}(W_J^+)$ for $k \geq 0$, it follows that $\mathrm{Exp}_k(W_J^+)$ is dense in $\mathcal{F}(W_J^+)$. We define a sesquilinear duality between $\mathrm{Exp}_k(W_J^+)$ and $\mathrm{Exp}_k(W_J^+)$ as follows:

$$(\chi, f)_B =: \int \bar{\chi} f \, d\nu$$

for $f \in \mathrm{Exp}_k(W_J^+)$ and $\chi \in \mathrm{Exp}_{-k}(W_J^+)$. It follows directly that:

$$|(\chi, f)_B| \leq \| f \|_k \, \| \chi \|_{-k} \, .$$

Note also that

$$\| f \|_k = \sup_{\substack{\chi \in \mathrm{Exp}_{-k}(W_J^+) \\ \| \chi \|_{-k} \leq 1}} |(\chi, f)_B|$$

Let $\mathrm{Exp}_\infty(W_J^+) =: \bigcap_{k \geq 0} \mathrm{Exp}_k(W_J^+)$. We say $\{f_n\}$ converges to 0 in $\mathrm{Exp}_\infty(W_J^+)$ if $\| f_n \|_k \to 0$ as $n \to \infty$ for all $k \geq 0$. It is clear that $\mathrm{Exp}_\infty(W_J^+)$ is a countable Hilbertian space. $\mathrm{Exp}_\infty(W_J^+)$ is dense in $\mathcal{F}(W_J^+)$ since it contains the polynomials. This space will serve as the domain for the quantized operators $Q(F)$.

A linear mapping $T : \mathrm{Exp}_\infty(W_J^+) \to \mathrm{Exp}_\infty(W_J^+)$ is continuous iff for each $k \geq 0$, there exists $k' \geq 0$, $K_k > 0$ such that $\| Tf \|_k \leq K_k \| f \|_{k'}$. A convenient way to define a continuous linear mapping on $\mathrm{Exp}_\infty(W_J^+)$ is via a sesquilinear form on $\mathrm{Exp}_{-\infty}(W_J^+) \times \mathrm{Exp}_\infty(W_J^+)$ where $\mathrm{Exp}_{-\infty}(W_J^+) =: \bigcup_{k>0} \mathrm{Exp}_{-k}(W_J^+)$. Recall that a form $[\, ,\,]$ is sesquilinear if it is conjugate linear in the first variable and linear in the second variable. We have the following simple but useful result.

THEOREM 3.4.1. *Suppose* $[\, ,\,]$ *is a sesquilinear form on* $\mathrm{Exp}_{-\infty}(W_J^+) \times \mathrm{Exp}_\infty(W_J^+)$ *such that for each* $k \geq 0$, *there exists* $k' \geq 0$, $K_k > 0$ *such that*

$$|[\chi, f]| \leq K_k \| f \|_{k'} \, \| \chi \|_{-k} \tag{3.4.3}$$

for $f \in \mathrm{Exp}_\infty(W_J^+)$, $\chi \in \mathrm{Exp}_{-k}(W_J^+)$. *Then there exists a unique continuous linear mapping* $T : \mathrm{Exp}_\infty(W_J^+) \to \mathrm{Exp}_\infty(W_J^+)$ *such that*

$$[\chi, f] = (\chi, Tf)_B. \tag{3.4.4}$$

PROOF. The existence and uniqueness of T satisfying (3.4.3) is direct. We only need to check continuity. Thus suppose $f_n \to 0$ in $\mathrm{Exp}_\infty(W_J^+)$. Let $k \geq 0$ and $\chi \in \mathrm{Exp}_{-k}(W_J^+)$ where k' is such that (3.4.3) holds, then, using (3.4.3 - 3.4.4), we have:

$$|(\chi, Tf_n)_B| \leq K_k \|f_n\|_{k'}$$

and therefore:

$$|Tf_n|_k \leq K_k |f_n|_{k'}$$

and thus $|Tf_n|_k \to 0$ as $n \to \infty$ and T is continuous.

We next need to show that $V(e^{\omega_+ + \omega_-})$ can be defined on $\mathrm{Exp}_\infty(W_J^+)$ and that it is a "weakly" holomorphic function on V_C. In fact, if $f \in \mathrm{Exp}_\infty(W_J^+)$, define:

$$V(e^{\omega_+ + \omega_-}) f(\omega_+') =: \exp[i(\tfrac{1}{2} <\omega_+, \omega_-> - <\omega_+', \omega_->)] f(\omega_+' - \omega_+) .$$

$V(e^{\omega_+ + \omega_-}) f$ is clearly in $\mathcal{B}(W_J^+)$. However, we have much more. The essential technical point is contained in the following lemma.

LEMMA 3.4.2. *For each* $k \geq 0$, $T > 0$, *there exists* $k'(k, T) \equiv k', K_{k,T} > 0$ *such that*

$$|V(e^{\omega_+ + \omega_-}) f|_k \leq K_{k,T} |f|_{k'} \tag{3.4.5}$$

for $\|\omega_+\|_+ \leq T$, $\|\omega_-\| =: \|\bar{\omega}_-\|_+ \leq T$ *and* $f \in \mathrm{Exp}_\infty(W_J^+)$.

PROOF. By making the change of variable $\omega_+' - \omega_+ \to \omega_+'$, we have:

$$\|V(e^{\omega_+ + \omega_-})f\|_k^2 = \int e^{k\|\omega'_+\|_+} \left| e^{i(\frac{1}{2}\langle\omega_+,\omega_-\rangle - \langle\omega'_+,\omega_-\rangle)} \right|^2 f(\omega'_+ - \omega_+)\, d\mu$$

$$= \left|\exp[-\tfrac{1}{2} i\langle\omega_+,\omega_-\rangle]\right|^2 \exp(-\|\omega_+\|_+^2)$$

$$\times \int \exp[k\|\omega'_+ + \omega_+\|_+ - 2\mathrm{Re}\{ i\langle\omega'_+,\bar{\omega}_+\rangle\}] \left|\exp[-i\tfrac{1}{2}\langle\omega'_+,\omega_-\rangle]\right|^2 |f|^2\, d\mu$$

$$\leq \exp[k\|\omega_+\|_+ - \mathrm{Re}(i\langle\omega_+,\omega_-\rangle) - \|\omega_+\|^2]$$

$$\times \int \exp[k\|\omega'_+\|_+ - \mathrm{Re}\{i\langle\omega'_+,\omega_-\rangle - 2i\langle\omega'_+,\bar{\omega}_+\rangle\}]\, |f|^2\, d\mu \; .$$

Thus if we let k' be such that $k' - k > 3T$, we have:

$$(k - k')\|\omega'_+\| - \mathrm{Re}\{i\langle\omega'_+,\omega_-\rangle - 2i\langle\omega'_+,\bar{\omega}_+\rangle\} < 0$$

and the last integral above is less than or equal to:

$$\int e^{k'\|\omega_+\|_+} |f(\omega')|^2\, d\mu = \|f\|_{k'}^2 \; .$$

If we let $K_{k,T} = \sup_{\substack{\|\omega_+\|_+ \leq T \\ \|\omega_-\|_+ \leq T}} (\exp[k\|\omega_+\|_+ - \mathrm{Re}\{i\langle\omega_+,\omega_-\rangle - \|\omega_+\|_+^2])^{\frac{1}{2}}$, (3.4.5) holds.

Corollary (of the Proof) 3.4.3. *Let* $\chi \in \mathrm{Exp}_{-\infty}(W_J^+)$ *and* $f \in \mathrm{Exp}_{\infty}(W_J^+)$ *and define* $\Theta_{\chi,f}(\omega)$ *by:*

$$\Theta_{\chi,f}(\omega) = \, : (\chi, V(e^{\omega}) f)_B \; .$$

Then $\Theta_{\chi,f}$ *is a holomorphic function on* V_C.

Proof. It follows from the proof of Lemma 3.4.2 that the integral $\int |\chi(\omega'_+)|\, |V(e^{\omega}) f(\omega'_+)|\, d\mu$ is locally bounded. If we choose a standard basis $\{\underline{a}^+, \underline{a}^-\}$ for V_C and identify it with $C^n \times C^n$, then $\Theta_{\chi,f}$ can be viewed as a function of the $2n$ complex variables (z, w). To show it is holomorphic, it is sufficient to show it is holomorphic in each variable separately. (For example, see [13], p. 28.)

Let Γ be a simple closed curve in C which we view as a path in say z_j space. Then by the local boundedness above, we have:

$$\int_\Gamma \int |\chi(z')|\,|V(z,w)f(z')|\,d\mu d|z_j| < \infty$$

where $V(z,w) =: \exp[w\cdot \underset{\sim}{A}^* - z\cdot \underset{\sim}{A}]$ and therefore by the Fubini-Tonelli theorem, we have

$$\int_\Gamma \Theta(z,w)\,dz_j = \iint_\Gamma \overline{\chi(z')}\,V(z,w)f(z')\,dz_j\,d\mu = 0\ .$$

$\Theta_{\chi,f}(z,w)$ is therefore analytic as a function of z_j by Morera's theorem. A similar statement holds for the w_j's and $\Theta_{\chi,f}$ is holomorphic on V_C.

If $F \in \mathrm{Exp}(V_C)$, we define a sesquilinear form $[\,,\,]_F$ on $\mathrm{Exp}_{-\infty}(W_J^+) \times \mathrm{Exp}_\infty(W_J^+)$ by:

$$[\chi,f]_F =: \nu_F((\chi, V(e^\omega)f)_B)\ .$$

THEOREM 3.4.4. *Given* $k>0$, *there exists* $k'\geq 0$, $K_k>0$ *such that*:

$$|[\chi,f]_F| \leq K_k \|\chi\|_{-k}\|f\|_{k'}$$

for $\chi \in \mathrm{Exp}_{-k}(W_J^+)$, $f\in \mathrm{Exp}(W_J^+)$ *and thus by Theorem 3.4.1, there exists a continuous linear mapping* $Q(F)$ *on* $\mathrm{Exp}_\infty(W_J^+)$ *such that*:

$$[\chi,f]_F = (\chi, Q(F)f)_B\ . \tag{3.4.6}$$

PROOF. It is a fact from the theory of analytic functionals on V_C that given $F\in \mathrm{Exp}(V_C)$, there exists $T_F \equiv T$ and $K_F>0$ such that:

$$|\nu_F(f) \leq K_F \sup_{\substack{\|\omega_+\|_+\leq T\\ \|\omega_-\|_+\leq T}} |\Theta(\omega_+,\omega_-)|$$

for $f \in Z(V_C)$ [24], p. 266. Therefore from Lemma 3.4.2 we have:

$$|[\chi, f]_F| = |\nu_F(\Theta_{\chi,f})| \le K_F K_{k,T} \|\chi\|_{-k} \|f\|_{k'}$$

for $\chi \in \mathrm{Exp}_{-k}(W_J^+)$, some $k' \ge 0$, $K_{k,T} > 0$. Thus by Theorem 3.4.1, there exists a unique $Q(F)$ such that (3.4.6) holds.

DEFINITION. The mapping $F \to Q(F)$ of $\mathrm{Exp}(V_C)$ into the continuous mappings on $\mathrm{Exp}_\infty(W_J^+)$ is called the *holomorphic quantization of* (V, J).

As we mentioned above, we will rigorously derive the properties of Q which we discussed only formally in Part B in a subsequent paper.

NOTES AND COMMENTS

Chapter II. For the material in §2.1, see [7].

Our derivation of the Schrödinger and B.F.S. realization follows the treatment of Lévy-Nahas in [7] very closely. This book gives a very nice treatment of the Kirilov [15] and Auslander and Kostant [1] treatments of the irreducible representations of nilpotent and type I solvable groups respectively. It should be pointed out that the authors of this book have made substantial contributions to the theory themselves.

The contents of §2.3 and §2.4 come directly from [2]. The characterization of $\mathcal{S}(\mathcal{F}(W_J^+))$ is given in [5] where rather detailed properties of the intertwining operators A are used. This leads to sharper results than we have obtained, but it requires considerably more hard analysis.

Chapter III. The material in §3.1 is from [3]. The material in §3.2 is from [2] and [5]. Many aspects of distribution theory are more transparent in $\mathcal{S}'(\mathcal{F}(W_J^+))$ than in $\mathcal{S}'(Q)$ as Bargmann's elegant paper illustrates. The material in §3.3 is from [8] and [9].

The holomorphic quantization we discuss seems to be somewhat related to hyperdifferential operators (see [16] and references given there). Formally the "normal ordering" quantization mapping Q_1, assigns a hyperdifferential operator Q_1 with symbol F in the terminology of [16]. Some other topics we did not have time to cover are: 1) the metaplectic representation of $Sp(n, R)$ and some of its subgroups [6], [14]; 2) differential equations with polynomial coefficients [18]; 3) the theory of tempered distributions in infinite dimensional spaces [10].

BIBLIOGRAPHY

[1] Auslander, L. and Kostant, B.: Polarizations and unitary representations of solvable Lie groups, Inventiones Mathematicae, *14*, 255-254 (1971).

[2] Bargmann, V.: On a Hilbert space of analytic functions and an associated integral transform – Part I, *Comm. on Pure and Applied Math.*, *14*, 187-214 (1961).

[3] ___________: On the representations of the rotation group, *Rev. of Modern Physics*, *34*, 829-845 (1962).

[4] ___________: Remarks on a Hilbert space of analytic functions, *Proc. N.A.S.*, *48*, 199-204 (1962).

[5] ___________: On a Hilbert space of analytic functions and an associated integral transform – Part II, *Comm. on Pure and Applied Math.*, *20*, 1-101 (1967).

[6] ___________: Group representation on Hilbert spaces of analytic functions, in *Analytic Methods in Mathematical Physics*, Ed. by R. P. Gilbert and R. G. Newton, Gordon and Breach (1968).

[7] Bernat, P., Conze, N., Duflo, M., Lévy-Nahas, M., Rais, M., Renouard, P., Vergne, M.: *Representation des Groupes de Lie Resolubles*, Dunod, Paris, 1972.

[8] Bruhat, F.: *Representations des Groupes Localement Compacts* – Lecture Notes for a course given at the Universite de Paris VII, 1969-1970.

[9] Cartier, P.: Quantum mechanical commutation relations and theta functions, in *Proceedings of Symposia in Pure Mathematics*, Vol. IX, A.M.S., Providence, (1966).

[10] Dwyer, T. A. III: Holomorphic Fock representations and partial differential equations on countably Hilbert spaces, *Bull. A.M.S.*, *79*, 1045-1050 (1973).

[11] Fock, V.: Verallgemeinerung und Lösung der Diracschen Statistischen Gleichung, *Z. Physik, 49*, 339-357 (1928).

[12] Gelfand, I. M. and Vilenkin, N.: *Generalized Functions*, Vol. IV, Academic Press, New York-London (1964).

[13] Hormander, L.: *An Introduction to Complex Analysis in Several Variables*, Van Nostrand, Princeton (1966).

[14] Itzykson, C.: Remarks on Boson commutation relations, *Comm. Math. Phys., 4*, 92-122 (1967).

[15] Kirilov, A. A.: Unitary representations of nilpotent Lie groups, *Uspekhi Mat. Nauk., 17*, 57-110 (1962) (in Russian); Trans. Russian Math. Surveys (1962).

[16] Miller, M. and Steinberg, S.: Applications of hyperdifferential operators to quantum mechanics, *Comm. Math. Phys. 24*, 40-60 (1971).

[17] Moyal, J.: Quantum Mechanics as a statistical theory, *Proc. Cambridge Phil. Soc.*, 99-124 (1949).

[18] Newman, D. and Shapiro, H.: Certain Hilbert spaces of entire functions, *Bull. A.M.S., 72*, 971-977 (1966).

[19] Pool, J. C. T.: Methematical aspects of the Weyl correspondence, *J. Math. Phys., 7*, 66-76 (1966).

[20] Reed, M. and Simon, B.: *Methods of Modern Mathematical Physics, I: Functional Analysis*, Academic Press, New York-London (1972).

[21] Segal, I. E.: Mathematical characterization of the physical vacuum for a linear Bose-Einstein field, *Ill. J. Math., 6*, 500-523 (1962).

[22] Stone, M. H.: Linear transformation in Hilbert space III. Operational Methods and group theory, *Proc. N.A.S., 16*, 172-175 (1930).

[23] Streater, R. F.: The representations of the oscillator group, *Comm. Math. Physics, 4*, 217-236 (1967).

[24] Treves, F.: *Topological Vector Spaces, Distributions and Kernels*, Academic Press, New York-London (1967).

[25] von Neumann, J.: Die Eindeutigkeit der Schrödingerschen Operatoren, *Math. Ann., 104*, 570-578 (1931).

[26] Warner, G.: *Harmonic Analysis on Semi-simple Lie Groups, I*, Springer-Verlag, New York-Heidelberg-Berlin (1972).

[27] Weyl, H.: *The Theory of Groups and Quantum Mechanics*, Methenen, London (1931).

SYMBOLS

Symbol	Page First Introduced
$\mathfrak{F}$	20
C.C.R.	20
B.F.S.	21
$\mathbf{C}$	21
$\mathbf{C}^n$	21
$z \cdot w$	21
$\bar{z}$	21
$\mathbf{R}$	21
$\mathbf{R}^n$	21
$\mathbf{Z}$	21
$\mathbf{N}$	21
$\mathbf{N}^n$	21
$\underset{\sim}{k}$	21
$\lvert\underset{\sim}{k}\rvert$	21
$\underset{\sim}{k}!$	21
$\underset{\sim}{k} \geq \underset{\sim}{\ell}$	21
$<,>$	21
$\mathbf{V}$	21
$<\cdot,\cdot>$	22
$\mathbf{N}(\mathbf{V}) = \mathbf{N}$	22
$\mathfrak{N}(\mathbf{V}) = \mathfrak{N}$	22
$\mathfrak{Z}$	22
$\mathbf{J}$	23
$(\,,\,)_J$	23
$\|\ \|_J$	23

Symbol	Page First Introduced
ker B_f	32
H	32
E^*	33
Q, P	33
$\mathfrak{H}_S$	33
$\mathfrak{H}_B$	33
$\rho(n)$	33
$\mathfrak{D}$	34
D	34
$C^\infty(N)$	34
χ	34
$d\alpha, dn, d\dot{n}$	34
$L^2(N/D, d\dot{n})$	34
$\mathcal{H}(\mathfrak{H})$	34
$\mathcal{H}_S = L^2(Q, d\underset{\sim}{q})$	35
$T_{Q,P} = T$	35
$T_{\alpha,\beta}$	35
$(\omega'_+, \omega_+)_+$	36
$\\|\cdot\\|_+$	36
$d\omega'$	36
$d\mu$	36
$\mathcal{H}_{B.F.S.} = \mathcal{F}(W_J^+, d\mu)$	36
$(\cdot\,,\cdot)_B$	36
$V_J = V$	36
$\mathcal{F}(W_J^+)$	36
$\mathcal{F}_n$	36
V_c	36
A	36
K	37
$A(\omega_+, q)$	37

Symbol	Page First Introduced
$\begin{pmatrix} m_1 & m_2 & m_3 \\ j_1 & j_2 & j_3 \end{pmatrix}$	56
F	61
$\hat{F}_t$	61
$\mathcal{F}_{2\pi}$	63
ω_J	63
$\mathcal{F}_\pi$	63
(J, τ)	64
$[\omega, \omega']$	66
$\hat{\nu}$	66
$\mathrm{Exp}(V_C)$	66
$A(V_C)$	66
$\delta^{(\underset{\sim}{n},\underset{\sim}{m})}$	67
ν_F	67
$Q(F)$	68
$F^*(z, w)$	68
$\nu_1 \boxed{X} \nu_2$	69
$F_1 \textcircled{M} F_2$	70
$\mathrm{Exp}_k(W_J^+)$	70
$\| \ \|_k$	70
$\mathrm{Exp}_\infty(W_J^+)$	71
$\mathrm{Exp}_{-\infty}(W_J^+)$	71
$\Theta_{\chi,f}(\omega)$	73
$[\chi, f]_F$	74

LOWER BOUND FOR THE GROUND STATE ENERGY OF THE SCHRÖDINGER EQUATION USING THE SHARP FORM OF YOUNG'S INEQUALITY

John F. Barnes
Herm Jan Brascamp*
Elliott H. Lieb*

We derive lower bounds for the ground state energy and free energy of a particle in a potential.

Professor Bargmann has always been interested in basic problems in quantum mechanics. In this paper, dedicated to him, we apply some modern technology to an old problem.

We shall be primarily interested in studying the lowest, or ground state, eigenvalue, E_0, of the Schrodinger operator $(\hbar^2/2m = 1)$

$$H = -\Delta + V(x) .$$

H acts on $L^2(R^n)$. Normally it is difficult to obtain a lower bound to E_0, although an upper bound can easily be obtained from the variational principle. We shall obtain the former by using a recent sharpening and generalization of Young's inequality, in the case that V satisfies the condition

$$I(\alpha) \equiv \int \exp[-\alpha\, V(x)]\, d^n x < \infty \tag{1}$$

for all $\alpha > 0$. We feel that one should be able to obtain a lower bound when (1) is not satisfied, for example when V(x) is a simple square well, but we have not done so. This is an open problem.

*Work partially supported by National Science Foundation Grant Number MPS 71-03375 A03.

Young's Inequality

The normal Young's inequality is that

$$\left| \iint f(x)\, g(x-y)\, h(y)\, d^n x\, d^n y \right| \le \|f\|_p \, \|g\|_q \, \|h\|_r \tag{2}$$

when $p^{-1} + q^{-1} + r^{-1} = 2$ and $p, q, r \ge 1$. Here $\|f\|_p \equiv [\int |f(x)|^p d^n x]^{1/p}$. Recently this inequality has been sharpened and generalized in two ways. The first is that

$$\left| \iint f(x)\, g(x-y)\, h(y)\, d^n x\, d^n y \right| \le [C_p C_q C_r]^n \|f\|_p \, \|g\|_q \, \|h\|_r \tag{3}$$

with

$$(C_p)^2 = p^{1/p} (p')^{-1/p'} \le 1$$

and $(p')^{-1} = 1 - p^{-1}$. A noteworthy point is that equality is reached in (3) when f, g and h are Gaussians which, apart from obvious scale factors, are

$$f(x) = \exp(-p'x^2), \quad g(x) = \exp(-q'x^2), \quad h(x) = \exp(-r'x^2) \,. \tag{4}$$

This result was obtained by W. Beckner [1] and by two of us [2]. The second generalization [2] is the following:

THEOREM 1. *Let* $f_1, \cdots, f_k$ *be complex valued functions on* R^1, *let* $a^1, \cdots, a^k$ *be vectors in* R^N, *let* B *be a real, symmetric, positive semi-definite* N*-square matrix of rank* m, *and let* $p_1, \cdots, p_k$ *satisfy* $p_i \ge 1$ *and* $N \ge \sum_1^k (p_i)^{-1} \ge N - m$. *Then*

$$\left| \int d^N x \exp[-\langle x, Bx\rangle] \prod_{j=1}^{k} f_j(\langle a^j, x\rangle) \right| \le D \prod_{j=1}^{k} \|f_j\|_{p_j} \tag{5}$$

where the optimal constant D, *which depends on the* $\{a^j\}$, $\{p_j\}$ *and* B, *but not on the* $\{f_j\}$, *is determined by restricting the* f_j's *to be Gaussians with mean zero.*

If $k = 3$, $N = 2$, $B = 0$ and $a^1 = (1,0)$, $a^2 = (0,1)$, $a^3 = (1,-1)$, then the previous inequality (3) is obtained when $n = 1$.

A generalization (but by no means the only possible one) of Theorem 1 to functions f_j from R^n to C, which we shall actually use, is the following: Let $x = \{y_1, \cdots, y_N\}$ with $y_i \in R^n$, $\langle a, x\rangle = \sum_1^N a_i y_i$, $\langle x, Bx\rangle = \sum\sum_1^N B_{ij}(y_i, y_j)$, with $(\cdot\,,\cdot)$ being the inner product in R^n, and $d^N x = d^n y_1 \cdots d^n y_N$. Then Theorem 1 is true with Gaussian taken to mean $f_j(y) = \exp(-z|y|^2)$. In particular, the R^n version of (3) results.

REMARK. If one substitutes Gaussians $f_j(x) = \exp(-z_j x^2)$ into (5) and differentiates (5) with respect to the z_j to determine D, the result is a complicated set of algebraic equations for the z_j. It can be shown [2] that *any* solution to these equations yields the optimum value of D.

The Partition Function

We wish to obtain an upper bound to

$$Z(\beta) = \operatorname{Tr} \exp(-\beta H), \qquad \beta > 0 . \tag{6}$$

A preliminary upper bound to $Z(\beta)$, which proves that $Z(\beta)$ is finite, is given by the Golden-Thompson-Symanzik inequality [3, 4, 5, 6, 7]

$$Z(\beta) \leq \operatorname{Tr} \exp(\beta\Delta)\exp(-\beta V) = (4\pi\beta)^{-n/2} \int \exp[-\beta V(x)]\, d^n x . \tag{7}$$

The Trotter product formula [8] states that

$$Z(\beta) = \lim_{N\to\infty} Z_N(\beta) , \tag{8}$$

$$\begin{aligned} Z_N(\beta) &= \operatorname{Tr}\{\exp(\beta\Delta/N)\exp(-\beta V/N)\}^N \\ &= C_N \int d^{nN}x \prod_{j=1}^{N} \exp[-N(x_{j-1}-x_j)^2(4\beta)^{-1} - \beta\, V(x_j)/N] \end{aligned} \tag{9}$$

with $x_j \in R^n$, $x_0 \equiv x_N$ and

$$C_N = (4\pi\beta/N)^{-nN/2} . \tag{10}$$

We shall use Theorem 1, generalized to R^n, to obtain an upper bound to Z_N. The first factor in (9) is the fixed Gaussian $\exp(-\langle x, Bx \rangle)$ of Theorem 1, (note that $\mathrm{rank}(B) = N-1$) and $f_j(x) = \exp(-\beta V(x)/N)$, $j = 1, \cdots, N$. Let $p_1, \cdots, p_N$ have a common value p such that

$$1 \leq N/p \leq N . \tag{11}$$

Replace $f_j(x)$ by $g_j(x) = \exp(-zx^2)$. (Note that the z_j will have a common value z by the remark and by the symmetry of Z_N.) Let $\omega^2 = Nz/\beta$. Then, by Theorem 1,

$$Z_N(\beta) \leq \max_{\omega \geq 0} Q_N(\omega, \beta) \{\|\exp(-\beta V/N)\|_p / \|\exp(-\beta\omega^2 x^2/N\|_p\}^N , \tag{12}$$

where $Q_N(\omega, \beta)$ is $Z_N(\beta)$ with V replaced $\omega^2 x^2$.

The equation determining p in terms of ω is $nN/p = -\omega\, d\ell n\, Q_N(\omega,\beta)/d\omega$. Instead of fixing p, we can and shall fix $\omega > 0$. Thus p depends on N. Now

$$\lim_{N\to\infty} Q_N(\omega, \beta) = [2 \sinh \beta\omega]^{-n} = Q(\omega, \beta) ,$$

the well known harmonic oscillator partition function. As $\ell n\, Q_N$ and $\ell n\, Q$ are convex and differentiable functions of ω^2,

$$\lim_{N\to\infty} d\, \ell n\, Q_N(\omega, \beta)/d\omega = d\, \ell n\, Q(\omega, \beta)/d\omega$$

for $\omega > 0$. Thus,

$$\lim_{N\to\infty} N/p = \beta\omega \coth \beta\omega > 1 .$$

Only the combination N/p appears in the last two factors of (12) and, therefore,

$$Z(\beta) \leq [2 \sinh \beta\omega]^{-n} [I(a)(\omega^2 a/\pi)^{n/2}]^{\beta/a} \tag{13}$$

where

$$a^{-1} = \omega \coth \beta\omega \ .$$

The inequality (13) is our main result. It is true for all $\omega > 0$. As $\omega \to 0$, we recover (7).

The right side of (13) simplifies as $\beta \to \infty$. One knows that

$$E_0 = -\lim_{\beta \to \infty} \beta^{-1} \ln Z(\beta)$$

and hence, using (13),

$$E_0 \geq \max_{\omega > 0} \omega[n + (n/2) \ln(\pi/\omega) - \ln I(1/\omega)] \ . \tag{14}$$

We have used the fact, which is a consequence of (1), that $a \to I(a)$ is continuous.

To test our bound (14), let us consider the potential

$$V(x) = v|x|^\gamma \ ,$$

with $v \geq 0$, $\gamma > 0$. Then

$$I(1/\omega) = \pi^{n/2}(\omega/v)^{n/\gamma}\Gamma(1+n/\gamma)/\Gamma(1+n/2) \ .$$

The right side of (14) can easily be maximized with respect to ω. The result is

$$E_0 \geq n\,v^{2/(\gamma+2)} \frac{\gamma+2}{2\gamma} \left[\frac{\Gamma(1+n/2)}{\Gamma(1+n/\gamma)}\right]^{2\gamma/(\gamma n+2n)} \exp\left(\frac{\gamma-2}{\gamma+2}\right) . \tag{15}$$

A Gaussian variational function yields the upper bound

$$E_0 \leq n\,v^{2/(\gamma+2)} \frac{\gamma+2}{2\gamma} \left[\frac{\gamma\Gamma((n+\gamma)/2))}{n\Gamma(n/2)}\right]^{2/(\gamma+2)} . \tag{16}$$

As $n \to \infty$, (15) and (16) agree with each other up to a relative error of order n^{-1}, i.e.

$$E_0 = [n^2/(2\gamma)]^{\gamma/(\gamma+2)}(\gamma/2+1)[1+O(n^{-1})]\, v^{2/(\gamma+2)} .$$

For a large class of V's, our lower bound (14) seems to be asymptotically exact as $n \to \infty$, but we shall not attempt to prove this here.

To compare (15) and (16) with the exact value of E_0 obtained numerically, we tabulate E_0 for various values of γ, with $v = 1$, and for $n = 1$ and 3. When $\gamma = 1, 2$ and ∞ the solution can be obtained analytically. It is to be noted that E_0 for $n = 3$ happens to be the first excited state, E_1, for $n = 1$. This is so because the ground state of the radial Schrödinger equation is the single node solution to the one-dimensional equation.

γ	eqn. (15)	E_0	eqn. (16)
0	1	1	1
1/4	0.9932	1.079542	1.0874
1/2	0.9908	1.059617	1.0688
1	0.9916	1.018793	1.0242
3/2	0.9955	1.001184	1.0026
2	1	1	1
3	1.0093	1.022948	1.0286
4	1.0158	1.060362	1.0817
5	1.0215	1.102298	1.1476
6	1.0262	1.144802	1.2209
10	1.0384	1.298844	1.5483
20	1.0508	1.560508	2.4346
50	1.0602	1.903191	5.1724
∞	1.0675	2.467401	∞

Table 1: The lower and upper bounds for E_0, (15) and (16) compared to the exact E_0 in one dimension.

γ	eqn. (15)	E_0	eqn. (16)
0	1	1	1
1/4	1.4412	1.492684	1.4981
1/2	1.7782	1.833394	1.8415
1	2.3067	2.338107	2.3448
3/2	2.6999	2.708092	2.7103
2	3	3	3
3	3.4218	3.450563	3.4621
4	3.6999	3.799673	3.8474
5	3.8950	4.089159	4.1968
6	4.0385	4.338599	4.5270
10	4.3610	5.097876	5.7681
20	4.6343	6.219360	8.7173
50	4.8096	7.610400	17.3037
∞	4.9296	9.869604	∞

Table 2: The lower and upper bounds for E_0, (15) and (16) compared to the exact E_0 in three dimensions. E_0 is also the one-dimensional first excited state energy.

JOHN F. BARNES
THEORETICAL DIVISION
LOS ALAMOS SCIENTIFIC LABORATORY
LOS ALAMOS, NEW MEXICO

HERM JAN BRASCAMP
DEPARTMENTS OF MATHEMATICS AND PHYSICS
PRINCETON UNIVERSITY
PRINCETON, NEW JERSEY

ELLIOTT H. LIEB
DEPARTMENTS OF MATHEMATICS AND PHYSICS
PRINCETON UNIVERSITY
PRINCETON, NEW JERSEY

REFERENCES

[1] W. Beckner, Inequalities in Fourier Analysis on R^n, Proc. Nat. Acad. Sci. (U.S.) *72* (1975), 638-641; Inequalities in Fourier Analysis, Ann. Math. *102* (1975), 159-182.

[2] H. J. Brascamp and E. H. Lieb, Best Constants in Young's Inequality, its Converse, and its Generalization to More Than Three Functions, Adv. in Math., in press.

[3] S. Golden, Lower Bounds for the Helmholtz Function, Phys. Rev. *137* (1965), B 1127-1128.

[4] C. J. Thompson, Inequality with Applications in Statistical Mechanics, J. Math. Phys. *6* (1965), 1812-1813.

[5] K. Symanzik, Proof and Refinements of an Inequality of Feynman, J. Math. Phys. *6* (1965), 1155-1156.

[6] M. B. Ruskai, Inequalities for Traces on Von Neumann Algebras, Comm. Math. Phys. *26* (1972), 280-289.

[7] M. Breitenecker and H. R. Gruemm, Note on Trace Inequalities, Comm. Math. Phys. *26* (1972), 276-279.

[8] H. F. Trotter, Approximation of Semigroups of Operators, Pacific J. Math. *8* (1958), 887-919.

ALTERNATIVE THEORIES OF GRAVITATION*

Peter G. Bergmann

Starting with a review of fiber bundle formulation of general relativity and its enrichment by the introduction of torsion, a further generalization of the geometric structure of Riemannian geometry is examined, whose relevance to physics remains, however, dubious. An attempt is made to delimit the probable usefulness of fiber bundles in the exploration of alternative theories of gravitation.

Introduction

Attempts to modify Einstein's general theory of gravitation go back at least to H. Weyl,[1] who replaced the Riemannian metric structure of the space-time manifold by a conformal metric so as to create room for a 1-form that might play the role of electromagnetic potentials. Since then, there have been numerous attempts to change the geometric structure that forms the basis of general relativity, as well as the assumed dynamical laws, so as to achieve a variety of purposes. One of these purposes has been to enrich the geometric structure in order to account for the variety of interactions observed in the physical universe, without sacrificing the unitary nature of that geometry. Other proposals have dealt with the perceived requirements of physics, without concentrating on an esthetically satisfying geometric structure.

To describe all these attempts by a single word, one might replace the traditional terms of "unified" or "unitary" field theories by one more neutral, perhaps "alternative" theories of gravitation and other force fields.[1-4]

*Research supported by the National Science Foundation under Grant GP43759X.

In this paper, which is dedicated to Valya Bargmann, I should like to comment on theories that are suggested by, or that lend themselves to a description by means of fiber bundles, and to contrast them to approaches for which this formalism appears less suitable. The quantum field theorist is apt to look at a field theory as a type of dynamics with an infinite number of degrees of freedom, whereas the differential geometer starts with the notion of space-time manifold, on which all kinds of structures are then grafted. It is the latter approach that lends itself most readily to the fiber bundle formalism. In this formalism all points of the space-time manifold are equivalent, that is to say, not only does the space-time admit arbitrary diffeomorphic mappings on itself, without change in the form of the dynamical laws, but the fibers are all isomorphic to each other, including the structure group that acts on each fiber separately. As a result, physical theories whose elements are presented in the form of fiber bundles will be manifestly covariant, both with respect to curvilinear coordinate transformations and to whatever additional gauge groups are assumed.

Fiber bundles are finite-dimensional manifolds; their topology is relatively manageable. A particular solution of the dynamical laws (the field equations) will constitute a cross section through the bundle.

There are some questions of concern to the physicist for whose exploration the fiber bundle formalism appears not the best method of approach. Any given physical theory will imply a definite law of causality, which is closely related to the Cauchy problem. If the theory is general-relativistic, or if it satisfies any kind of gauge invariance, even a well-set initial-value problem cannot have a formally unique solution; at best all solutions that belong to one set of data are equivalent to each other. Hence all solutions that transform into each other under the invariance group can be collected into so-called equivalence classes; this term is also applied to the collection of all sets of initial-value data that lead to equivalent solutions.

Finally, there are mappings of the equivalence classes on themselves, to be discussed in the final section, that do not belong to the initially intended invariance group but which must be considered to belong to that group. These transformations, foremost among them canonical transformations, do not map fiber on fiber, or world point on world point. The projective mapping that relates the bundle manifold to the base manifold is lost; under these circumstances the fiber bundle becomes unwieldy. The three sections that follow next are, however, cast in the spirit of bundle formalism. They are concerned with geometries that generalize Riemannian geometry along lines that might be expected to lend themselves to the incorporation of additional physical fields into the geometric base.

Weyl-Utiyama Presentation of General Relativity

Consider a differentiable manifold with its tangent bundle, and adopt instead of the usual basis for the latter an arbitrary basis of linearly independent covariant vectors at each point of the manifold, $h^m_\mu(x)$. With their help any vector U^μ may be represented by n numbers

$$U^m = h^m_\mu U^\mu \ . \tag{1}$$

This representation will be bi-unique if the ranges of the two kinds of indices are equal; in this case the contravariant vectors $h^\mu_m(x)$ will be uniquely determined by the covariant basis,

$$h^m_\mu h^\nu_m = \delta^\nu_\mu \ . \tag{2}$$

As the choice of basis vectors at each point of the underlying manifold is arbitrary, all such choices are related to each other by elements of the group $GL(n, R)$, which forms the structure group acting on the fiber whose elements are n-ples of the real numbers U^m.

To relate fibers at neighboring points x and $(x+dx)$ to each other, we introduce the notion of "horizontality," which is equivalent to the concept of parallel transfer. We consider two n-ples of numbers U^m at these two points horizontally related to each other if they obey the equations

$$d_H U^m = -A^m_{\ell\rho}(x) U^\ell dx^\rho , \tag{3}$$

where the coefficients $A^m_{\ell\rho}(x)$ form a structure imposed on our fiber bundle. The expression (3) may be integrated along any curve connecting two points $\bar{x}$ and $\bar{\bar{x}}$ with each other, resulting in a linear mapping of the fiber at $\bar{x}$ on the fiber at $\bar{\bar{x}}$. This mapping will in general depend on the curve chosen to connect the two points with each other. If it does not, the horizontal connection $\underline{A}$ is said to be holonomic (= integrable).

Suppose we require that parallel transfer of the vector components U^μ leads to the same result as horizontal mappings of the fiber. If so, the components of the affine connection are related to the coefficients $\underline{A}$ by the requirement:

$$h^m_{\mu,\rho} + A^m_{\ell\rho} h^\ell_\mu - \Gamma^\nu_{\mu\rho} h^m_\nu = 0 . \tag{4}$$

In what follows it will be assumed that this relation holds.

On the basis of horizontality one can define a covariant derivative with respect to vector and tensor components with respect to the chosen basis vectors. We shall introduce the notation

$$U^m{}_{|\rho} \equiv U^m{}_{,\rho} + A^m{}_{\ell\rho} U^\ell . \tag{5}$$

With its help we can express the special condition that an affine connection Γ be symmetric entirely in terms of the properties of A-coefficients and basis co-vectors:

$$h^m_{\mu|\nu} - h^m_{\nu|\mu} = 0 . \tag{6}$$

Parenthetically it might be remarked that the terms on the left of Equation (6) individually have no covariant significance with respect to curvilinear coordinate transformations, but the difference does, being the exterior derivative of a 1-form.

Suppose that the n-dimensional manifold that serves as our base manifold possesses a Riemannian (indefinite) metric. In that event the basis vectors may be chosen orthonormal, "tetrads" if the base space is four-dimensional. The remaining choice corresponds to the appropriate orthonormal group, the Lorentz group $0(3,1)$ in the case of space-time. This new and smaller structure group replaces $GL(n,R)$. The tetrad components of the metric tensor thereby reduce to invariant numerics, η_{mn}. With respect to horizontal transfers these components will be preserved (the connection will be "metric") if

$$\begin{aligned} A^{mn}{}_{\rho} + A^{nm}{}_{\rho} &= 0 \,, \\ A^{mn}{}_{\rho} &\equiv A^{m}{}_{\ell\rho}\,\eta^{\ell n} \,. \end{aligned} \tag{7}$$

The connection will be Riemannian if both conditions (6) and (7) are satisfied. Together they are equivalent to setting the components of the affine connection equal to the Christoffel symbols.

The quantities $A^{k}{}_{\ell\sigma}$ are the components of a "matrix-valued 1-form." Their transformation law under Lorentz transformations involves the derivatives of the Lorentz matrix. One obtains an object that transforms linearly by forming the commutator, a matrix-valued 2-form:

$$R_{\iota\kappa\ell}{}^{m} = A^{m}{}_{\ell\iota,\kappa} - A^{m}{}_{\ell\kappa,\iota} - A^{m}{}_{s\iota}A^{s}{}_{\ell\kappa} + A^{m}{}_{s\kappa}A^{s}{}_{\ell\iota} \,. \tag{8}$$

With its help, and using the tetrad components, one can form a scalar, and a scalar density; the latter is formally suitable as the integrand of an invariant action principle. The resulting theory is equivalent to conventional general relativity, of which it is well known that one can obtain the usual field equations, if desired, by treating the metric tensor and the components of the affine connection as initially independent variables, according to Palatini.

The formulation of general relativity by way of fiber bundles does, however, suggest generalizations, and these will be discussed next.

Poincaré-Invariant Fibration

A possible generalization of a fiber bundle whose structure group is the (homogeneous) Lorentz group is one whose fibers are acted on by the Poincaré group. We introduce a four-dimensional fiber, whose coordinates may be designated by $u^k (k = 1, 2, 3, 0)$, and which differs from the fiber of the Weyl-Utiyama formalism merely by the prescription that the permitted mappings of the fiber on itself are elements of the Poincaré group,

$$U^{\ell'} = L^{\ell}{}_{s} U^{s} + a^{\ell} . \tag{9}$$

This fiber is no longer a linear vector space, in that its origin may be mapped on any other point of the fiber.[5-9] If we wish to introduce an appropriate rule for the horizontal mapping of one fiber on the fiber at an adjacent world point, this rule should be of the form:

$$du^{\ell} = -[\varepsilon^{\ell}{}_{s\rho}(x) u^{s} + a^{\ell}{}_{\rho}(x)]\, dx^{\rho} . \tag{10}$$

There is, of course, a corresponding commutator, a 2-form:

$$du^{n} = (R^{n}{}_{\ell\iota\kappa} u^{\ell} + S^{n}{}_{\iota\kappa})\, dx^{\iota} \wedge dx^{\kappa} ,$$

$$R^{n}{}_{\ell\iota\kappa} = \varepsilon^{n}{}_{\ell\iota,\kappa} - \varepsilon^{n}{}_{\ell\kappa,\iota} - \varepsilon^{n}{}_{s\iota} \varepsilon^{s}{}_{\ell\kappa} + \varepsilon^{n}{}_{s\kappa} \varepsilon^{s}{}_{\ell\iota} , \tag{11}$$

$$S^{n}{}_{\iota\kappa} = a^{n}{}_{\iota,\kappa} - a^{n}{}_{\kappa,\iota} - \varepsilon^{n}{}_{s\iota} a^{s}{}_{\kappa} + \varepsilon^{n}{}_{s\kappa} a^{s}{}_{\iota} .$$

The two terms of the 2-form (11) are separable, that is to say, not only does each term transform by itself under Lorentz transformations, but $R^{n}{}_{\ell\iota\kappa}$ transforms by itself even under Poincaré transformations. This is because there is a homomorphism between the Poincaré group and the Lorentz group. Given any two objects that transform as "affine-vectors" under Poincaré transformations, i.e. according to Equation (9), their difference is an (ordinary) vector, which is invariant with respect to translations. The obvious law for the parallel transfer of such a vector about an infinitesimal closed loop in space-time involves only the first term of Equation (11), hence the coefficient $R^n{}_{\ell\iota\kappa}$ must be a tensor even under Poincaré transformations.

If we introduce, in analogy to the previous construction, tetrad components, we can form a scalar, and a scalar density, only if these tetrads are invariant with respect to translations. An action density obtained in this manner will be based on the first term of the commutator expression (11) exclusively, with no role given to the form $S^n{}_{\iota\kappa}$. This latter form can be related to source fields, and this route has in fact been proposed by advocates of the importance of "torsion."[5-9]

In these papers it is assumed, as before, that the tetrad components are covariantly constant, in the sense of Equation (4). Hence it is possible to convert the one tetrad index of $S^n{}_{\iota\kappa}$ into a coordinate superscript. There must exist a class of geometric objects analogous to the affine-vectors introduced in Equation (9), say u^κ, for which the law of parallel transport is:

$$du^\kappa = -(\Gamma^\kappa_{\lambda\rho} u^\lambda + a^\kappa_\rho) dx^\rho \ . \tag{12}$$

If the definition of horizontality adopted restricts the a-variables to

$$\begin{aligned} a^\kappa_\rho &= -\delta^\kappa_\rho \ , \\ a^\ell_\rho &= -h^\ell_\rho \ , \end{aligned} \tag{13}$$

then the u^κ might be called radius vectors,[9] as they change on horizontal displacements like the rectilinear coordinates do in a Euclidean space. With the adoption of the restriction (13), one obtains the equality:

$$S^\nu{}_{\iota\kappa} = \Gamma^\nu_{\kappa\iota} - \Gamma^\nu_{\iota\kappa} \ . \tag{14}$$

Formally, this generalization of Riemannian geometry, and of Einstein's theory of gravitation, is very slight, indeed. Though the affine connection is assumed to be asymmetric, its skew-symmetric part, the torsion, can be split off invariantly, and the enrichment of the geometry amounts in effect to the addition of the tensor (14) to the machinery of unmodified Riemannian geometry.

This additional geometric object might eventually turn out to be essential, as the depository of a special coupling of the spin density of matter to the gravitational field. But in the present form of the theory the torsion is coupled to the local spin density algebraically, and it does not propagate beyond the extension of its sources. That is to say, there is no distant spin-spin coupling, except by way of the gravitational potentials that form the basis of conventional gravitational theory. Very possibly, with the same geometric building blocks, the field equations of the modified theory might be constructed so as to permit the propagation of torsion beyond its sources; this remains to be seen.

The Conformal Group

As long as the structure group of a fiber bundle is homomorphic with the Lorentz group, it is reasonable that geometric objects decompose into those encountered in Riemannian geometry and additional tensorial structures. In this section an alternative approach will be described that leads to an interesting geometry, but which has so far failed to yield a physical field theory.

The point of departure was an attempt to retain within the structure group the Poincaré group as a subgroup, but without the possibility of a homomorphic reduction. The group adopted here is the so-called conformal group, which is isomorphic to $O(4,2)$. It is a 15-parametric Lie group, which incorporates, in addition to the usual elements of the Poincaré group, an isotropic change of scale (or "inflation") and mappings in which the two frames of reference are "accelerated" with respect to each other. The infinitesimal mappings that can be performed on a four-dimensional fiber coordinated by u^k are:

$$du^k = a^k + b^k{}_s u^s + cu^k + e_s\left(u^k u^s - \frac{1}{2}\eta^{ks} u^r u_r\right).$$
$$b_{k\ell} + b_{\ell k} = 0 \tag{15}$$

These mappings are not even affine. A corresponding condition of horizontality will be:

$$du^k = -\left[\alpha^k{}_\rho + \beta^k{}_{s\rho} u^s + \gamma_\rho u^k + \varepsilon_{s\rho}\left(u^k u^s - \frac{1}{2}\eta^{ks} u^r u_r\right)\right] dx^\rho, \tag{16}$$

with a similar commutator two-form,

$$du^n = -\Bigg[S^n{}_{\iota\kappa} + R^n{}_{\ell\iota\kappa} u^\ell + T_{\iota\kappa} u^n + + V_{s\iota\kappa}\left(u^n u^s - \frac{1}{2}\eta^{ns} u^r u_r\right)\Bigg] dx^\iota \wedge dx^\kappa. \tag{17}$$

Perhaps of some interest is the appearance of a scalar-valued 1-form in (16), and a scalar-valued 2-form in the commutator (17), which, optimistically, could be related to the electromagnetic field. However, under the assumed structure group the various 2-forms appearing in (17) are not separate geometric objects, but will combine with each other.

There is a "fiber metric," η_{mn}, which will be form-invariant if it is transformed according to the rule:

$$\delta\eta_{mn} = -(du^s)_{,m}\eta_{sn} - (du^s)_{,n}\eta_{sm} + \frac{1}{2}(du^s)_{,s}\eta_{mn} = 0. \tag{18}$$

That is to say, the transformation law is that of a tensor density of weight $-\frac{1}{2}$.

As long as we are concerned only with the definition of horizontality and the structure of the associated holonomy group, it does not really matter whether we adopt the structure group itself as the model for the fiber (so-called principal bundles) or whether the fiber has a dimensionality of its own. Even the existence of an invariant, such as the fiber bundle metric, has essentially only heuristic significance. The actual structure of the fiber enters substantively when we adopt the point of view of Yang and Mills,[3] and define a field over the base manifold. The field is then a cross section through the bundle, and the covariant derivative of a field is defined as the difference between the partial derivative of the field (which depends also on the variability of the gauge frame employed) and the change caused by a horizontal displacement in the bundle. The components of the field itself must form a geometric object, but though the transformation law of any geometric object is a representation of the gauge group (or the structure group, as we have called it up to now), different representations correspond to fields with different properties.

A possible object might be a tetrad, h^n_ν, which would make it possible, as before, to convert a world vector U^ν into four scalars, U^n. These four scalars will not only be functions of the space-time coordinates but will also depend on the fiber coordinates u^k. This dependence will change under a mapping of the type (15); it is desirable to assign to the tetrad components, and hence to U^n, transformation properties that will help to preserve the general form of that u-dependence. This aim will be achieved if the h^n_ν are assumed to transform as vectors (with respect to the index n) under fiber mappings of the type (15), and if initially they are assumed to have a u-dependence of the same kind as the right-hand side of Equation (15). Because the infinitesimal mappings of type (15) form a Lie algebra, this u-dependence will not change in kind. With these assumptions U^n are, and remain, second-degree polynomials in the fiber coordinates u^k, whose coefficients may be considered to be the 15 components of an x-field that transform together when acted on by the conformal structure group.

With the help of the conformal metric of the fiber, η_{mn}, and of the tetrad components h^n_ν one can construct an ordinary metric tensor, or, more precisely, a conformal metric tensor density. In one gauge frame, one can choose tetrad components that are u-independent, which means that in any gauge frame their Lie derivatives with respect to each other must vanish:

$$\frac{\partial h^m_\mu}{\partial u^s} h^s_\nu - \frac{\partial h^m_\nu}{\partial u^s} h^s_\mu = 0 . \tag{19}$$

The conformal metric of the fiber is form-invariant under the transformation law (18), hence a gauge-invariant metric can be put together only by the formula:

$$g_{\mu\nu} = h^m_\mu h^n_\nu (\det h)^{-\frac{1}{2}} \eta_{mn} . \tag{20}$$

This is a tensor density of weight $\left(-\frac{1}{2}\right)$, no matter what the assumed coordinate-weight of the tetrad component is, and its own determinant is -1. To make a real metric out of the expression (20) one would have to introduce a scalar density as an additional building block.

This is not the only, or even the worst difficulty that must be overcome before a new (speculative) generalized theory of gravitation can be constructed. An action density that is linear in the "curvature," Equation (17), cannot be constructed with the help of the tetrads because the number of indices of the 2-form is odd. True, the second and the third term within the square bracket on the right have coefficients with even numbers of indices, but by themselves these terms are not gauge-invariant. One might attempt to construct an action density quadratic in the curvature; the physical implications of the resulting dynamical laws are obscure. An action density obtained by this, or any similar route, will be u-independent only if the u-Lie derivatives of the various factors with respect to the tetrad components vanish. None of these problems has been solved.

There is an alternative procedure available. The conformal group will act linearly on a six-dimensional vector space, or on its non-linear subspace that consists of null lines through the origin. Suppose Latin indices are permitted to assume the values 0, 1, 2, 3, 5, 6, with 5 and 6 being null directions, so that the fiber metric has the form

$$\mathcal{H} = \left(\begin{array}{c|c} \eta_0 & 0 \\ \hline 0 & \begin{array}{c|c} 0 & 1 \\ \hline 1 & 0 \end{array} \end{array}\right) . \tag{21}$$

Null directions are then characterized by the condition

$$\sum_{0,\cdots,3} \eta_{\mu\nu} X^{\mu} X^{\nu} + 2 X^5 X^6 = 0 . \qquad (22)$$

In this formalism there will be a condition of horizontality, and a resulting curvature, thus:

$$dX^n = R^n{}_{\ell\iota\kappa} X^{\ell} dx^{\iota} \wedge dx^{\kappa} . \qquad (23)$$

The coefficients $R^n{}_{\ell\iota\kappa}$ will be functions of the space-time coordinates, but independent of the fiber coordinates X^m. If we introduce tetrad components h^n_{ν}, which also are to depend only on the space-time coordinates, we can construct from them an ordinary metric tensor,

$$g_{\mu\nu} = \mathcal{H}_{mn} h^m_{\mu} h^n_{\nu} , \qquad (24)$$

but the tetrad tensor

$$g^{mn} = h^m_{\mu} h^n_{\nu} g^{\mu\nu}$$

will be a singular matrix, and hence not equal to the fiber metric. The symbols

$$g^b_a \equiv \mathcal{H}_{as} g^{sb} \qquad (26)$$

are idempotent matrices.

If the horizontal mappings and the tetrads introduced are to be attuned to each other, it is reasonable to require that both the fiber metric and the fiber tensor g^{mn} be covariantly constant. In that case, the subspaces belonging to the eigenvalues 1 and 0 of the matrix (26) are invariant under the holonomy group. Accordingly, the truly irreducible structure groups of the theory are the subgroups of the conformal group that map these subspaces on themselves. And as one of these subspaces is four-dimensional, the apparent generalization falls to the ground.

Non-Local Transformations

Ordinarily general-relativistic theories involve various structures (fields) over differentiable manifolds. Their dynamical laws have the form of differential equations interconnecting the field. The theory as a whole must be form-invariant with respect to diffeomorphic mappings of the space-time manifold on itself, with the fields at each world point determining those at the image point.

There is at least one important formulation of the theory of general relativity involving more general mappings than space-time diffeomorphisms; this is Dirac's Hamiltonian version of the theory.[10] In the pure theory of gravitation the Ricci-flat manifolds among all Lorentz-Riemannian metric fields obey the dynamical laws; similar statements can be made of more comprehensive models. Though these formulations are valid, they are not the best for studying the dynamic structure of a given physical theory. One would like to know, for instance, whether the theory is (locally) Cauchy, and if so, what data precisely must be given in a three-dimensional space-like domain to provide for unique propagation off that hypersurface.

Dirac has identified as a possible, and very intuitive set of such data the internal geometry of the hypersurface, $g_{mn}(m, n = 1, 2, 3)$, and the Lie derivatives of that metric with respect to a unit vector field perpendicular to the hypersurface, v_{mn}. (These have simple algebraic relations to the canonical momentum densities actually used by Dirac, his p^{mn}.) It turns out that the field equations impose on these variables at each 3-point four differential conditions, the so-called Hamiltonian constraints. The propagation off the initial hypersurface is unique, except for four-dimensional diffeomorphisms that map the initial hypersurface on itself but otherwise are not restricted.

Consider now two (four-dimensional) vector fields representing infinitesimal diffeomorphisms and form their commutator, that is to say, the Lie derivative of one with respect to the other. To do so on a three-dimensional hypersurface, one will need the off-surface derivatives of the vectors just in order to form the commutator itself on the hypersurface, and generally the (i+1)th order off-surface derivatives in order to form the i-th derivatives of the commutator field. Accordingly, infinitesimal coordinate transformations of this type, and the associated infinitesimal changes of the field variables, cannot be set up as canonical transformations of the Cauchy data, whose Lie algebra would have to correspond to Poisson brackets. Even if the generator of an infinitesimal coordinate transformation is to be some functional over the Cauchy hypersurface, involving the vector fields that represent the infinitesimal mappings, there is no way by which Poisson brackets between two such functionals can yield off-surface derivatives of (arbitrary) space-time functions that are of higher order than those involved in the functionals themselves.

Nevertheless Dirac succeeded in constructing a complete Hamiltonian formalism. His procedure works because there are mappings with different, and acceptable properties. These are diffeomorphisms of the space-time manifold that involve in some way the metric (or other field variables) present. It is not the bare space-time manifold that is mapped on itself, but Dirac's phase space. The "points" of this phase space represent (acceptable) sets of Cauchy data on three-dimensional hypersurfaces. There are phase space mappings corresponding to coordinate transformations. But the image of a given world point depends both on the point itself and on the metric of that point's four-dimensional neighborhood.[11]

There is no simple relationship between Dirac's phase space and fiber bundles over the space-time manifold. Canonical fields are defined on three-dimensional hypersurfaces, whereas the standard bundles of relativistic theories are erected over four-dimensional base spaces. A particular four-dimensional Ricci-flat space-time that is coordinated can be represented in phase space as a parametrized curve. Each point on the curve represents one time-slice, and the parametrization corresponds to the chosen time coordinate. Without a specific coordinatization a particular space-time will be represented by an infinite-dimensional region of phase space, a so-called equivalence class, which comprises all the possible curves obtained from various coordinatizations. That same Ricci-flat space time will be represented by a family of cross sections through the bundle of metrics. A world point will be represented by a single fiber, which, by definition, is vertical.

Under an infinitesimal diffeomorphism of the usual kind, a curve within the equivalence class of Dirac's phase space will be mapped on a neighboring curve within the same equivalence class. Similarly, a cross section through the fiber bundle will be mapped on a slightly different cross section, and a fiber on a neighboring fiber. For the bundle as a whole, the mapping is point-to-point, whereas for the phase space it is not.

Now consider the kind of infinitesimal coordinate transformation that might be called "canonical": *For any set of complete and permissible* data on a space-like hypersurface, there is an infinitesimal mapping to a slightly different set of data, generated by a specified Hamiltonian generator (the integral over a linear combination of the Hamiltonian constraints, with specified coefficients). In the phase space this mapping is point-to-point, and a trajectory again is mapped on a new trajectory within the same equivalence class. For the bundle, the mapping is no longer point-to-point. Depending on the generality of the canonical transformation considered (e.g. whether the coefficients of the constraints in the generator are numerics, or depend on the canonical variables as well), the map of a given point in the bundle will depend on additional properties of the chosen cross section through that point. A cross section will be mapped on a cross section, but a fiber will not be mapped on a fiber, nor necessarily on a domain having the same dimension as the original fiber. And whereas the projection of the fiber on the base manifold is, by definition, a point, the projection of the fiber's image on the base space is not. Infinitesimal canonical coordinate transformations do not preserve the integrity of a world point, though they do preserve the integrity of the equivalence classes, which are in fact the orbits of a permissible point in phase space under the totality of all coordinate transformations of whatever kind.

In summary, the notion of fiber bundle appears appropriate, and it may indeed be a powerful tool, in the construction and analysis of classical field theories. Its usefulness becomes questionable in the consideration of dynamical properties (particularly Cauchy problems), and in the speculative exploration of conceptual structures in which the concepts of world point and of space-time manifold are to be modified.

Acknowledgments

The author wishes to acknowledge with gratitude helpful discussions with J. N. Goldberg, D. Lerner, A. Trautman, and J. Weinberg.

PETER G. BERGMANN
DEPARTMENT OF PHYSICS
SYRACUSE UNIVERSITY
SYRACUSE, NEW YORK

REFERENCES

[1] H. Weyl, Preuss. Akad. Wiss. Ber. 1918, p. 465; Ann. d. Physik *59*, 101 (1919).

[2] M. A. Tonnelat, Les Théories Unitaires de l'Electromagnétisme et de la Gravitation, Gauthier-Villars, Paris, 1965. (Extensive bibliography).

[3] C. N. Yang and R. Mills, Phys. Rev. *96*, 191 (1954).

[4] H. Weyl, Phys. Rev. *77*, 699 (1950); R. Utiyama, Phys. Rev. *101*, 1597 (1956).

[5] E. Cartan, Ann. Ec. Norm. *40*, 325 (1923).

[6] T. W. B. Kibble, J. Math. Phys. *2*, 212 (1961).

[7] R. Kerner, Ann. Inst. H. Poincaré *9*, 143 (1968).

[8] F. W. Hehl, Gen. Rel. Grav. *4*, 333 (1973), *5*, 491 (1974).

[9] A. Trautman, Reports Math. Phys. *1*, 29 (1970); Ist. Naz. Alta Matematica Symposia Mathematica *12*, 139 (1973). (Extensive bibliography).

[10] P. A. M. Dirac, Roy. Soc. London Proc. *A 246*, 333 (1958); Phys. Rev. *114*, 924 (1959).

[11] P. G. Bergmann and A. Komar, Intl. J. Theoret. Phys. *5*, 15 (1972).

GENERALIZED WRONSKIAN RELATIONS: A NOVEL APPROACH TO BARGMANN-EQUIVALENT AND PHASE-EQUIVALENT POTENTIALS

F. Calogero

Generalized wronskian-type equations are derived, yielding relations between the values that the solutions of a linear second order differential equation and their first derivatives take at the two ends of an interval, and integrals over the same interval of the solutions themselves times appropriate (nonlinear) combinations of the functions entering as coefficients in the differential equation. These relations are then used to derive a number of results in potential scattering theory, including a prescription yielding pairs of Bargmann-equivalent and phase-equivalent potentials.

1. *Introduction*

In this paper we derive a generalization of the usual wronskian theorem, providing relations between the values of the solutions of linear second-order differential equations, and of their first derivatives, evaluated at the two ends of an interval of values of the independent variable, and an integral, over the same interval, of the solutions themselves multiplied by appropriate coefficients derived directly from the differential equations. These relations are then shown to provide a flexible and powerful tool to extract information about the solutions of the differential equation, by using them in the context of the nonrelativistic quantal scattering problem based on the Schroedinger equation.

The generalized wronskian relations are derived in Section 2. In Section 3 they are applied to the conventional potential scattering problem, i.e. to the radial Schroedinger equation; the main results obtained, some of them quite new, relate to Bargmann-equivalent and phase-equivalent potentials. It is also possible to apply the same approach to the one-dimensional scattering and bound-state problem; the corresponding results, while interesting *per se*, are especially important in connection with the recently discovered possibility to solve certain nonlinear evolution equations by the so called "inverse scattering method." But these results fall outside the scope of the present paper and will therefore be published elsewhere.

2. *Generalized wronskian relations*

Let

(2.1) $$\psi_j''(x) = f_j(x)\psi_j(x), \quad j = 1, 2 ,$$

and set, for notational convenience

(2.2) $$s(x) = f_1(x) + f_2(x), \quad d(x) = f_1(x) - f_2(x) .$$

It is then well known that the wronskian of the functions $\psi_1(x)$ and $\psi_2(x)$ satisfies the relation

(2.3) $$\{\psi_1'(x)\psi_2(x) - \psi_1(x)\psi_2'(x)\}\Big|_{x_1}^{x_2} = \int_{x_1}^{x_2} dx\, \psi_1(x)\psi_2(x)\, d(x) .$$

Another, more general, wronskian-type relation may be obtained as follows. Let

(2.4) $$\tilde{\psi}(x) = \phi(x)\psi_2'(x)$$

and note that, as a consequence of (2.1), this function satisfies the differential equation

(2.5) $$\tilde{\psi}''(x) = (2f_2\phi' + f_2'\phi)\psi_2 + (\phi'' + f_2\phi)\psi_2' .$$

Consider next the wronskian between the functions $\psi_1(x)$ and $\tilde{\psi}(x)$, and using equations (2.1) and (2.5), obtain

(2.6)
$$\{\psi_1'(x)\tilde{\psi}(x)-\psi_1(x)\tilde{\psi}'(x)\}\Big|_{x_1}^{x_2}$$
$$=\int_{x_1}^{x_2} dx\,\{\psi_1(x)\psi_2(x)H(x)+\psi_1(x)\psi_2'(x)L(x)\}$$

with

(2.7)
$$H(x) = -2f_2\phi' - f_2'\phi \ ,$$

(2.8)
$$L(x) = -\phi'' + (f_1-f_2)\phi \ .$$

Focus now attention on the second term in the right-hand side of equatior (2.6), and by partial integration cast it in a form similar to the first term. This can be done as follows. First note that

(2.9)
$$\frac{1}{2}\int_{x_1}^{x_2} dx\,\psi_1\psi_2'\,L = \frac{1}{2}\{\psi_1(x)\psi_2(x)L(x)\}\Big|_{x_1}^{x_2}$$
$$-\frac{1}{2}\int_{x_1}^{x_2} dx\,\psi_1\psi_2 L' - \frac{1}{2}\int_{x_1}^{x_2} dx\,\psi_1'\psi_2\,L \ ,$$

so that

(2.10)
$$\int_{x_1}^{x_2} dx\,\psi_1\psi_2'\,L = \frac{1}{2}\{\psi_1(x)\psi_2(x)L(x)\}\Big|_{x_1}^{x_2}$$
$$-\frac{1}{2}\int_{x_1}^{x_2} dx\,\psi_1\psi_2 L' - \frac{1}{2}\int_{x_1}^{x_2} dx(\psi_1'\psi_2-\psi_1\psi_2')L \, .$$

Then introduce the quantity

(2.11)
$$M(x) = \int^{x} dx'\ L(x')$$

to integrate once more by parts the last integral in (2.10). Using equations (2.1), this yields

$$(2.12)\quad \int_{x_1}^{x_2} dx\,(\psi_1'\psi_2 - \psi_1\psi_2')L = \{[\psi_1'(x)\psi_2(x) - \psi_1(x)\psi_2'(x)]M(x)\}\Big|_{x_1}^{x_2} - \int_{x_1}^{x_2} dx\,\psi_1\psi_2(f_1 - f_2)M\,.$$

We can now rewrite equation (2.6) using (2.10) and (2.12), as well as the definitions (2.2) and (2.4) and the equations (2.1) and (2.3). We thus obtain our final result, that reads

$$(2.13)\quad \Big\{\psi_1(x)\psi_2(x)[\phi''(x) - s(x)\phi(x)] + 2\psi_1'(x)\psi_2'(x)\phi(x) + -[\psi_1'(x)\psi_2(x) + \psi_1(x)\psi_2'(x)]\,\phi'(x) + [\psi_1'(x)\psi_2(x) - \psi_1(x)\psi_2'(x)]\int_{x_0}^{x} dx'\,d(x')\phi(x')\Big\}\Big|_{x_1}^{x_2}$$
$$= \int_{x_1}^{x_2} dx\,\psi_1(x)\psi_2(x)\Big\{\phi'''(x) - 2s(x)\phi'(x) - s'(x)\phi + d(x)\int_{x_0}^{x} dx'\,d(x')\phi(x')\Big\}\,.$$

It should be emphasized that the function $\phi(x)$ that enters this equation is arbitrary, except for the properties required to justify the various integrations by parts that have been performed. It is clearly sufficient that $\phi(x)$ be thrice differentiable in the interval (x_1, x_2), with $\phi'''(x)$ finite-valued. Some obvious regularity requirements on $f_j(x)$, or equivalently $s(x)$ and $d(x)$, are of course also required.

The validity of (2.13) is a consequence of the differential equations (2.1), and in view of the flexibility implied by the presence of the arbitrary function $\phi(x)$, may be used to extract a lot of information about their solutions $\psi_j(x)$. It is remarkable that the two functions $f_j(x)$ enter this relation naturally in the combinations $s(x)$ and $d(x)$, equations (2.2).

The bulk of this paper is devoted to indicate how equation (2.13), together with the usual wronskian relation (2.3), may be exploited in the context of nonrelativistic scattering theory. The simplicity with which a number of results, some of them new, are yielded by this approach, suggests that this method might be useful also in other problems of applied mathematics, where Sturm-Liouville type equations play a rôle.

We are not aware of other investigations based on such an approach, nor have we seen any previous display of equation (2.13), that may, however, have been considered sufficiently clumsy not to deserve special notice; indeed its merit rests in its usefulness rather than in its elegance. A special case of this equation had been used to derive integral expressions for the difference between scattering phase shifts corresponding to different angular momenta.[1]

It appears that equations (2.3) and (2.13) are the only ones of this kind that can be obtained. It is remarkable that equation (2.13) is *symmetrical* in ψ_1 and ψ_2, even though its derivation given above did not treat these two functions symmetrically; indeed, analogous treatments with different starting points are possible, but they eventually reproduce equation (2.13), or yield trivial identities, or give equation (2.3). In contrast to equation (2.13), equation (2.3) (that has of course been known and used for more than a century) contains no arbitrary function and is *antisymmetrical* in ψ_1 and ψ_2. The more general nature of equation (2.13) relative to equation (2.3) is evidenced by the fact that the former implies the latter, due to the arbitrariness in the choice of x_0. A less direct way to obtain the same result obtains setting $\phi(x) = \psi_1(x)\psi_2(x)$ in equation (2.13) and integrating the right-hand side of equation (2.13), getting

$$(2.14a)\quad \left\{\psi_1'(x)\psi_2(x)-\psi_1(x)\psi_2'(x) - \int_{x_0}^{x} dx'\psi_1(x')\psi_2(x')d(x')\right\}^2 \Bigg|_{x_1}^{x_2} = 0$$

or, equivalently,

$$\left\{ [\psi_1'(x)\psi_2(x) - \psi_1(x)\psi_2'(x)]\Big|_{x_1}^{x_2} - \int_{x_1}^{x_2} dx\, \psi_1(x)\psi_2(x)\, d(x) \right\} \tag{2.14b}$$

$$\cdot \left\{ [\psi_1'(x_2)\psi_2(x_2) - \psi_1(x_2)\psi_2'(x_2)] + [\psi_1'(x_1(x_1)\psi_2(x_1) - \psi_1(x_1)\psi_2'(x_1)] \right.$$

$$\left. + \int_{x_1}^{x_0} dx\, \psi_1(x)\psi_2(x) d(x) + \int_{x_2}^{x_0} dx\, \psi_1(x)\psi_2(x) d(x) \right\} = 0 \ ,$$

that, in view of the arbitrariness of x_0, implies of course equation (2.3). To obtain equation (2.14a), we have used the fact that, for $\phi(x) = \psi_1(x)\psi_2(x)$, the right-hand side of equation (2.13) is exactly integrable, as implied by the possibility to rewrite it, quite generally, in the form

$$\int_{x_1}^{x_2} dx\, [\psi_1(x)\psi_2(x)/\phi(x)]$$

$$\cdot \left\{ \phi''(x)\phi(x) - \frac{1}{2}[\phi'(x)]^2 - s(x)\phi^2(x) + \frac{1}{2}\left[\int_{x_0}^{x} dx' d(x')\phi(x')\right]^2 \right\} .$$

In the special case $\psi_1(x) = \psi_2(x) = \psi(x)$, $f_1(x) = f_2(x) = f(x)$, both sides of equation (2.3) vanish, while equation (2.13) yields the nontrivial relation

$$\left\{ \psi^2(x)[\phi''(x) - 2f(x)\phi(x)] + 2[\psi'(x)]^2\phi(x) \right. \tag{2.15}$$

$$\left. - 2\psi'(x)\psi(x)\phi'(x) \right\}\Big|_{x_1}^{x_2}$$

$$= \int_{x_1}^{x_2} dx\, [\psi(x)]^2 \{\phi'''(x) - 4f(x)\phi'(x) - 2f'(x)\phi(x)\} \ .$$

However, also from equation (2.3), a nontrivial (and well-known) relation can be obtained, setting $f_1(x) = f(x, t+dt)$, $f_2(x) = f(x,t)$, $dt \to 0$. We thus get (indicating for short with a dot partial differentiation with respect to t, and as usual with a dash partial differentiation with respect to x):

$$\{\psi'(x,t)\dot{\psi}(x,t) - \psi(x,t)\dot{\psi}'(x,t)\}\Big|_{x_1}^{x_2} = \int_{x_1}^{x_2} dx\,[\psi(x,t)]^2\,\dot{f}(x,t) \ , \tag{2.16}$$

where, of course, $\psi(x,t)$ satisfies now the differential equation

$$\psi''(x,t) = f(x,t)\psi(x,t) \ . \tag{2.17}$$

3. *Applications to the radial Schroedinger equation*

As is well known, the 3-dimensional Schroedinger equation can be reduced to an infinite sequence of radial equations in the single variable r – each of these equations being a second order linear differential equation, disconnected from the others if, as we assume hereafter, the potential in the original Schroedinger equation depends only on the radial variable r (central problems). In this section we discuss a number of results that can be obtained applying the generalized wronskian technique of the preceding section to the radial Schroedinger equation. Although in the first subsection below we review tersely some results associated with the radial Schroedinger equation, we do assume the reader to be already familiar with this problem.

3.1. *Notation and preliminaries*

In appropriate units $(\hbar^2/2m = 1)$, the radial Schroedinger equation for energy κ^2 reads

$$\psi_j''(\kappa,r) = [V_j(r) - \kappa^2]\psi_j(\kappa,r), \quad r \geq 0, \quad j = 1,2 \ , \tag{3.1.1}$$

and is associated with the boundary conditions

(3.1.2) $$\psi_j(\kappa, 0) = 0$$

(3.1.3) $$\lim_{r \to \infty} \{\psi_j(\kappa, r) - \sin[\kappa r + \eta_j(\kappa)]\} = 0 .$$

Equation (3.1.2) characterizes, up to a multiplicative constant, the "physical" solutions $\psi_j(\kappa, r)$; their detailed behavior as $r \to 0$ depends on the behavior at the origin of the functions $V_j(r)$ (that are termed "potentials" hereafter, even though they are generally the sum of the actual potentials $v_j(r)$ and of the centrifugal terms:

(3.1.4) $$V_j(r) = v_j(r) + \ell_j(\ell_j + 1)/r^2) .$$

Specifically, if the potential is less singular at the origin than $1/r^2$,

(3.1.5) $$\lim_{r \to 0} [r^{2-\varepsilon} V_j(r)] = 0, \quad \varepsilon > 0 ,$$

then

(3.1.6) $$\lim_{r \to 0} [\psi_j(\kappa, r)/r] = \psi_j'(\kappa, 0) \neq 0 ;$$

if the potential is proportional to r^{-2} (as is the case when the centrifugal term is present and dominant as $r \to 0$),

(3.1.7) $$\lim_{r \to 0} [r^2 V_j(r)] = \ell_j(\ell_j + 1), \quad \ell_j > 0 ,$$

then

(3.1.8) $$\lim_{r \to 0} [\psi_j(\kappa, r)/r^{\ell_j+1}] = \text{const} \neq 0 ;$$

and if the potential is more singular at the origin than r^{-2},

(3.1.9) $$\lim_{r \to 0} [r^m V_j(r)] = c_j > 0, \quad m > 2 ,$$

then $\psi_j(\kappa, r)$ vanishes faster than any power of r as $r \to 0$,

(3.1.10) $$\lim_{r \to 0} [r^{-n} \psi_j(\kappa, r)] = 0 \ .$$

The first case obtains for s-wave scattering on regular potentials, the second for higher-wave scattering on regular potentials, and the last for "singular" potentials (that make physical sense only if they are repulsive at the origin).

The asymptotic boundary condition (3.1.3) characterizes uniquely the solution $\psi_j(\kappa, r)$, by specifying its normalization, and defines the "phase shift" $\eta_j(\kappa)$ corresponding to the potential $V_j(r)$ (the actual scattering phase shift obtains adding $\ell_j \pi/2$ to $\eta_j(\kappa)$). Note that, in writing equation (3.1.3), we have assumed that the potentials $V_j(r)$ vanish asymptotically faster than $1/r$,

(3.1.11) $$\lim_{r \to \infty} [r^{1+\varepsilon} V_j(r)] = 0 , \quad \varepsilon > 0 \ ;$$

this assumption will be retained hereafter. We moreover assume the potentials to be finite-valued for $r > 0$.

We define the "S-matrix" corresponding to the potential $V_j(r)$ to be the quantity

(3.1.12) $$S_j(\kappa) = \exp[2i\, \eta_j(\kappa)] \ .$$

If $V_j(r)$ is real, $\eta_j(\kappa)$ is also real, and $S_j(\kappa)$ is unitary:

(3.1.13) $$|S_j(\kappa)| = 1 \ .$$

We also recall the well-known formal symmetry property $\eta_j(-\kappa) = -\eta_j(\kappa)$, that yields

(3.1.14) $$S(\kappa)\, S(-\kappa) = 1 \ .$$

In addition to the positive energy solutions corresponding to the scattering regime, the radial Schroedinger equation (3.1.1) (with (3.1.2)) may possess, for some negative values of the energy, normalizable solutions, characterized by the asymptotic behavior

(3.1.15) $$\psi_j(i p_{jn}, r) \underset{r\to\infty}{\longrightarrow} C_{jn} \exp(-p_{jn} r), \quad p_{jn} > 0 .$$

The index j of p_{jn} is a reminder that the binding energies p_{jn}^2 depend generally on the potential; the index n labels different bound states of the same potential, and will be omitted whenever possible. The hypotheses that we have made on the potentials imply that only a finite number of bound states may occur. To the (eventual) presence of bound states corresponds the well-known property of the functions $S_j(\kappa)$, to have simple poles for $\kappa = i p_j$, $p_j > 0$.

Hereafter we use the notation

(3.1.16a) $$S(r) = V_1(r) + V_2(r), \quad D(r) = V_1(r) - V_2(r)$$

implying

(3.1.16b) $$V_1(r) = \frac{1}{2}[S(r)+D(r)], \quad V_2(r) = \frac{1}{2}[S(r)-D(r)] .$$

The different contexts in which the S-matrix functions $S_j(\kappa)$, and the sum of the two potentials S(r) appear, should eliminate any ambiguity that might possibly arise from the use of the same letter to denote these quantities.

3.2. *Expressions for phase shift differences*

We now apply the wronskian relations of Section 2 to the potential scattering problem. This will be done identifying the differential equation (2.1) with the radial Schroedinger equation (3.1.1), i.e. setting $x \equiv r$ and

(3.2.1) $$f_j = V_j - \kappa^2 ,$$

considering the wronskian relations (2.3) and (2.13) with $x_1 = 0$ and $x_2 = \infty$ and using the boundary conditions (3.1.2) and (3.1.3). We get in this manner, from (2.3), the well-known relation

$$\kappa \sin[\eta_1(\kappa) - \eta_2(\kappa)] = -\int_0^\infty dr\, \psi_1(\kappa, r)\psi_2(\kappa, r) D(r) , \tag{3.2.2}$$

and from (2.13), the two equations

$$\kappa^2 \cos[\eta_1(\kappa) - \eta_2(\kappa)] = \lim_{r\to 0} [\psi_1'(\kappa, r)\psi_2'(\kappa, r)] - \frac{1}{2}\int_0^\infty dr\, \psi_1(\kappa, r)\psi_2(\kappa, r)\left\{S'(r) + D(r)\int_r^\infty dr'\, D(r')\right\} , \tag{3.2.3}$$

$$\begin{aligned} -4\kappa^2 \int_0^\infty dr\, \psi_1(\kappa, r)\psi_2(\kappa, r) F'(r) &= 2 \lim_{r\to 0} [\psi_1'(\kappa, r)\psi_2'(\kappa, r) F(r)] \\ &+ \int_0^\infty dr\, \psi_1(\kappa, r)\psi_2(\kappa, r)\left\{F'''(r) - 2S(r) F'(r) - S'(r) F(r) - D(r)\int_r^\infty dr'\, D(r') F(r')\right\} . \end{aligned} \tag{3.2.4}$$

The first one obtains setting $\phi = 1$ in (2.13); the second, setting $\phi = F$, and assuming the (otherwise arbitrary) function $F(r)$ to satisfy the asymptotic conditions

$$F(\infty) = F'(\infty) = F''(\infty) = 0 \tag{3.2.5}$$

and to be finite (nondivergent) at $r = 0$ (this last condition is sufficient, but not necessary; see below).

The possibility to choose $F'(r)$ in the left-hand side of (3.2.4) to coincide with the factor multiplying $\psi_1\psi_2$ in the right-hand side of (3.2.2) or (3.2.3) yields two new expressions for the sine and the cosine of $\eta_1 - \eta_2$; and the iteration of this procedure yields the formulas

(3.2.6)
$$\kappa(-4\kappa^2)^n \sin[\eta_1(\kappa)-\eta_2(\kappa)]$$
$$= 2 \lim_{r\to 0}\left[\psi_1'(\kappa,r)\psi_2'(\kappa,r)\sum_{m=0}^{n-1}(-4\kappa^2)^{n-m-1}\sigma_m(r)\right]$$
$$+\int_0^\infty dr\, \psi_1(\kappa,r)\psi_2(\kappa,r)\sigma_n'(r), \quad n = 0,1,2,\cdots,$$

(3.2.7)
$$(-4\kappa^2)^n \cos[\eta_1(\kappa)-\eta_2(\kappa)]$$
$$= 2 \lim_{r\to 0}\left[\psi_1'(\kappa,r)\psi_2'(\kappa,r)\sum_{m=0}^{n-1}(-4\kappa^2)^{n-m-1}\gamma_m(r)\right]$$
$$+\int_0^\infty dr\, \psi_1(\kappa,r)\psi_2(\kappa,r)\gamma_n'(r), \quad n = 1,2,3,\cdots,$$

where $\sigma_n(r)$ and $\gamma_n(r)$ are defined by the recurrence relation (here $a_n(r)$ stands for $\sigma_n(r)$ or $\gamma_n(r)$)

(3.2.8a)
$$a_{n+1}'(r) = a_n'''(r) - 2S(r)a_n'(r) - S'(r)a_n(r)$$
$$+ D(r)\int_r^\infty dr'\, D(r')a_n(r'), \quad a_n(\infty) = 0, \quad n = 0,1,2,3,\cdots,$$

or, equivalently,

(3.2.8b)
$$\alpha_{n+1}(r) = \alpha''_n(r) - S(r)\alpha_n(r)$$

$$+\int_r^\infty dr' \left[S(r')\alpha'_n(r') - D(r') \int_{r'}^\infty dr'' \, D(r'')\alpha_n(r'') \right], \quad n = 0, 1, 2, \cdots,$$

with

(3.2.9)
$$\sigma_0(r) = \int_r^\infty dr' \, D(r') \ ,$$

(3.2.10)
$$\gamma_0(r) = -2 \ .$$

These equations imply:

(3.2.11)
$$\sigma'_0(r) = -D(r) \ ,$$

(3.2.12)
$$\gamma'_1(r) = 2\left\{ S'(r) + D(r) \int_r^\infty dr' \, D(r') \right\} ,$$

(3.2.13)
$$\sigma'_1(r) = -D''(r) + 2D(r)S(r)$$

$$-S'(r)\int_r^\infty dr' \, D(r') - \frac{1}{2} D(r) \left[\int_r^\infty dr' \, D(r') \right]^2 ,$$

(3.2.14)
$$\gamma'_2(r) = 2\left\{ S'''(r) - 3S'(r)S(r) \right.$$

$$-3D'(r)D(r) + [D''(r) - 2D(r)S(r)] \int_r^\infty dr' \, D(r')$$

$$-D(r)\int_r^\infty dr' \, D(r')S(r') + \frac{1}{2} S'(n) \left[\int_r^\infty dr' \, D(r') \right]^2$$

$$\left. + \frac{1}{6} D(r) \left[\int_r^\infty dr' \, D(r') \right]^3 \right\} ,$$

and so on.

Clearly the expressions (3.2.6) and (3.2.7) are not the only ones that can be obtained for the sine and cosine of the phase shift difference, from (3.2.2-4); for instance, one could choose $F(r)$ in (3.2.4) so that $F'''(r)$, or $S'(r)F(r)$, coincide with the factors multiplying $\psi_1\psi_2$ in the right-hand side of (3.2.2) or (3.2.3). The equations we have displayed, (3.2.6-7), are singled out because the factors multiplying $\psi_1\psi_2$ within the integrals are independent of κ; it is this property, as we shall see in the following, that makes them particularly suitable to derive interesting results.

In deriving equations (3.2.6-7), we have ignored the legitimacy of setting $x_1 = 0$ in equations (2.3) and (2.13), which depends on the behavior of all quantities at the origin. Let us clarify this point now.

If both potentials $V_j(r)$, and therefore also $S(r)$ and $D(r)$, are holomorphic at $r = 0$, there is of course no problem, and the whole infinite sequence of equations (3.2.6-7) hold. In this case all the quantities $\psi_1'(\kappa,0)\psi_2'(\kappa,0)$, $\sigma_n(0)$ and $\gamma_n(0)$ exist; and generally only the very first equation (3.2.2) does not contain in the right-hand side an additional contribution besides the integral.

It is also easy to show that the whole infinite sequence of equations hold as well in the case of singular potentials; moreover in this case the equations (3.2.6-7) simplify, because then the contributions at the origin vanish, as implied by equation (3.1.10), and only the integrals remain in the right-hand side.[2]

The other case that deserves a special mention is when the potentials diverge at the origin proportionally to r^{-2}, equation (3.1.7). In this case the very validity of equations (3.2.6-7) is questionable, since their derivation violates the condition stated after equation (3.2.5). But using the fact that the recursion relations (3.2.28) imply that, if both $S(r)$ and $D(r)$ are proportional at the origin to r^{-2}, then as $r \to 0$

$$\sigma_n(r) \sim r^{-(1+2n)} \tag{3.2.15}$$

(3.2.16) $$\gamma_n(r) \sim r^{-2n} ,$$

while, of course,

(3.2.17) $$\psi_j(\kappa, r) \sim r^{\ell_j+1}$$

with ℓ_j defined by equation (3.1.7), it is easy to prove that equations (3.2.6) hold only(3) for $n < \nu_s = (\ell_1 + \ell_2 + 1)/2$ and equations (3.2.7) only(3) for $n < \nu_c = \nu_s + \frac{1}{2}$. Moreover, in all these cases, the first term in the right-hand side is missing, as in the case of singular potentials.

The analysis of other cases is easily done along similar lines, and is left to the interested reader.

If $V_1(r) = V_2(r) = V(r)$, $\psi_1(\kappa, r) = \psi_2(\kappa, r) = \psi(\kappa, r)$, $\eta_1(\kappa) = \eta_2(\kappa) = \eta(\kappa)$, the equations (3.2.6) become the trivial identity $0 = 0$, while the equations (3.2.7) yield the sum rules

(3.2.18) $$(-4\kappa^2)^n = 2 \lim_{r\to 0} \left\{ [\psi'(\kappa, r)]^2 \sum_{m=0}^{n-1} (-4\kappa^2)^{n-m-1} \gamma_m(r) \right\}$$

$$+ \int_0^\infty dr\, [\psi(\kappa, r)]^2\, \gamma_n'(r) , \qquad n = 1, 2, 3, \cdots ,$$

with the quantities $\gamma_n(r)$ given by equations (3.2.8), (3.2.10) (and (3.2.12), (3.2.14)), but now with $D(r) = 0$, $S(r) = 2V(r)$. But it is also possible to obtain nontrivial relationship from the "sine" equations, introducing the potential $V(r, \rho)$ that depends on the parameter ρ. The corresponding Schroedinger equation reads, of course,

(3.2.19) $$\psi''(\kappa, r, \rho) = [-\kappa^2 + V(r, \rho)]\psi(\kappa, r, \rho) ,$$

with

(3.2.20) $$\psi(\kappa, 0, \rho) = 0 ,$$

(3.2.21) $$\psi(\kappa, r, \rho) \xrightarrow[r\to\infty]{} \sin[\kappa r + \eta(\kappa, \rho)] .$$

We then get, setting, as at the end of Section 2, $V_1(r) = V(r, \rho + d\rho)$, $V_2(r) = V(r, \rho)$, $d\rho \to 0$,

$$(3.2.22) \quad \dot{\eta}(\kappa, r) = 2 \lim_{r \to 0} \left\{ [\psi'(\kappa, r, \rho)]^2 \sum_{m=0}^{n-1} (-4\kappa^2)^{n-m-1} \sigma_m(r, \rho) \right\}$$

$$+ \int_0^\infty dr \, [\psi(\kappa, r, \rho)]^2 \, \sigma_n'(r, \rho), \quad n = 0, 1, 2, \cdots,$$

with $\sigma_n(r, \rho)$ defined now by

$$(3.2.23) \qquad \sigma_0(r, \rho) = \int_r^\infty dr' \, \dot{V}(r', \rho) \; ,$$

$$(3.2.24) \qquad \sigma'_{n+s}(r, \rho) = \sigma'''_n(r, \rho) - 4V(r, \rho)\sigma'_n(r, \rho)$$

$$-2V'(r, \rho)\sigma_n(r, \rho), \quad \sigma_n(\infty) = 0, \quad n = 0, 1, 2, \cdots .$$

Here dots indicate partial differentiation with respect to ρ, while primes indicate as usual partial differentiation with respect to r. The expressions for $\sigma_n(r, \rho)$ yielded by these formulas obtain from those given above setting $S(r) = 2V(r, \rho)$ and $D(r) = \dot{V}(r, \rho)$ or $D(r) = 0$ according to whether $D(r)$ appears linearly or nonlinearly.

The first of the equations (3.2.22) in the well-known formula

$$(3.2.25) \qquad \kappa\dot{\eta}(\kappa, \rho) = - \int_0^\infty dr \, [\psi(\kappa, r, \rho)]^2 \, \dot{V}(r, \rho) \; .$$

With the special choice

$$(3.2.26) \qquad V(r, \rho) = V(r)\theta(\rho - r)$$

that implies, through equation (3.2.21),

$$(3.2.27) \qquad \psi(\kappa, r, \rho) = \sin\left[\kappa r + \eta(\kappa, \rho)\right] \quad \text{for} \quad r \geq \rho \; ,$$

and also

(3.2.28) $$\eta(\kappa, 0) = 0$$

and

(3.2.29) $$\eta(\kappa, \infty) = \eta(\kappa)$$

(since obviously (3.2.26) implies $V(r, 0) = 0$, $V(r, \infty) = V(r)$; $\eta(\kappa)$ is clearly the phase shift produced by the potential $V(r)$), we get the variable phase equation[4]

(3.2.30) $$\dot{\eta}(\kappa, \rho) = -\kappa^{-1} V(\rho) \sin^2[\kappa\rho + \eta(\kappa, \rho)] .$$

Note however that this elegant derivation of the phase equation exploits only the "usual" wronskian theorem, since it is this that yields equation (3.2.25).

3.3. *Bargmann-equivalent and phase-equivalent potentials*

We define the potentials $V_1(r)$ and $V_2(r)$ to be *Bargmann-equivalent* if the corresponding S-matrices are related by the equation

(3.3.1) $$S_2(\kappa) = S_1(\kappa) R(\kappa)$$

$R(\kappa)$ *being a rational function.* We shall in particular focus attention on the case when

(3.3.2a) $$R(\kappa) = (a - ib\kappa + c\kappa^2)/(a + ib\kappa + c\kappa^2)$$

(3.3.2b) $$= (\alpha - i\beta\kappa + \kappa^2)/(\alpha + i\beta\kappa + \kappa^2) ,$$

with a, b, and c (and $\alpha = a/c$, $\beta = b/c$) real. It is clear that this structure of $R(\kappa)$ automatically implies that $S_2(\kappa)$ satisfies the two conditions (3.1.13) and (3.1.14) if $S_1(\kappa)$ satisfies them; and vice versa. Moreover, with this choice of $R(\kappa)$, $S_1(\kappa)$ and $S_2(\kappa)$ coincide at $\kappa = 0$ and at $\kappa \to \infty$.

In the special case $\beta = 0$, the two S-matrices coincide; the two potentials $V_1(r)$ and $V_2(r)$ are then called "phase-equivalent."

In his classic paper of 1949 "On the Connection between Phase Shifts and Scattering Potential"[5] Bargmann constructed classes of solvable potentials, that produce S-matrices of rational type. Such potentials are now generally called "Bargmann potentials";[6] they were used by Bargmann to manufacture explicit examples of phase-equivalent potentials.[5] Subsequently a general procedure was given to construct the potential whose S-matrix $S_2(\kappa)$ is related by an equation such as (3.3.1) to the S-matrix $S_1(\kappa)$ produced by any potential $V_1(r)$ such that the solutions of the corresponding Schroedinger equation are explicitly known (for all values of κ, including those corresponding to bound states, if any).[6] In the special case $V_1(r) \equiv 0$, $S_1(\kappa) = 1$, this procedure reproduces of course the Bargmann potentials.

We now show how in general, given a rational function $R(\kappa)$ of type (3.3.2), pairs of Bargmann-equivalent potentials can be obtained. The simplicity of the procedure, based on the results of the previous section, is remarkable, as well as the fact that it is possible in this manner to obtain pairs of Bargmann-equivalent potentials even though the Schroedinger equations for neither one of them is solvable, nor the S-matrices corresponding to either one of them known. Indeed the method yields also pairs of Bargmann-equivalent potentials that are singular, and as it is well known, there is no example of singular potentials whose exact S-matrix is known for all κ.[7] In the special case $b = \beta = 0$ (see equation (3.3.2)), the procedure yields of course phase-equivalent potentials.

A drawback of this approach is, that it does not allow to compute explicitly the potentials $V_2(r)$ that are Bargmann-equivalent to a given potential $V_1(r)$, although it provides the nonlinear differential equation that in principle determines them. However, as we shall presently see, this differential equation can be solved in closed form for $S(r) = V_1(r) + V_2(r)$ if $D(r) = V_1(r) - V_2(r)$ (or rather its integral $\sigma(r)$; see below) is given, so that any (reasonable) choice of such a function $\sigma(r)$ yields in explicit form a pair of Bargmann-equivalent potentials.

The starting point of our analysis are the first three equations of the sequence (3.2.6-7), that we rewrite here:

(3.3.3) $$\kappa \sin[\eta_1(\kappa)-\eta_2(\kappa)] = \int_0^\infty dr\, \psi_1(\kappa,r)\psi_2(\kappa,r)\sigma'(r) ,$$

(3.3.4) $$\kappa^2 \cos[\eta_1(\kappa)-\eta_2(\kappa)] = z(\kappa) - \frac{1}{2}\int_0^\infty dr\, \psi_1(\kappa,r)\psi_2(\kappa,r)\{S'(r)-\sigma'(r)\sigma(r)\} ,$$

(3.3.5) $$\kappa^3 \sin[\eta_1(\kappa)-\eta_2(\kappa)] = -\frac{1}{2} z(\kappa)\sigma(0) + \frac{1}{4}\int_0^\infty dr\, \psi_1(\kappa,r)\psi_2(\kappa,r)\Big\{-\sigma'''(r) + 2\sigma'(r)S(r)+\sigma(r)S'(r)-\frac{1}{2}\sigma'(r)\sigma^2(r)\Big\} .$$

Here

(3.3.6) $$z(\kappa) = \psi_1'(\kappa,0)\psi_2'(\kappa,0)$$

and (see (3.2.11), and note that we have dropped the subscript)

(3.3.7) $$\sigma(r) = \int_r^\infty dr'\, D(r')$$

so that

(3.3.8) $$D(r) = -\sigma'(r) .$$

Note moreover that, as implied by the analysis of the previous subsection, *the terms with* $z(\kappa)$ *are present only if the potentials* $V_j(r)$ *are less singular than* r^{-2} *at the origin*[8] (this case is hereafter referred to, for short, as the case of "regular potentials").

Multiply now the three equations respectively by a, b and c, and sum them. It is then clear that if the potentials $V_j(r)$, or rather the two functions $S(r)$ and $D(r)$, satisfy the equation

$$a\,\sigma'(r) - \frac{1}{2}\,b\{S'(r) - \sigma'(r)\sigma(r)\} + \frac{1}{4}\,c\left\{-\sigma'''(r) + 2\sigma'(r)\,S(r) + \sigma(r)\,S'(r) - \frac{1}{2}\,\sigma'(r)\sigma^2(r)\right\} = 0 \,, \tag{3.3.9}$$

and in addition (in the case of regular potentials) the condition

$$b - \frac{1}{2}\,c\,\sigma(0) = 0 \,, \tag{3.3.10}$$

then for the phase-shift difference $\eta_1(\kappa) - \eta_2(\kappa)$ we get the relation

$$(a + c\kappa^2)\,\sin[\eta_1(\kappa) - \eta_2(\kappa)] + b\kappa\,\cos[\eta_1(\kappa) - \eta_2(\kappa)] = 0 \,. \tag{3.3.11}$$

It is immediately seen, through the definition (3.1.12) of the S-matrix, that this last equation coincides with equations (3.3.1) and (3.3.2). Thus we may conclude that the potentials $V_1(r)$ and $V_2(r)$ are Bargmann-equivalent (in the sense of equations (3.3.1) and (3.3.2)) if they satisfy the equations (3.3.9) and (3.3.10); the second condition, namely (see (3.3.2))

$$\sigma(0) = 2\beta \,, \tag{3.3.12}$$

is required *only for regular potentials.*

The equation (3.3.9) is easily solved for $S(r)$:

$$S(r) = [\sigma(r) - 2\beta]^{-2}\left\{\frac{1}{8}\,\sigma^4(r) - \beta\sigma^3(r) + 2(\beta^2 - a)\sigma^2(r) + 8a\beta\,\sigma(r) - \frac{1}{2}\,[\sigma'(r)]^2 + \sigma''(r)\,[\sigma(r) - 2\beta]\right\}. \tag{3.3.13}$$

To obtain this result we have assumed $\beta \neq 0$ (the special case $\beta = 0$, corresponding to phase-equivalent potentials, is considered below), and we have eliminated the integration constant by exploiting the asymptotic vanishing of $S(r)$ and of $\sigma(r)$ with its derivatives; indeed this equation and (3.3.8) imply that if $\sigma(r)$ vanishes asymptotically as an exponential, so do $S(r)$ and $D(r)$,

$$\sigma(r) \approx g \exp(-\mu r) , \tag{3.3.14}$$

$$D(r) \approx g\mu \exp(-\mu r) , \tag{3.3.15}$$

$$S(r) \approx [(4a-\mu^2)/(2\beta)]\, g \exp(-\mu r)\{1+0[\exp(-\mu r)]\} , \tag{3.3.16}$$

and if $\sigma(r)$ vanishes asymptotically as a power, so do $D(r)$ and $S(r)$,

$$\sigma(r) \approx g r^{-p}, \quad p > 1 , \tag{3.3.17}$$

$$D(r) \approx g p r^{-p-1} , \tag{3.3.18}$$

$$S(r) \approx 2(a/\beta) r^{-p} . \tag{3.3.19}$$

We may now choose $\sigma(r)$ with large arbitrariness, and through equations (3.3.8) and (3.3.13), we then get explicit expressions for $S(r)$ and $D(r)$ (and therefore also for $V_1(r)$ and $V_2(r)$; see (3.1.16b)). Before giving some examples, let us investigate what are the restrictions that $\sigma(r)$ must satisfy for the consistency of the approach, that requires the potentials $V_j(r)$ to obey the conditions they were assumed to satisfy to begin with.

Consider first the case of singular potentials. Then equation (3.3.12) can be ignored, and one only needs a function that be itself singular at $r = 0$, that be finite-valued for $r > 0$, that vanish asymptotically, and such that[9]

$$\sigma(r) - 2\beta \neq 0 \tag{3.3.20}$$

for all (positive) values of r. Specifically if $\sigma(r)$ diverges as a power as $r \to 0$, so does $S(r)$, and we get:

$$\sigma(r) \approx g r^{-m} , \tag{3.3.21}$$

(3.3.22) $$D(r) \approx g\,m\,r^{-m-1} ,$$

(3.3.23a) $$S(r) \approx \frac{1}{8} g^2 r^{-2m}, \quad m > 1 ,$$

(3.3.23b) $$S(r) \approx \left(\frac{1}{8} g^2 + \frac{3}{2}\right) r^{-2}, \quad m = 1 .$$

Thus, for $m = 1$, we have the restriction $g \leq 2$ or $g \geq 6$, while if $m > 1$, we have no additional restriction besides (3.3.20),[9] since in this case the behavior of $S(r)$ at the origin is repulsive and dominant over that of $D(r)$, so that both potentials $V_j(r)$ are always physically sound (we assume of course g to be real).

There is, of course, a great variety of possible choices of functions $\sigma(r)$ that satisfy all these conditions, for instance

(3.3.24) $$\sigma(r) = g\,r^{-m} \exp(-\mu r), \quad m > 1, \quad \mu > 0 ,$$

with g an arbitrary constant having the same sign as β. Every such choice produces an explicit pair of Bargmann-equivalent potentials.

Let us now consider the cases when the function $\sigma(r)$ is chosen so as to satisfy equation (3.3.12), as it is required for regular potentials (and it is, of course, also permissible for singular potentials). There are then two possibilities. If

(3.3.25) $$a = \frac{1}{4} [\sigma'(0)/\sigma(0)]^2 - \frac{1}{16} [\sigma(0)]^2$$

and

(3.3.26) $$\sigma''(0) \neq 0 ,$$

then the function $S(r)$, as well as the function $D(r)$, are generally regular at $r = 0$,[10] and equation (3.3.13) can be rewritten in the form

(3.3.27) $$S(r) = \frac{1}{8} \sigma(r)[\sigma(r) - 2\sigma(0)] - \frac{1}{2} [\sigma'(0)/\sigma(0)]^2$$

$$- \frac{1}{2} [\sigma'(r) + \sigma'(0)][\sigma'(r) - \sigma'(0)]/[\sigma(r) - \sigma(0)]^2 + \sigma''(r)/[\sigma(r) - \sigma(0)] .$$

Note that, even though the last two terms here are separately singular, their singularities cancel out, so that generally

(3.3.28a) $$S(0) = -\frac{1}{8}\sigma^2(0) - \frac{1}{2}[\sigma'(0)/\sigma(0)]^2 + \frac{1}{2}[\sigma'''(0)/\sigma'(0)]$$

is finite, except in the special case $\sigma'(0) = 0$, $\sigma'''(0) \neq 0$, when instead, as $r \to 0$,

(3.3.28b) $$S(r) \approx \frac{2}{3}[\sigma'''(0)/\sigma''(0)]\, r^{-1} \; .$$

Thus in general this situation yields pairs of regular potentials.

If instead the condition (3.3.25) is not satisfied, then $S(r)$ is generally proportional to r^{-2} as $r \to 0$,

(3.3.29) $$S(r) \approx 2[\sigma(0)/\sigma'(0)]^2 \cdot \left\{ a - \frac{1}{4}[\sigma'(0)/\sigma(0)]^2 + \frac{1}{16}[\sigma(0)]^2 \right\} r^{-2} \; ,$$

and the only additional condition required to guarantee the physical soundness of the corresponding potentials is positivity as $r \to 0$, namely validity of the inequality

(3.3.30) $$a > \frac{1}{4}[\sigma'(0)/\sigma(0)]^2 - \frac{1}{16}[\sigma(0)]^2 \; .$$

In deriving equation (3.3.29), we have assumed $\sigma'(0) \neq 0$; if instead

(3.3.31) $$d^p\sigma(r)/dr^p\big|_{r=0} = 0 \quad \text{for} \quad p = 1, 2, \cdots, n-1 \; ;$$

$$d^n\sigma(r)/dr^n\big|_{r=0} \neq 0, \quad n \geq 2 \; ,$$

then in place of equation (3.3.29), one gets

(3.3.32) $$S(r) \approx 2[n!\, \sigma(0)/\sigma^{(n)}(0)]^2 \left\{ a + \frac{1}{16}[\sigma(0)^2 \right\} r^{-2n} \; ,$$

and in place of the condition (3.3.30),

(3.3.33) $$a > -\frac{1}{16}[\sigma(0)]^2 .$$

We have thus seen that any function $\sigma(r)$ that is holomorphic at $r = 0$, vanishes as $r \to \infty$, does not vanish at $r = 0$ and satisfies the condition (3.3.26), and is such that[11]

(3.3.34) $$\sigma(r) - \sigma(0) \neq 0 \quad \text{for} \quad r > 0 ,$$

yields through the equations (3.3.27) and (3.3.8) (together with (3.1.16)) two regular potentials $V_j(r)$ such that the corresponding S-matrices are related by the equations (3.3.1) and (3.3.2), with a and β given by equations (3.3.25) and (3.3.12). These two potentials are holomorphic, or proportional to r^{-1} (but with their difference holomorphic), at $r = 0$, depending whether $\sigma'(0)$ is different or equal to zero.[12] Their asymptotic behavior is simply related to that of $\sigma(r)$ (see equations (3.3.14-19)).

A particularly simple example obtains choosing $\sigma(r) = g \exp(-\mu r), \mu > 0$, to which there corresponds

(3.3.35a) $$\beta = \frac{1}{2} g , \quad a = \frac{1}{4}\mu^2 - \frac{1}{16} g^2$$

and

(3.3.35b) $$V_{1,2}(r) = \frac{1}{4} g^2 \exp(-2\mu r) - \frac{1}{2} g(g \mp \mu) \exp(-\mu r) .$$

Thus far we have considered the case $\beta \neq 0$. If instead

(3.3.36) $$\beta = 0 ,$$

the two potentials are *phase-equivalent*, and it is easily seen that in this case, in place of equation (3.3.13), we get

(3.3.37) $$S(r) = \frac{1}{8}\sigma^2(r) + \sigma''(r)/\sigma(r) - \frac{1}{2}[\sigma'(r)/\sigma(r)]^2 - 2a .$$

The constant α is now fixed by the condition that $S(r)$ vanish asymptotically. For instance if, as $r \to \infty$,

$$\sigma(r) \approx g\, r^{-p} \exp(-\mu r) \ , \tag{3.3.38}$$

one must choose

$$\alpha = \frac{1}{4} \mu^2 \ , \tag{3.3.39}$$

and then

$$S(r) \approx -3p\mu/r + p(p+1)/r^2 + \frac{1}{8} g^2 r^{-2p} \exp(-2\mu r) \ ; \tag{3.3.40}$$

thus only if either p or μ vanishes does $rS(r)$ vanish asymptotically, and only if p vanishes, the potentials $V_j(r)$ are short-ranged (indeed, if

$$\sigma(r) \approx g \exp(-\mu r) \{1 + 0[\exp(-\lambda r)]\} \ , \tag{3.3.41}$$

then

$$S(r) \approx 0\{\exp[-r \min(\lambda, 2\mu)]\} \ ). \tag{3.3.42}$$

Note that now $\sigma(r)$ should not vanish asymptotically faster than exponentially, because in such a case $S(r)$ would diverge.

Now, if $\sigma(r)$ diverges as $r \to 0$, say

$$\sigma(r) \approx g\, r^{-m}, \quad m \geq 1 \ , \tag{3.3.43}$$

then

$$S(r) \approx \frac{1}{8} g^2 r^{-2m} + \frac{1}{2} m(m+2) r^{-2} \tag{3.3.44}$$

and the corresponding potentials are singular. In this case, no additional condition[13] on $\sigma(r)$ is required, except the nonvanishing of $\sigma(r)$, and its finite-valuedness, for $r > 0$.[14]

If instead $\sigma(r)$ is holomorphic at $r = 0$, then it is required that

$$\sigma(0) = \sigma'(0) = 0, \quad \sigma''(0) \neq 0 \ . \tag{3.3.45}$$

The potentials thus produced are regular, since as $r \to 0$

(3.3.46) $$S(r) \approx \frac{2}{3}\,[\sigma'''(0)/\sigma''(0)]\,r^{-1} \;;$$

indeed, if $\sigma'''(0) = 0$, $S(0)$ is finite.

In conclusion, we thus see that the choice of a largely arbitrary function $\sigma(r)$ yields, through equations (3.3.37) and (3.3.38) (with (3.1.16)), a pair of phase-equivalent potentials. All that is required of the function $\sigma(r)$ is that it be finite-valued and not vanish for $r > 0$;[14] if singular potentials are sought, $\sigma(r)$ should be singular at $r = 0$ (see equations (3.3.43) and (3.3.44)); if regular potentials are instead sought, $\sigma(r)$ should be holomorphic at $r = 0$, and it should moreover satisfy the additional conditions (3.3.45).

A simple example yielding singular phase-equivalent potentials is

(3.3.47) $$\sigma(r) = g\,r^{-2\ell}$$

which, when inserted in (3.3.37) (with $\alpha = 0$) and (3.3.8), yields

(3.3.48) $$V_{1,2}(r) = \frac{1}{16}\,g^2\,r^{-4\ell} + \ell(\ell+1)\,r^{-2} \pm g\ell\,r^{-2\ell-1} \;.$$

It is easily seen that this result holds for $\ell \geq \frac{1}{2}$, with no restriction on g. It should be emphasized that the equality of the phases produced by the potentials $V_j(r)$ of equation (3.3.48) holds mod (π), as the derivation of this result implies.

A simple example yielding regular phase-equivalent potentials is

(3.3.49) $$\sigma(r) = g\,\sinh^2(\mu r)/\cosh(\nu r)\,, \qquad \nu > 2\mu > 0 \;.$$

We then obtain (with $\alpha = (\mu - \frac{1}{2}\,\nu)^2$),

(3.3.50) $$D(r) = g\,[\sinh(\mu r)/\cosh^2(\nu r)]$$

$$\cdot\,[\nu\,\sinh(\mu r)\,\sinh(\nu r) - 2\mu\,\cosh(\mu r)\,\cosh(\nu r)] \;,$$

(3.3.51) $$S(r) = \frac{1}{8} g^2 \sinh^4(\mu r)/\cosh^2(\nu r)$$

$$+ 2\mu\nu [1 - \operatorname{cotgh}(\mu r) \operatorname{tgh}(\nu r)] - \frac{3}{2} \nu^2/\cosh^2(\nu r) .$$

In this case, of course, $D(0)$ vanishes, and

(3.3.52) $$S(0) = -\frac{1}{2} \nu(7\nu - 4\mu)$$

is finite (and negative).

Another example yielding regular phase-equivalent potentials obtains from

(3.3.53) $$\sigma(r) = g r^2 \exp(-\mu r), \quad \mu > 0 .$$

This yields (with $d = \frac{1}{4} \mu^2$)[15]

(3.3.54) $$D(r) = -g r(2 - \mu r) \exp(-\mu r) ,$$

(3.3.55) $$S(r) = \frac{1}{8} g^2 r^4 \exp(-2\mu r) - 2\mu/r .$$

It should be noted that any two phase-equivalent potentials obtained by this procedure and holomorphic at $r = 0$ do satisfy the conditions

(3.3.56) $$V_1(0) = V_2(0), \quad V_1''(0) = V_2''(0) .$$

Indeed these conditions are known to hold quite generally for any phase-equivalent potentials that are holomorphic at $r = 0$.[16]

As is now well known, the existence of phase-equivalent potentials is associated with the presence of bound states.[17] The corresponding binding energies are easily identified by noting that the equations (3.3.1) and (3.3.2) imply that, barring accidental cancellations, the S-matrix $S_2(\kappa)$ has two poles at

(3.3.57) $$\kappa = i[-\beta \pm (\beta^2 + \alpha)^{\frac{1}{2}}] ,$$

while $S_1(\kappa)$ has two poles at

(3.3.58) $$\kappa = i\,[\beta \pm (\beta^2+\alpha)^{\frac{1}{2}}] \; .$$

These poles correspond generally to bound states if they occur on the positive imaginary axis in the complex κ plane. We thus see that, if $\alpha > 0$, both $V_1(r)$ and $V_2(r)$ have (at least) one bound state; if $\alpha < -\frac{1}{4}\beta^2$, neither $V_1(r)$ nor $V_2(r)$ need have any bound state; if $-\frac{1}{4}\beta^2 < \alpha < 0$, $V_1(r)$ need have no bound state while $V_2(r)$ has at least two if $\beta > 0$, and vice versa if $\beta < 0$. In the special case $\beta = 0$, corresponding to phase-equivalent potentials, it is moreover easily seen that the requirement that $S(r)$, equation (3.3.37), vanish asymptotically, always yields a nonnegative value for α. If this value is actually positive (as is always the case for short-ranged potentials), then by a continuity argument in the limit $\beta \to 0$, we may expect both phase-equivalent potentials to possess a bound state with binding energy $p^2 = \alpha$ (whose value is thus associated in a straightforward manner to the asymptotic behavior of the potentials; see (3.3.39)). If instead α vanishes (as is the case when $\sigma(r)$ behaves asymptotically as a power, yielding potentials that are asymptotically proportional to r^{-2}, see (3.3.40)), then generally only one of the two S-matrices $S_j(\kappa)$ has a pole on the positive imaginary axis, that approaches the origin as $\beta \to 0$. Thus by the same continuity argument, we expect in this case that only one of the two phase-equivalent potentials $V_j(r)$ possess a (zero-energy) bound state. Indeed for instance in the example of equation (3.3.48), one of the two potentials is everywhere repulsive and therefore cannot certainly support any bound state.[18]

In conclusion we see that any two short-ranged phase-equivalent potentials obtained with this procedure generally possess a bound state with the same binding energy, whose value is simply related to their asymptotic behavior. If instead the potentials vanish asymptotically only proportionally to r^{-2}, it happens generally that of two phase-equivalent potentials, only one has a (zero-energy) bound state. These findings were first obtained, on the basis of specific examples, by Bargmann;[5] they were successively understood in a more general context through the analysis of the inverse scattering problem;[17] note however that such an analysis, as well as Bargmann examples, only refer to regular potentials, while the treatment given here also applies to the singular case.

It should be emphasized that the identification of the bound states has been based here on the examination of the poles that the presence in equation (3.3.1) of the rational function $R(\kappa)$ of equation (3.3.2) implies for the S-matrices $S_j(\kappa)$. But in the following section we shall show how the very approach used here, based on the generalized wronskian relations, and applied to the bound-state regime, confirms this interpretation.

We end this subsection considering the special case with $c = 0$, namely when the function $R(\kappa)$ of equations (3.3.1-2) becomes

(3.3.59a) $$R(\kappa) = (a - ib\kappa)/(a + ib\kappa)$$

(3.3.59b) $$= (i\gamma + \kappa)/(i\gamma - \kappa), \quad \gamma = a/b .$$

Note that this implies $S_1(0) = S_2(0)$, but

(3.3.60) $$\lim_{\kappa\to\infty} [S_1(\kappa)/S_2(\kappa)] = -1 .$$

In this case we expect $V_2(r)$ resp. $V_1(r)$ to possess a bound state with binding energy $p^2 = \gamma^2$, depending whether γ is positive resp. negative.

Now the same treatment yields, in place of equations (3.3.13) or (3.3.37), the simpler equation

(3.3.61) $$S(r) = \frac{1}{2}\,\sigma(r)[\sigma(r) + 4\gamma] ,$$

with $D(r)$ always related to $\sigma(r)$ by equation (3.3.8). It is, however, applicable *only in the case of singular potentials*, because otherwise the term $z(\kappa)$ in the right-hand side of equation (3.3.4) cannot be gotten rid of.[8] This restriction is consistent with equation (3.3.60) (for regular potentials, both $S_j(\kappa) \to 1$ as $\kappa \to \infty$).

A simple example of Bargmann-equivalent potentials whose S-matrices are related through (3.3.1) with (3.3.59) obtains setting

(3.3.62) $$\sigma(r) = g\,r^{-2L}$$

in equations (3.3.61) and (3.3.8), yielding

(3.3.63) $$V_{1,2}(r) = \frac{1}{4}\,g^2\,r^{-4L} + g\gamma\,r^{-2L} \pm g\,L\,r^{-2L-1} \ .$$

It is easily seen that this applies for $L \geq \frac{1}{2}$; for $L = \frac{1}{2}$ it yields, through the positions $g = 2\ell$, $\gamma = C/(2\ell)$, the result $\operatorname{tg}(\delta_\ell - \delta_{\ell-1}) = -2\ell\kappa/C$, where δ_ℓ is the phase shift corresponding to angular momentum ℓ (and linear momentum k) produced by the Coulomb potential C/r (although this case falls, strictly speaking, outside of the class of potentials considered here, it is easily seen, by going through its derivation, that the result holds; its validity may be verified using the explicit expression of the Coulomb scattering phase shifts). If $C < 0$, in this potential there is, moreover, a bound state with angular momentum ℓ and energy $E_\ell = -\frac{1}{4}\,(C/\ell)^2$. If $C > 0$, there is of course no bound state; this shows that a pole, or a zero, of $R(\kappa)$, does not necessarily imply the occurrence of a pole of $S_2(\kappa)$ or $S_1(\kappa)$, since, for instance, a pole of $R(\kappa)$ might be cancelled by a zero of $S_1(\kappa)$. This is indeed what happens in this case. (And there is, of course, also the possibility that, to a pole of the S-matrix, there correspond no bound state; in the language of Jost functions, this occurs if the pole of the S-matrix is due to a pole of the Jost function in the numerator rather than to a zero of that in the denominator.)

In the special case $\gamma = 0$, the two potentials are anti-phase-equivalent, namely

(3.3.64) $$\eta_2(\kappa) = \eta_1(\kappa) + \pi/2 \quad (\text{mod } \pi) \ ,$$

as directly implied by equation (3.3.61), that now reads

(3.3.65) $$S(r) = \frac{1}{2}\,\sigma^2(r)$$

and (3.3.4), that now reads

(3.3.66) $$\cos\,[\eta_1(\kappa) - \eta_2(\kappa)] = 0 \ .$$

The usual analysis suggests that only one of the two potentials $V_j(r)$ has a (zero-energy) bound state. The simpler example of such potentials obtains setting $\gamma = 0$ in the right-hand side of equation (3.3.63).

We finally note that, while the analysis given in this subsection has been based on the first 3 equations of the sequence (3.2.6-7), a similar, but more general, analysis, could be based on a larger set of these relations. This would, however, yield equations, linking $S(r)$ and $D(r)$, that are more complicated and do not allow an explicit solution.

3.4. *Bound states*

In this subsection we apply the generalized wronskian technique of Section 2 to the (radial) Schroedinger problem for bound states.

We assume the potentials $V_j(r)$ to satisfy the conditions of subsection 3.1, and moreover, we assume one of the potentials, say $V_1(r)$, to possess a bound state with energy $E = -p^2$, so that[19]

$$\psi_1(ip, r) \xrightarrow[r \to \infty]{} \exp(-pr) \quad . \tag{3.4.1}$$

Here, and always in the following, we assume p to be positive. This equation replaces (3.1.3), and identifies $\psi_1(ip, r)$. The potential $V_2(r)$ does not generally possess a bound state at the same energy, so that the solution $\psi_2(ip, r)$ of the corresponding Schroedinger equation (at the same energy $-p^2$) may be characterized by the asymptotic boundary condition[19]

$$\psi_2(ip, r) \xrightarrow[r \to \infty]{} \exp(pr) \quad . \tag{3.4.2}$$

This condition is, of course, applicable only if the potential $V_2(r)$ *does not* possess a bound state at the energy $-p^2$ (see below).

We now follow the procedure of Subsection 3.2 and we thus get the equations

$$2p = -\int_0^\infty dr\, \psi_1(ip, r)\psi_2(ip, r) D(r) \quad , \tag{3.4.3}$$

(3.4.4)
$$-2p^2 = \lim_{r\to 0} [\psi_1'(ip, r)\psi_2'(ip, r)]$$

$$-\frac{1}{2}\int_0^\infty dr\, \psi_1(ip, r)\psi_2(ip, r)\left\{S'(r) + D(r)\int_r^\infty dr'\, D(r')\right\},$$

(3.4.5)
$$4p^2\int_0^\infty dr\, \psi_1(ip, r)\psi_2(ip, r)\, F'(r)$$

$$= 2\lim_{r\to 0} [\psi_1'(ip, r)\psi_2'(ip, r)\, F(r)]$$

$$+\int_0^\infty dr\, \psi_1(ip, r)\psi_2(ip, r)\left\{F'''(r) - 2S(r)\, F'(r)\right.$$

$$\left. - S'(r)\, F(r) - D(r)\int_r^\infty dr'\, D(r')\, F(r')\right\},$$

with (3.2.5) as the only condition on $F(r)$ required for the validity of the last equation.

From these equations, again following Subsection 3.2, one also gets the sequence of relations

(3.4.6)
$$(2p)^{2n+1} = 2\lim_{r\to 0}\left[\psi_1'(ip, r)\psi_2'(ip, r)\sum_{m=0}^{n-1}(2p)^{2(n-m-1)}\sigma_m(r)\right]$$

$$+\int_0^\infty dr\, \psi_1(ip, r)\psi_2(ip, r)\sigma_n'(r), \qquad n = 0, 1, 2, \cdots,$$

(3.4.7)
$$2(2p)^{2n} = 2\lim_{r\to 0}\left[\psi_1'(ip, r)\psi_2'(ip, r)\sum_{m=0}^{n-1}(2p)^{2(n-m-1)}\gamma_m(r)\right]$$

$$+\int_0^\infty dr\, \psi_1(ip, r)\psi_2(ip, r)\gamma_n'(r), \qquad n = 1, 2, 3, \cdots,$$

with $\sigma_n(r)$ and $\gamma_n(r)$ defined as in Subsection 3.2, equations (3.2.8-14).

We may now compare these results with those of Subsection 3.3. It is then immediately seen that the validity of equations (3.3.9) (with, if need be, (3.3.10)), implies now, in place of (3.3.11), the equation

$$a + bp - cp^2 = 0 \tag{3.4.8}$$

or equivalently

$$R(ip) = 0 \tag{3.4.9}$$

with $R(\kappa)$ defined by (3.3.2). We thus see that the identification of the bound state of $V_1(r)$ with the pole of $S_1(\kappa)$ is confirmed through equation (3.3.1).

It is easily seen that, if the bound state with energy $-p^2$ were possessed by $V_2(r)$ (and $V_1(r)$ had no bound state at the same energy), all the previous equations would hold with p replaced by $-p$, confirming that in this case $S_2(\kappa)$ rather than $S_1(\kappa)$ would possess a pole at $\kappa = ip$.

We have already called attention to the requirement, for the validity of all these equations, that $V_2(r)$ not possess a bound state at the same energy $-p^2$ as $V_1(r)$. Indeed, if this were to happen, the equations (3.4.3), (3.4.4) (and, more generally, (3.4.6) and (3.4.7)) could be replaced by other relations with the same right-hand sides but with vanishing left-hand sides, and with $\psi_2(ip, r)$ characterized by the boundary condition

$$\psi_2(ip, r) \xrightarrow[r \to \infty]{} \exp(-pr) \tag{3.4.10}$$

in place of equation (3.4.2). It is, of course, the abrupt change in this boundary condition that accounts for the (abrupt) change in the left-hand side of these equations; if instead $V_2(r)$ were changed continuously until it has a bound state at energy $-p^2$, then all the equations (3.4.3-4), (3.4.6-7), would continue to hold as they stand, but with their right-hand side requiring a limiting definition, since the wave-function $\psi_2(ip, r)$, that is characterized by the boundary condition (3.4.2) as long as $V_2(r)$ has no bound state at the energy $-p^2$ and by continuity as $V_2(r)$

changes so as to acquire such a bound state, develops a diverging normalization factor (corresponding to the transition between the asymptotic behavior (3.4.2) and (3.4.10)).

In the special case $V_1(r) = V_2(r) = V(r)$, when, of course, the two coincident potentials $V_j(r) = V(r)$ necessarily have bound states with the *same* energy, the equation (3.4.3) (and, more generally, (3.4.6)), become the trivial identity $0 = 0$ (see, however, below), while the equations (3.4.7) yield the family of sum rules

$$0 = 2 \lim_{r \to 0} \left\{ [\psi'(ip, r)]^2 \sum_{m=0}^{n-1} (-4E)^{n-m-1} \gamma_m(r) \right\} + \int_0^\infty dr\, [\psi(ip, r)]^2\, \gamma_n'(r), \quad n = 1, 2, 3, \cdots, \tag{3.4.11}$$

where $E = -p^2$ is the energy of the bound state, $\psi(ip, r)$ is the corresponding wave function, and $\gamma_n(r)$ is defined as in Subsection 3.2, but now for the special case $D(r) = 0$, $S(r) = 2V(r)$. The first of these sum rules,

$$0 = -\lim_{r \to 0} \{[\psi'(ip, r)]^2\} + \int_0^\infty dr\, [\psi(ip, r)]^2\, V'(r) \tag{3.4.12}$$

was already known.[20] The next one reads

$$0 = \lim_{r \to 0} \{[\psi'(ip, r)]^2 [V(r) + 2E]\} + \frac{1}{2} \int_0^\infty dr\, [\psi(ip, r)]^2 [V'''(r) - 6V'(r)\, V(r)] \, . \tag{3.4.13}$$

Multiplying the first of these sum rules by the constant a and adding the second, we immediately infer that, if the regular potential $V(r)$ satisfies the nonlinear equation

(3.4.14) $$aV'(r) = V'''(r) - 6V'(r)V(r)$$

and has a bound state, then the energy of this bound state is given by the formula

(3.4.15) $$E = -\frac{1}{2} V(0) - \frac{1}{4} a .$$

Indeed the differential equation (3.4.14) is easily integrated,[21] yielding

(3.4.16) $$V(r) = -\frac{1}{2} \mu^2/\cosh^2\left[\frac{1}{2} \mu(r-r_o)\right] .$$

Here we have set $a = \mu^2$, and of the 3 integration constants, only r_o survives, the other two being fixed by the requirement of asymptotic vanishing. And it is easily verified[22] that the potential (3.4.16) has one bound state if r_o is positive, whose binding energy is given by equation (3.4.15), namely

(3.4.17) $$E = -\frac{1}{4} \mu^2 \operatorname{tgh}^2\left(\frac{1}{2} \mu r_o\right) .$$

If instead r_o is negative, the potential (3.4.16) supports no bound state.[22]

The wronskian relations (2.3) and (2.13) may also be applied, in the bound-state case, with a more general choice for f_j and ψ_j. Specifically, let

(3.4.18) $$f_j = -E_j + V_j(r), \quad E_j = -p_j^2, \quad p_j > 0 ,$$

with E_j the energy of a bound state of the potential $V_j(r)$, and let $\psi_j(ip_j, r)$ be the corresponding bound-state wave function, characterized by the asymptotic behavior

(3.4.19) $$\psi_j(ip_j, r) \xrightarrow[r \to \infty]{} \exp(-p_j r)$$

and vanishing at $r = 0$ (see equation (3.1.2), and specifically equations (3.1.5-10)). We then get:

(3.4.20) $$0 = \int_0^\infty dr\, \psi_1(ip_1, r)\psi_2(ip_2, r)[E_2 - E_1 + D(r)]$$

(3.4.21) $$0 = -\lim_{r\to 0} [\psi_1'(ip_1, r)\psi_2'(ip_2, r)] + \int_0^\infty dr\, \psi_1(ip_1, r)\psi_2(ip_2, r) \cdot \left\{ S'(r) + [E_2 - E_1 + D(r)] \left[(E_1 - E_2) r + \int_r^\infty dr'\, D(r') \right] \right\},$$

(3.4.22) $$-2(E_1 + E_2) \int_0^\infty dr\, \psi_1(ip_1, r)\psi_2(ip_2, r) F'(r)$$
$$= 2 \lim_{r\to 0} [\psi_1'(ip_1, r)\psi_2'(ip_2, r) F(r)] + \int_0^\infty dr\, \psi_1(ip_1, r)\psi_2(ip_2, r) \left\{ F'''(r) - 2S(r) F'(r) - S'(r) F(r) - [E_2 - E_1 + D(r)] \int_r^R dr'\, [E_2 - E_1 + D(r')] F(r') \right\}.$$

For the validity of the last equation, it is sufficient that

(3.4.23) $$\lim_{r\to\infty} \{F(r) \exp[-(p_1 + p_2) r]\} = 0 ;$$

the constant R can be chosen arbitrarily (as implied by (3.4.20)), including $R = \infty$ if the last integral remains convergent (note, however, that R is a constant; the choice $R = r$, that would eliminate the last term in the right-hand side of equation (3.4.22), is not permitted).

Actually, for the validity of the 3 equations (3.1.20-22), it is not necessary that both E_1 and E_2 be the energies of bound states of the potentials $V_1(r)$ and $V_2(r)$ respectively; it is sufficient that only one, say E_1, be the energy of a bound state, provided the inequality[(23)]

(3.4.24a) $$E_2 > E_1$$

or, equivalently,

(3.4.24b) $$p_1 > \mathrm{Re}\, p_2$$

holds. In this case $\psi_2(ip_2, r)$, that is always characterized as the solution of the Schroedinger equation with potential $V_2(r)$ and energy E_2 that vanishes at the origin, behaves asymptotically proportionally to $\exp(p_2 r)$ (if p_2 is real), or $\sin[\kappa_2 r + \eta_2(\kappa_2)]$ if $\kappa_2 = -ip_2$ is real (i.e., if E_2 is positive rather than negative). In this case, the condition (3.4.23) is replaced by the more stringent condition(23)

(3.4.25) $$\lim_{r\to\infty} \{F(r) \exp[-p_1 r]\} = 0 \ .$$

Note, however, that this condition is less stringent than the usual requirement of asymptotic vanishing of $F(r)$, $F'(r)$ and $F''(r)$, equations (3.2.5).

By the usual iterative technique, one gets, from these equations, the sequence of relations

(3.4.26) $$0 = 2 \lim_{r\to 0} \left\{ \psi_1'(ip_1,r)\psi_2'(ip_2,r) \sum_{m=0}^{n-1} [-2(E_1+E_2)]^{n-m-1} \tilde{\sigma}_m(r) \right\}$$

$$+ \int_0^\infty dr\, \psi_1(ip_1,r)\psi_2(ip_2,r)\, \tilde{\sigma}_n'(r), \qquad n = 0,1,2,\cdots,$$

(3.4.27) $$0 = \lim_{r\to 0} \left\{ \psi_1'(ip_1,r)\psi_2'(ip_2,r) \sum_{m=0}^{n-1} [-2(E_1+E_2)]^{n-m-1} \tilde{\gamma}_m(r) \right\}$$

$$+ \int_0^\infty dr\, \psi_1(ip_1,r)\psi_2(ip_2,r)\, \tilde{\gamma}_n'(r), \qquad n = 1,2,3,\cdots,$$

with

(3.4.28) $$\tilde{\sigma}_0(r) = (E_1 - E_2)r + \int_r^\infty dr'\, D(r') \ ,$$

(3.4.29) $$\tilde{\gamma}_0(r) = -2 \ ,$$

and (here $\tilde{a}_n(r)$ stands for $\tilde{\sigma}_n(r)$ or $\tilde{\gamma}_n(r)$)

$$\tilde{a}'_{n+1}(r) = \tilde{a}'''_n(r) - 2S(r)\tilde{a}'_n(r) - S'(r)\tilde{a}_n(r) - [D(r) + E_2 - E_1] \int_r^{R_n} dr'[D(r') + E_2 - E_1]\, \tilde{a}_n(r') \; , \tag{3.4.30}$$

where the quantities R_n are arbitrary constants and the functions $a_n(r)$ are characterized by the asymptotic boundary conditions

$$\lim_{r\to\infty} [\tilde{a}_n(r) \exp(-\varepsilon r)] = 0 \, , \quad \varepsilon > 0 \, , \quad n = 1, 2, 3, \cdots \; . \tag{3.4.31}$$

Let us reiterate that a sufficient condition for the validity of these relations is either that both $E_j = -p_j^2$, $p_j > 0$ be the energies of bound states in the potentials $V_j(r)$ and $\psi_j(ip, r)$ the corresponding wave functions, or that only one, say, E_1, be the energy of a bound state (and $\psi_1(ip_1, r)$ the corresponding wave function), while E_2 (that can be negative or positive) satisfies the strict inequality (3.4.24). The conditions of Subsection 3.1 on the potentials $V_j(r)$ are also assumed; and the discussion given at the end of Subsection 3.2 applies in this case without any change, implying certain restrictions on the validity of these equations if the potentials $V_j(r)$ are neither holomorphic at $r = 0$ nor singular there more than r^{-2} (in this latter case the equations simplify due to the vanishing of all contributions at $r = 0$).

In deriving these relations, we have assumed the potentials $V_j(r)$ to vanish asymptotically.(24) It is easily seen that the same approach is also applicable to potentials that diverge asymptotically (to positive infinity), and that possess therefore only a discrete spectrum. Then the equations (3.4.3-4), (3.4.6-7) hold, but with vanishing left-hand sides, and the equations (3.4.11-13), (3.4.20-22) and (3.4.26-30) hold as they stand;(25) but for the validity of the last equations, it is now required that both E_1 resp. E_2 be the energies of bound states in the potentials $V_1(r)$ resp. $V_2(r)$.

It is worthwhile to display the first few of the equations (3.4.26-27). In writing them, we assume again the potentials to vanish at infinity. The first two equations of the sequence are (3.4.20) and (3.4.21); the following two read

$$0 = -\lim_{r\to 0}\left\{\psi_1'(ip_1,r)\psi_2'(ip_2,r)\int_r^\infty dr'\; D(r')\right\} \tag{3.4.32}$$
$$+\frac{1}{2}\int_0^\infty dr\,\psi_1(ip_1,r)\psi_2(ip_2,r)\;\cdot$$
$$\cdot\left\{D''(r) + 2S(r)[E_1 - E_2 - D(r)] + S'(r)\left[(E_1 - E_2)r + \int_r^\infty dr'\; D(r')\right]\right.$$
$$\left.-\frac{1}{2}[E_1 - E_2 - D(r)]\left[(E_1 - E_2)r + \int_r^\infty dr'\; D(r')\right]^2\right\},$$

$$0 = \lim_{r\to 0}\left\{\psi_1'(ip_1,r)\psi_2'(ip_2,r)\right. \tag{3.4.33}$$
$$\left.\cdot\left[E_1 + E_2 + \frac{1}{2}S(r) - \frac{1}{4}\left(\int_r^\infty dr'\; D(r')\right)^2\right]\right\}$$
$$+\frac{1}{2}\int_0^\infty dr\,\psi_1(ip_1,r)\psi_2(ip_2,r)\left\{\frac{1}{2}S'''(r) - \frac{3}{2}S'(r)S(r)\right.$$
$$-\frac{1}{2}[E_2 - E_1 + D(r)]\left[3D'(r) + \int_r^\infty dr'\;\{S(r')[E_2 - E_1 + D(r')]\}\right]$$
$$+\frac{1}{2}\left[(E_1 - E_2)r + \int_r^\infty dr'\; D(r')\right][D''(r) - 2S(r)\{E_2 - E_1 + D(r)\}]$$
$$+\frac{1}{4}S'(r)\left[(E_1 - E_2)r + \int_r^\infty dr'\; D(r')\right]^2$$
$$\left.+\frac{1}{12}[E_2 - E_1 + D(r)]\left[(E_1 - E_2) + \int_r^\infty dr'\; D(r')\right]^3\right\}.$$

It is also of some interest to display these equations for the special case $D(r) = 0$, $S(r) = 2V(r)$. Consider first the case with $E_1 \neq E_2$. We then get:

(3.4.34)
$$0 = \int_0^\infty dr\, \psi(ip_1, r)\psi(ip_2, r) ,$$

(3.4.35)
$$0 = - \lim_{r\to 0} [\psi'(ip_1, r)\psi'(ip_2, r)] + \int_0^\infty dr\, \psi(ip_1, r)\psi(ip_2, r)\left\{V'(r) - \frac{1}{2}(E_1 - E_2) r\right\} ,$$

(3.4.36)
$$0 = \int_0^\infty dr\, \psi(ip_1, r)\psi(ip_2, r)\left\{4V(r) + 2r\, V'(r) - \frac{1}{2}(E_1 - E_2)^2 r^2\right\} ,$$

(3.4.37)
$$0 = \lim_{r\to 0} \{\psi'(ip_1, r)\psi'(ip_2, r)[E_1 + E_2 + V(r)]\} + \frac{1}{2}\int_0^\infty dr\, \psi(ip_1, r)\psi(ip_2, r)\Big\{V'''(r) - 6V'(r)\, V(r) + (E_1 - E_2)^2 \left[2r\, V(r) + \frac{1}{2} r^2 V'(r) - \int_r^\infty dr' V(r')\right] - \frac{1}{12}(E_1 - E_2)^3 r^3\Big\} .$$

Here we have eliminated the subscript in the wave functions, since they are solutions of the same Schroedinger equation (i.e., with the same potential) for different energies. The first of these equations is, of course, just the orthogonality condition.

If we consider instead the case with $D(r) = 0$, $S(r) = 2V(r)$ and moreover $E_1 = E_2$, the second and fourth of these equations reproduce equations (3.4.12-13), while the first and third become the identity $0 = 0$, but can be manipulated to yield (using the same method as at the end of *Section 2*)

(3.4.38) $$0 = \int_0^\infty dr\, [\psi(r,\rho)]^2\, [\dot{V}(r,\rho) - \dot{E}(\rho)] \ ,$$

(3.4.39) $$0 = -\lim_{r\to 0} \left\{ [\psi'(r,\rho)]^2 \int_r^\infty dr'\, \dot{V}(r,\rho) \right\}$$
$$+ \frac{1}{2}\int_0^\infty dr\, [\psi(r,\rho)]^2 \left\{ \dot{V}''(r,\rho) + 4V(r,\rho)\,[\dot{E}(\rho) - \dot{V}(r,\rho)] \right.$$
$$\left. + 2V'(r,\rho) \left[r\,\dot{E}(\rho) + \int_r^\infty dr'\, \dot{V}(r',\rho) \right] \right\} .$$

Here we have indicated by a dot partial differentiation with respect to ρ, while primes indicate as usual partial differentiation with respect to r. $E(\rho)$ is the bound-state energy in the potential $V(r,\rho)$, and $\psi(r,\rho)$ the corresponding wave function, that satisfies the Schroedinger equation

(3.4.40) $$\psi''(r,\rho) = [V(r,\rho) - E(\rho)]\,\psi(r,\rho) \ .$$

The sequence of equations (3.4.26-27) have been derived using equation (3.4.22), in conjunction with (3.4.20) and (3.4.21), in the same manner as in the scattering case. These equations involve, however, the energies in a more complicated way than in the scattering case. Thus there is less justification for singling out these sequences of equations out of the many that can be obtained with other choices of F(r). We leave the derivation of such relations to the interested reader.

From these relationships it is possible to derive a number of properties of bound states; in some cases it is even possible to determine their energies and/or angular momenta, although the typical results are bounds on these quantities. The techniques that yield such results (and also similar results for Regge poles and complex poles of the S-matrix) are displayed in the paper of Reference (18) and in the example given above.

F. CALOGERO
ISTITUTO DI FISICA, UNIVERSITÀ DI ROMA, 00185 ROMA, ITALY
ISTITUTO NAZIONALE DI FISICA NUCLEARE, SEZIONE DI ROMA

REFERENCES AND FOOTNOTES

(1) D. M. Fradkin and F. Calogero, Nuclear Phys. *75*, 475 (1966).

(2) Although for simplicity of presentation, we are limiting consideration to singular potentials that diverge to positive infinity as powers, equation (3.1.9), this statement would hold *a fortiori* for more singular potentials, that diverge to positive infinity as $r \to 0$ faster than any power.

(3) In the marginal cases (equations (3.2.6) with $n = \nu_s$ or (3.2.7) with $n = \nu_c$, that may of course occur only if either ν_s or ν_c are integral) a significant relation of type (3.2.6) or (3.2.7) can also be obtained; but it has in the right-hand side, in addition to the integral, a nonvanishing contribution from the origin, whose evaluation can be easily done (although it requires the use of the more complete version of equation (3.2.4), that obtains from (2.13) when the simplification implied by the condition stated after equation (3.2.5) does not apply). We leave to the interested reader the derivation of such a result. It should be noted that in this case the logarithmic divergence in the integral, that might be expected counting powers on the basis of equations (3.2.15-17), does not occur due to a cancellation.

(4) F. Calogero, *Variable Phase Approach to Potential Scattering*, Academic Press, New York, 1967.

(5) V. Bargmann, Rev. Mod. Phys. *21*, 488 (1949).

(6) See R. G. Newton, "*Scattering Theory of Waves and Particles*," McGraw-Hill, 1966, Section 14.7.

(7) The only possible exception is the potential $V(r) = g^2/r^4$ whose Schroedinger equation can be reduced to the Mathieu equation (but the Mathieu functions are quite complicated, and their asymptotic behavior far from simple). See: E. Vogt and G. H. Wannier, Phys. Rev. *95*, 1190 (1954); R. M. Spector, J. Math. Phys. *5*, 1185 (1964).

(8) More precisely, it is sufficient, in order for these terms to be absent, that only one of the potentials be more singular than r^{-2}, in the sense of equation (3.1.9). If both potentials behave as r^{-2} at the origin (see equation (3.1.7)), then these terms are absent if $\ell_1 + \ell_2 > 1$ (in the case $\ell_1 = \ell_2 = \ell$, it is sufficient that $\ell > 0$).

(9) Actually $\sigma(\bar{r}) = 2\beta$ is permitted, provided $4\alpha + \beta^2 > 0$ and $\sigma'(\bar{r}) = 2\beta(4\alpha + \beta^2)^{\frac{1}{2}}$, because then also the numerator in the right-hand side of (3.3.13) has a double zero at $r = \bar{r}$.

(10) We are assuming, for simplicity, $\sigma(r)$ to be holomorphic at $r = 0$.

(11) Actually $\sigma(\bar{r}) = \sigma(0)$ is permitted, provided $\sigma'(\bar{r}) = \pm\sigma'(0)$ and $\sigma''(\bar{r}) = 0$, because then the right-hand side of equation (3.3.27) is not singular at $r = \bar{r}$.

(12) In the second case $\sigma'''(0) \neq 0$ is also required.

(13) Except for $m = 1$, when the additional condition $g \geq 6$ or $g \leq 2$ is also required.

(14) Actually $\sigma(\bar{r}) = 0$ is permitted, provided also $\sigma'(\bar{r}) = \sigma''(\bar{r}) = 0$.

(15) In this case, strictly speaking, the potentials $V_j(r)$ are outside the class we have considered, since they both contain, in addition to a short-ranged part, the (same) attractive Coulomb contribution $-\mu/r$.

(16) F. Calogero and A. Degasperis, J. Math. Phys. 9, 90 (1968).

(17) See any treatment of the inverse scattering problem, for instance that given in Section 20.2 of Reference 6.

(18) Indeed for the special example (3.3.48) the zero-energy Schroedinger equation can be solved exactly, and it is easily seen that the potential $V_2(r)$ (if $g > 0$) or $V_1(r)$ (if $g < 0$) has just one bound state, with vanishing binding energy. See, for instance: F. Calogero and G. Cosenza, Nuovo Cimento 45, 867 (1966). (Note, however, that this paper is marred by several misprints: in equation (2.20) $(\ell + 1)$ should be replaced by $\ell(\ell + 1)$, in equation (2.23) the last $-$ should be $+$, in equations (2.24) and (A.7) the 3 should disappear, in equation (A.4) the first exponent should be $-\ell$, and the quantity $-b$ in the fifth line from the bottom on p. 880 should read $-b + 1$.)

(19) Note that this function is not the analytic continuation in κ of the corresponding scattering wave function $\psi_j(\kappa, r)$, since a comparison with equation (3.1.3) implies a renormalization.

(20) See the paper by Calogero and Cosenza quoted above (Reference 18).

(21) Integrate once, then multiply by $V'(r)$ and integrate again.

(22) See, for instance, problem 39 in S. Flügge, *Practical Quantum Mechanics*, Springer, 1971, vol. I.

(23) Throughout this discussion we consider E_j to be real, and therefore p_j either real (and positive) or pure imaginary.

(24) Indeed even an asymptotic vanishing that is slower than required by (3.1.11) is sufficient for the validity of all these equations. In particular they apply if Coulomb tails are present.

(25) Except for the replacement of an arbitrary finite constant in place of ∞ as the upper limit of the integral $\int_r^\infty dr' D(r')$ appearing in equations (3.4.20) and (3.4.27); note that this can be done generally even when the integral is convergent at infinity.

OLD AND NEW APPROACHES TO THE INVERSE SCATTERING PROBLEM

Freeman J. Dyson

I. *Historical*

When I agreed to write a chapter about the inverse scattering problem, I thought I would produce a systematic review article describing the historical development of the subject from its origins in the two classic papers of Bargmann.[1,2] However, I soon discovered that the review article which I wanted to write had already been written. The article of Faddeev,[3] written in 1959, is a masterpiece of exposition, and it is fortunately available in a good English translation by Seckler.[4] So far as I can discover, the Diplomarbeit of Steinmann,[5] on the inverse scattering problem for periodic potentials, is the only significant paper in the literature up to 1959 that Faddeev missed. After 1959 the subject went into a long sleep. The main problem had been solved, and Faddeev had tied the loose ends together so skillfully that he left hardly any openings for further investigation. Faddeev's article is one of those select few that effectively kill a whole area of research (such articles are usually written by very great mathematicians, for example Von Neumann[6] on the mathematical foundations of quantum mechanics, or occasionally by great physicists, for example Bohr and Rosenfeld[7] on the problem of measurement of electromagnetic fields). Recently there has been a revival of interest in the inverse scattering problem, since Gardner, Greene, Kruskal and Miura[8] discovered a remarkable connection between this problem and the solution of certain nonlinear partial differential equations. Another recent development is the beautiful solution by Case and Kac[9] of a discrete version of the inverse scattering problem, with a difference equation replacing the differential (Schrödinger) equation. But these new discoveries have not led to any major change in the formulation of the inverse scattering theory, which remains essentially as Faddeev left it. Confronted with this situation, and having no desire either to plagiarize or to paraphrase Faddeev, I decided to abandon the idea of writing a comprehensive review. Instead, I shall give an elementary account of certain aspects of the inverse scattering problem that happen to interest me personally. I shall not assume that the reader is familiar with the Faddeev article. On the contrary, I hope that this chapter may serve as an appetizer, encouraging any reader who finds the subject attractive to go to Faddeev for a solid meal.

I begin by stating what we mean by the inverse scattering problem. In elementary quantum mechanics we consider a particle with positive energy E moving in a spherically symmetric potential $V(r)$ with a wave-function $\psi(r, E)$ which is also spherically symmetric. The Schrödinger equation is then

$$[(d^2/dr^2) - V(r) + E]\, u(r, E) = 0 \ , \tag{1}$$

valid for $0 < r < \infty$, where

$$u(r, E) = r\psi(r, E) \tag{2}$$

satisfies the boundary condition

$$u(0, E) = 0 \ . \tag{3}$$

The scattering of the particle by the potential is described by the behavior of $u(r, E)$ as $r \to \infty$. If the potential decreases more rapidly than r^{-1}, or more precisely if

$$\int_0^\infty |V(r)|\, dr < \infty \ , \tag{4}$$

then the asymptotic behavior of the wave-function is given by

$$u(r, E) \sim \sin(kr - \eta(E)), \quad k = E^{\frac{1}{2}} , \tag{5}$$

as $r \to \infty$. The function $\eta(E)$ appearing in Equation (5) is the phase-shift for s-wave scattering at energy E, and is a directly measurable quantity. Historically, for the first twenty years (1926 - 1946) after the discovery of wave-mechanics, physicists were concerned with the "direct scattering problem," which is the problem of calculating the phase-shift $\eta(E)$ produced by a given potential $V(r)$. Potentials were either deduced from atomic or nuclear models or simply guessed, and the resulting phase-shifts were compared with experiment. About 1946 it became clear to the leading potential-guessers that many different potentials were consistent with the experimental data, so that the choice of a potential, for example in the case of neutron-proton scattering, was determined more by personal taste than by the experiments. To bring some order into a chaotic situation, people began to ask the question, "Is there a way to solve the problem backwards, to begin with a phase-shift $\eta(E)$, supposedly measured for all energies E, and to calculate by a strict mathematical procedure the potential $V(r)$ that would produce this phase-shift?" Thus, the inverse scattering problem was born.

II. *The Bargmann Potentials*

After a few abortive attempts[10,11] had been made to solve the inverse problem by series expansions which turned out to be in general divergent, Bargmann[1] threw his bombshell. He showed that the two potentials

$$V_1(r) = -6\lambda^2 e^{-\lambda r}(1+e^{-\lambda r})^{-2} , \tag{6}$$

$$V_2(r) = -24\lambda^2 e^{-2\lambda r}(1+3e^{-\lambda r})^{-2} , \tag{7}$$

give identical phase-shifts, with

$$3\lambda \; k \cot \eta(E) = 2k^2 - \lambda^2 \tag{8}$$

for all values of $E = k^2$. This single example doomed all attempts to determine V from η by straightforward calculation, and immediately raised the relationship between V and η to the status of a famous unsolved problem. It also explained why the efforts to determine V from experimental scattering data had led to such inconsistent and confusing answers.

Bargmann discovered the phase-equivalent potentials V_1 and V_2 using only elementary calculus and native ingenuity. His starting-point, following an earlier analysis of the scattering problem by Jost,[12] was to look for solutions of the Schrödinger equation (1) of the form

$$u(r,E) = e^{ikr}[4k^2 + 2ika(r) + b(r)] . \qquad (9)$$

The functions $a(r)$, $b(r)$ then satisfy ordinary differential equations which can be solved exactly. The solutions are rational combinations of exponentials similar to those which appear in Equations (6) and (7). The potential is

$$V(r) = (da/dr) , \qquad (10)$$

and the phase-shift is

$$\eta(E) = \mathrm{Arg}[4k^2 + 2ika(0) + b(0)] - \mathrm{Arg}[4k^2 + 2ika(\infty) + b(\infty)] . \qquad (11)$$

This means that $(k \cot \eta)$ is a rational function of E, depending only on the four coefficients $a(0)$, $b(0)$, $a(\infty)$, $b(\infty)$. If in particular we choose $a(0) = 0$, then Equation (11) implies

$$2a(\infty)\, k \cot \eta = -(4E + b(\infty)) , \qquad (12)$$

so that the phase-shift depends only on two parameters $a(\infty)$, $b(\infty)$ and is independent of $b(0)$. But the potential defined by Equation (10) still contains $b(0)$ as a free parameter. In this way Bargmann[2] found several continuous families of potentials giving identical phase-shifts. The examples (6) and (7) are only a special case of a more general phenomenon.

It soon became clear that the existence of phase-equivalent potentials is closely connected with the existence of bound states. A bound state here means a solution of Equation (1) with negative E and with $u(r, E)$ tending to zero exponentially as $r \to \infty$. Levinson[13] proved that the potential is uniquely determined if the phase-shift is given and if no bound state is allowed. Both of Bargmann's potentials (6) and (7) possess a bound state. But it is not true that the potential is uniquely determined even when the phase-shift and the energies of all bound states are given. In fact the potentials (6) and (7) each have only a single bound state with the same energy $E = -\frac{1}{4}\lambda^2$. Bargmann's second paper[2] exhibited a variety of examples of phase-equivalent potentials, some with equal and some with unequal bound-state energies. These examples posed an irresistible challenge to mathematicians and physicists all over the world, to discover the causal connections underlying the mysteriously ambiguous relationship between phase-shift and potential.

The inverse scattering problem was attacked by at least three groups working more-or-less independently of each other. These were Gelfand and Levitan,[14] Marchenko,[15] and Jost and Kohn.[16] The solution found by Gelfand and Levitan was more complete and mathematically rigorous than the others, so that it has come to be accepted as the definitive solution. During this period Bargmann corresponded with Jost[17] and made a substantial contribution to the Jost-Kohn formulation of the problem. It was typical of Bargmann that, seeing the problem satisfactorily solved by others, he did not rush into print with his own very elegant analysis. The essential result at which all three groups arrived was the following. The potential $V(r)$ is uniquely determined only when three sets of quantities are given:

(i) The phase-shift $\eta(E)$ for $0 < E < \infty$,

(ii) the number n and energies $E_1, \cdots, E_n$ of the bound states,

(iii) a set of positive real parameters $c_1, \cdots, c_n$, one corresponding to each bound state.

The parameter c_j is connected with the normalized bound-state wave-function $u(r, E_j)$ by the relation

$$c_j = [(d/dr)\, u(r, E_j)]^2_{r=0} \ . \tag{13}$$

Furthermore, if certain mathematical conditions are satisfied, a potential will exist corresponding to any given set of quantities $\eta(E)$, E_j, c_j.

Since Bargmann did not publish his solution,[17] and since it appears[18] in Faddeev's review[3] in a somewhat disguised form, I here place on record Bargmann's equations as he wrote them down in 1952. Let $V(r)$ and $V_0(r)$ be two phase-equivalent potentials, each having n bound-states with the same energies $E_1, \cdots, E_n$. Let $u^0(r, E_j)$ be the normalized wave-functions of the bound-states in the potential V_0, and let

$$F_{jk}(r) = \int_r^\infty u^0(r', E_j)\, u^0(r', E_k)\, dr' \ , \tag{14}$$

$$c_j^0 = [(d/dr)\, u^0(r, E_j)]^2_{r=0} \ . \tag{15}$$

The quantities F_{jk}, c_j^0 refer only to the potential V_0. The potential V is then given explicitly by the formula

$$V(r) = V_0(r) - 2(d^2/dr^2) \log D_n(r) \ , \tag{16}$$

where D_n is the determinant

$$D_n(r) = \operatorname{Det}[\delta_{jk} + \mu_j F_{jk}(r)]_{j,k=1,\cdots,n} \ , \tag{17}$$

and the parameters μ_j are related to the c_j by

$$(1+\mu_j)\, c_j = c_j^0 \ . \tag{18}$$

These equations of Bargmann were suggested by the 1952 papers of Jost and Kohn[16] and were in turn generalized by Jost and Kohn in their 1953 paper.

III. The Gelfand-Levitan Construction by a Variational Method

Bargmann's equation (16) gives the simplest and most direct solution of the problem with which Bargmann was concerned, namely to characterize precisely the manifold of potentials $V(r)$ which possess the same scattering phase-shifts and bound-state energies as a given potential $V_0(r)$. He assumed that one potential V_0 in the manifold was already known. Gelfand and Levitan[14] solved the more difficult problem of determining the manifold of V when no V_0 is given. They showed how a potential V could be constructed explicitly from a given phase-shift $\eta(E)$ and given bound-state parameters $E_1, \cdots, E_n$, $c_1, \cdots, c_n$. They proved the existence and uniqueness of the potential for any $\eta(E)$, E_j and c_j satisfying a set of physically reasonable conditions. In the following paragraphs I sketch the basic idea of the Gelfand-Levitan construction, borrowing details freely from Marchenko.[15]

Bargmann's formula (16) was obtained by comparing the potential V with another potential V_0 having the same phase-shifts. Instead of this, Gelfand and Levitan compare V with the potential $V_0 = 0$ which gives zero phase-shifts. In both cases, the essential idea is to compare the solutions of the Schrödinger equation (1) with the two potentials V and V_0. For Bargmann the transformation from V_0 to V was independent of the phase-shifts, which entered the solution only implicitly through the wave-functions $u^0(r, E)$ appearing in Equations (14) and (15). For Gelfand and Levitan the wave-functions u^0 are elementary functions (sines and exponentials), but the transformation from V_0 to V is correspondingly more complicated and depends explicitly on the phase-shifts.

We define the comparison wave-function for $E > 0$ by

$$u^0(r, E) = \sin(kr - \eta(E)), \quad k = E^{\frac{1}{2}}, \tag{19}$$

the same expression which appears on the right side of Equation (5). Thus $u^0(r, E)$ satisfies Equation (1) with potential $V_0 = 0$, and coincides with $u(r, E)$ asymptotically as $r \to \infty$. Corresponding to each bound-state energy E_j, we define a comparison wave-function

$$u^0(r, E_j) = m_j \exp(-\kappa_j r), \quad \kappa_j = (-E_j)^{\frac{1}{2}}, \tag{20}$$

where m_j is a free positive real parameter. We suppose that the phase-shift $\eta(E)$ and the parameters $E_1, \cdots, E_n$, $m_1, \cdots, m_n$ are given, and our objective is to construct the potential $V(r)$ which will produce scattering wave-functions $u(r, E)$ for $E > 0$ and bound-state wave-functions $u(r, E_j)$ for $E_j < 0$ coinciding asymptotically with Equations (19) and (20) as $r \to \infty$. It is only a matter of convenience that we now use the m_j as free parameters instead of the c_j defined by Equation (13).

We define the symmetric kernel f on the domain $0 < r < \infty$, $0 < s < \infty$, by

$$\delta(r-s) + f(r,s) = (2/\pi) \int_0^\infty u^0(r,k^2)\, u^0(s,k^2)\, dk + \sum_j u^0(r,E_j)\, u^0(s,E_j). \tag{21}$$

Since

$$\delta(r-s) = (2/\pi) \int_0^\infty \sin kr \sin ks\, dk, \tag{22}$$

Equations (19), (20), (21) imply

$$f(r,s) = f(r+s) = (2/\pi) \int_0^\infty [\cos k(r+s) - \cos(k(r+s) - 2\eta(k^2))]\, dk + \sum_j m_j^2 \exp(-\kappa_j(r+s)). \tag{23}$$

Thus f is completely determined by the quantities $\eta(E)$, $E_1, \cdots, E_n$, $m_1, \cdots, m_n$. We call f the "spectral function" because it summarizes all information concerning the spectrum which we desire to obtain for the Schrödinger equation (1). The Gelfand-Levitan construction will deduce the potential V from a knowledge of f alone.

We use operator notation, so that any kernel $K(r,s)$ is regarded as an operator K operating on the vector-space $L_2[0,\infty]$. The left side of Equation (21) is then the operator $(1+f)$. For any vector v, Equation (21) implies

$$(v,(1+f)v) > 0 \ , \tag{24}$$

so that $(1+f)$ is positive definite. It is easy to prove that $(1+f)$ has an inverse which we denote by $(1+F)$, so that

$$f = -F(1+F)^{-1} \ , \quad F = -f(1+f)^{-1} \ . \tag{25}$$

Let $P(t)$ be the diagonal kernel

$$P(t,r,s) = \delta(r-s) \ , \quad r,s > t \ , \tag{26}$$

$$P(t,r,s) = 0 \ , \quad r,s < t \ . \tag{27}$$

We consider the quantity

$$Q(t,K) = \mathrm{Tr}[P(t)(KK^\dagger + (1+K)f(1+K^\dagger))] \tag{28}$$

which is a functional bilinear in the kernel K and its adjoint $K^\dagger$. A little algebra gives

$$Q(t,K) = \mathrm{Tr}[P(t)(-F + (K-F)(1+f)(K^\dagger-F))] \geq -\mathrm{Tr}[P(t)F] \ , \tag{29}$$

so that Q is bounded below for all K. Let now K belong to the class of causal kernels, that is to say

$$K(r,s) = 0 \quad \text{for} \quad r > s \ . \tag{30}$$

"Causal" means that K propagates influence only inwards, from larger values of r to smaller values. Within this class, $Q(t,K)$ is still bounded below and attains its minimum value

$$M(t) = Q(t,A) \tag{31}$$

for a particular causal kernel A which is independent of t. Taking the variation of Equation (28) with respect to K, we find the Marchenko[15] equation

$$[A+f+Af](r,s) = 0 \quad \text{for} \quad r > s \ , \tag{32}$$

a linear integral equation of Fredholm type which determines the kernel A uniquely. Equation (32) is equivalent to the statement that

$$A + f + Af = B^{\dagger} \tag{33}$$

where B is another causal kernel. From Equation (33) it follows that

$$(1+B^{\dagger})(1+A^{\dagger}) = (1+A)(1+f)(1+A^{\dagger}) \ . \tag{34}$$

But the right side of Equation (34) is self-adjoint while the left side is the unit operator plus an anti-causal kernel. Therefore both sides of Equation (34) are equal to unity, and we have

$$(1+A)(1+B) = 1 \ . \tag{35}$$

Combining Equation (35) with (25) and (33), we find

$$1 + f = (1+B)(1+B^{\dagger}) \ , \tag{36}$$

$$1 + F = (1+A^{\dagger})(1+A) \ . \tag{37}$$

Equation (37) is a nonlinear integral equation for A, sometimes called the nonlinear Gelfand-Levitan equation.

We use Equations (33), (36), (37) to rewrite Equation (28) in the form

$$Q(t,K) = \mathrm{Tr}[P(t)(B+B^{\dagger} + (K-A)(1+f)(K^{\dagger} - A^{\dagger}))] \ , \tag{38}$$

valid for any causal kernel K. In deducing Equation (38), we also used the fact that

$$\mathrm{Tr}[P(t)KB] = \mathrm{Tr}[P(t)AB] = 0 \ , \tag{39}$$

since P(t) is diagonal while K, A, B are causal. Equation (38) shows explicitly that the minimum value

$$M(t) = \mathrm{Tr}[P(t)(B+B^{\dagger})] \tag{40}$$

of Q(t, K) is attained at K = A for every value of t.

The result (40) allows us to obtain a direct relation between the potential V and the spectral function f. The explicit representation of B in terms of the wave-functions $u(r,E)$ with potential and $u^0(r,E)$ without potential is

$$\delta(r-s) + B(r,s) = (2/\pi)\int_0^\infty u^0(r,k^2)\,u(s,k^2)\,dk + \sum_j u^0(r,E_j)\,u(s,E_j) . \qquad (41)$$

Since the wave-functions $u(s,E)$ are a complete orthonormal set, a comparison of Equation (41) with (21) shows that B defined by (41) satisfies Equation (36), and it is easy to prove that Equation (36) with the property of causality defines B uniquely. Since $u^0(r,E)$ and $u(s,E)$ satisfy the Schrödinger equation without and with potential, Equation (41) implies

$$[(\delta^2/\delta s^2) - (\delta^2/\delta r^2)]\,B(r,s) = V(s)[\delta(r-s) + B(r,s)] . \qquad (42)$$

The left side of this equation has a singular term proportional to $\delta(r-s)$ due to the discontinuity of B at $r = s$. Equating the singular terms on both sides gives the Gelfand-Levitan formula for the potential

$$V(r) = 2\,(d/dr)\,B(r,r) , \qquad (43)$$

where $B(r,r)$ means the limit of $B(r,s)$ as $s \to r$ from above. Now Equation (40) implies

$$B(r,r) = -(d/dr)\,M(r) , \qquad (44)$$

and therefore

$$V(r) = -2(d^2/dr^2)\,M(r) . \qquad (45)$$

We wish finally to eliminate the auxiliary kernel B from Equation (40) and obtain an expression for $M(t)$ in terms of f alone. Since B is causal we can transform Equation (40) into a logarithm of a Fredholm determinant

$$\begin{aligned} M(t) &= \mathrm{Tr}\,\log(1+P(t)B) + \mathrm{Tr}\,\log(1+B^\dagger P(t)) \\ &= \log\,\mathrm{Det}\,(1+P(t)B) + \log\,\mathrm{Det}\,(1+B^\dagger P(t)) \qquad (46) \\ &= \log\,\mathrm{Det}\,[1+P(t)B + B^\dagger P(t) + P(t)BB^\dagger P(t)] . \end{aligned}$$

But since B is causal,

$$P(t)B = P(t)BP(t), \quad B^{\dagger}P(t) = P(t)B^{\dagger}P(t) , \tag{47}$$

and Equations (36) and (46) give

$$\begin{aligned} M(t) &= \log \operatorname{Det}[1 + P(t) f P(t)] \\ &= \log \operatorname{Det}[P(t)(1+f)P(t) + (1-P(t))] . \end{aligned} \tag{48}$$

Thus M(t) is simply the logarithm of the Fredholm determinant of the spectral function f(r, s) restricted to the interval $t < r, \quad s < \infty$. Our final formula for the potential is

$$V(t) = -2(d^2/dt^2) \log \operatorname{Det}[\delta(r-s) + f(r,s)]_t^{\infty} . \tag{49}$$

This result of the Gelfand-Levitan analysis has the advantage of avoiding the necessity of solving the integral equations for the kernels A and B. There is a close formal similarity between Equation (49) and Bargmann's Equation (16). Both equations relate the potential to the second derivative of the logarithm of a determinant, in one case continuous and in the other discrete.

I have been unable to find Equation (49) stated explicitly in the published literature. It appears in some unpublished lecture notes of Jost written in 1954.

IV. *Motivation*

The foregoing presentation of the Gelfand-Levitan construction has one serious defect. We began with a variational principle, minimizing the quantity Q(t, K) given by Equation (28), but we never provided any physical motivation for considering this quantity. The choice of Q appeared simply as a mathematical trick justified by its results. I do not know of any satisfactory motivation, within the framework of scattering theory, for making Q a minimum. In fact, my motivation for approaching the theory in this fashion came from a totally different direction, from a problem in optics[19] which seems at first sight to have nothing whatever to do with inverse scattering. I will now briefly describe how the optics problem leads to a variational principle of the form (28).

In the optics problem[19] we are concerned with an unknown disturbance (atmospheric noise) which is a random function $a(t)$ of the time t. The statistical behavior of the disturbance is described by the atmospheric autocorrelation function

$$f(t,t') = \langle a(t)\, a(t') \rangle \ . \tag{50}$$

We attempt to compensate the disturbance by subtracting from it a servo-signal, so that the residual disturbance is

$$s(t) = a(t) - \int_{-\infty}^{t} dt' \; L(t,t')\, I(t') \ , \tag{51}$$

where $L(t,t')$ is a causal feed-back kernel (propagating influence only from earlier to later times), and $I(t')$ is the output of an optical sensor which instantaneously monitors the value of $s(t')$. The response of the optical sensor is on the average

$$\langle I(t) \rangle = b\,s(t) + c \ , \tag{52}$$

with positive coefficients b and c, but $I(t)$ is a photon-count output and is subject to photon-noise fluctuations. The object of the game is to choose the feed-back kernel L so as to make the mean-square residual disturbance

$$D = \int \langle (s(t))^2 \rangle \, dt \ , \tag{53}$$

averaged over both atmospheric and photon fluctuations, as small as possible. The time-integration in Equation (53) extends over a certain finite interval of observation, $T_1 < t < T_2$. When the equations (50) to (52) are solved, taking the quantum nature of light properly into account, the figure of merit D becomes

$$D = \mathrm{Tr}[(P(T_1) - P(T_2))(1+bL)^{-1}(f + LcL^{\dagger})(1+bL^{\dagger})^{-1}], \quad (54)$$

with $P(t)$ defined by Equations (26) and (27). In Equation (54) the term involving f represents atmospheric noise and the term involving $LL^{\dagger}$ represents photon noise, both terms being modified by factors $(1+bL)^{-1}$ which represent the effect of feedback. Now we define the causal kernel K by

$$K = -bL(1+bL)^{-1} \quad (55)$$

and substitute into Equation (54). The result is

$$D = \mathrm{Tr}[(P(T_1) - P(T_2))(c/b^2)KK^{\dagger} + (1+K)f(1+K^{\dagger}))], \quad (56)$$

an expression identical with Equation (28) except for trivial numerical factors. If we choose the unit of measurement of $a(t)$ and $s(t)$ so as to make $(c/b^2) = 1$, we have precisely

$$D = Q(T_1, K) - Q(T_2, K). \quad (57)$$

We find that D is minimized for all values of T_1 and T_2 if we make for the feedback kernel the particular choice

$$K = A, \quad bL = B, \quad (58)$$

where A and B are the Gelfand-Levitan kernels defined by Equations (33), (36) and (37). Note that, to make the definition of causality consistent between Equations (30) and (51), we must identify the space-coordinate r of the scattering problem with minus the time-coordinate t of the optics problem. Causal influence propagates from $r = +\infty$ in the one case and from $t = -\infty$ in the other. The identification (58) is then causally consistent, and leads to the minimum value of D given by Equation (40), namely

$$D = \mathrm{Tr}[(P(T_1) - P(T_2))(B + B^{\dagger})] = \int_{T_1}^{T_2} B(t,t)\,dt. \quad (59)$$

The relation between the inverse scattering problem and the optics problem is a peculiar one. Although the two problems have much in common, essential features of each are absent in the other. The potential, which provides the whole raison d'être of the inverse scattering problem, does not appear at all in the optics problem. And the minimum principle, which appears in a completely natural way in the optics problem, becomes only an artificial mathematical device in the inverse scattering problem. The optics problem is a special case of the general problem of optimization of a linear control system. The general theory of optimum control systems is summarized in an excellent historical review by Thomas Kailath.[20] In Kailath's article it is easy to find places (for example his equations (10), (112) and (122)) where the formal analogies between control theory and the inverse scattering problem become obvious. But the formal analogies remain mysterious so long as no physical connection is found between control theory and the Gelfand-Levitan analysis. It is likely that much still remains to be learned by exploring on a deeper level the connections between these two problems, arising from two areas of physics apparently so remote from one another.

In the practical context where the optics problem arose,[19] I was dealing with a multichannel rather than a single-channel situation. The atmospheric disturbance $a(t)$ and the optical detector output $I(t)$ were not scalar quantities but vectors belonging to different vector-spaces. In fact, $a(t)$ was a function of space as well as time, while $I(t)$ was a vector with one component corresponding to each photon detector. These complications do not change in any essential way the results of the analysis. Equations formally analogous to Equations (50) to (59) still hold. The Gelfand-Levitan construction has also been extended by Newton and Jost[21] to the case of multichannel scattering. The correspondence between the inverse scattering problem and the optics problem remains valid in the multichannel case. In particular, the optics problem provides a natural variational principle generalizing Equation (28) to the multichannel situation, and from this variational principle the solution of the inverse scattering problem can be obtained just as I described it in Section III for a single channel. The natural variational principle for the multichannel optics problem is to minimize the statistician's parameter chi-squared, the sum over detector channels of the mean square signal fluctuation divided by the mean counting-rate in that channel. It turns out that the chi-squared variational principle is precisely the one which gives at the minimum a logarithm of a Fredholm determinant as in Equations (48) and (49). But this chapter is already long enough, so I will refrain from spelling out any further details of the extension of the analysis to the multichannel case. I urge any reader who is seriously interested to work out the details for himself.

Acknowledgement

I am grateful to Res Jost, Tullio Regge and Valya Bargmann for help in the preparation of this chapter. They are, of course, not responsible for any technical or historical errors that it may contain.

FREEMAN J. DYSON (on leave of absence from the Institute for Advanced Study, Princeton, N. J., U.S.A.).
MAX-PLANCK-INSTITUT FÜR PHYSIK UND ASTROPHYSIK
MÜNCHEN, GERMANY

REFERENCES

1) V. Bargmann, Phys. Rev. *75*, 301 (1949).

2) V. Bargmann, Rev. Mod. Phys. *21*, 488 (1949).

3) L. D. Faddeev, Usp. Mat. Nauk *14*, 57 (1959).

4) B. Seckler, J. Math. Phys. *4*, 72 (1963).

5) O. Steinmann, Helv. Phys. Acta *30*, 515 (1957).

6) J. von Neumann, *Mathematische Grundlagen der Quantenmechanik*, (Springer, Berlin, 1932), English translation by R. T. Beyer, (Princeton, 1955).

7) N. Bohr and L. Rosenfeld, Kgl. Danske Vidensk. Selsk. Mat.-fys. Medd. *12*, No. 8 (1933).

8) C. S. Gardner, J. M. Greene, M. D. Kruskal and R. M. Miura, Phys. Rev. Letters *19*, 1095 (1967); for later developments see M. J. Ablowitz, D. J. Kaup, A. C. Newell and H. Segur, Phys. Rev. Letters *31*, 125 (1973), and H. H. Chen, Phys. Rev. Letters *33*, 925 (1974).

9) K. M. Case and M. Kac, J. Math. Phys. *14*, 594 (1973); K. M. Case, J. Math. Phys. *14*, 916 (1973).

10) C. E. Fröberg, Phys. Rev. *72*, 519 (1947).

11) E. A. Hylleraas, Phys. Rev. *74*, 48 (1948).

12) R. Jost, Helv. Phys. Acta *22*, 256 (1947).

13) N. Levinson, Kgl. Danske Vidensk. Selsk. Mat.-fys. Medd. *25*, No. 9 (1949).

14) I. M. Gelfand and B. M. Levitan, Izv. Akad. Nauk SSSR, Ser. Mat. *15*, 309 (1951); English translation in Am. Math. Soc. Translations (2), *1*, 253 (1955).

15) V. A. Marchenko, Dokl. Akad. Nauk SSSR, *72*, 457 (1950) and *104*, 695 (1955).

16) R. Jost and W. Kohn, Phys. Rev. *87*, 979 (1952), Phys. Rev. *88*, 382 (1952), and Kgl. Danske Vidensk. Selsk. Mat.-fys. Medd. *27*, No. 9 (1953).

17) Unpublished letters from V. Bargmann to R. Jost, August 2 and August 31, 1952; I am indebted to R. Jost for copies of these letters.

18) In Section 13 of Reference 3, especially Equation (13.25).

19) F. J. Dyson, J. Opt. Soc. Am. *65*, 551 (1975).

20) T. Kailath, "A View of Three Decades of Linear Filtering Theory," IEEE Transactions on Information Theory, Vol. IT-20, No. 2 pp. 145-181 (1974).

21) R. G. Newton and R. Jost, Nuovo Cim. *1*, 590 (1955).

A FAMILY OF OPTIMAL CONDITIONS FOR THE ABSENCE OF BOUND STATES IN A POTENTIAL

V. Glaser and A. Martin
H. Grosse and W. Thirring

We derive the optimal condition

$$I_p(V) = \frac{(p-1)^{p-1}\Gamma(2p)}{p^p\Gamma^2(p)} \int \frac{d^3x}{4\pi} |x|^{2p-3} \left|\frac{2m}{\hbar^2} V^-(x)\right|^p < 1$$

for the absence of bound states in a potential. The condition is valid

(i) for arbitrary V if $p \geq \frac{3}{2}$;

(ii) for spherically symmetric V if $1 \leq p \leq \frac{3}{2}$.

In the special case of spherical symmetry, the number of bound states with angular momentum ℓ is less than $(2\ell+1)^{1-2p} I_p(V)$. An application for mesic atoms is presented.

1. *Introduction*

A standard problem in Schrödinger's theory is to obtain conditions on a potential in order to guarantee that this potential has at most one or n bound states. For instance, Jost and Pais,(1) Bargmann,(2) Birman and Schwinger(3) have shown that a spherically symmetric potential such that

$$\int V^-(r)\, r\, dr < 1 \tag{1}$$

where $V^- > 0$ is the attractive part of the potential, with units such that $2m/\hbar^2 = 1$, has no bound states. Similarly, it is also known(4) that if

$$\sup r^2 V^-(r) < 1/4 \tag{2}$$

there is no bound state. Bargmann has also obtained that in the state of angular momentum ℓ, the number of bound states ν_ℓ (counted without the $2\ell + 1$ multiplicity factor) is such that

$$(2\ell+1)\nu_\ell < \int r\, V^-(r)\, dr \ . \tag{3}$$

On the other hand, Faris[5] has obtained an inequality of the type

$$\int d^3x\, |V^-(x)|^{3/2} < C \tag{4}$$

for arbitrary potentials, not necessarily spherically symmetric, and one of us,[6] in the special limit of a fixed shape potential, has obtained the asymptotic estimate, for $V = \lambda v$, $\lambda \to \infty$

$$N \sim \frac{1}{6\pi^2} \int d^3x\, |V^-(x)|^{3/2} \ , \tag{5}$$

where N is the total number of bound states.

What we want to do here is to find a complete family of optimal inequalities, incorporating inequalities (1) and (2) as extreme cases. Formulae (1) and (2) are already optimal, in the sense that the numerical constants they contain cannot be improved. Inequality (4), on the other hand, is not yet optimal. The inequalities we shall obtain in Section 2 involve arbitrary powers of the potential $[V^-(r)]^p$, $1 \le p \le \infty$ and are all optimal, as shown in Section 3. Comparison with standard potentials shows that for a convenient p, they give excellent results. In Section 4, an application to muonic atoms is given, which gives relatively tight bounds on the ground-state energy levels.

For monotonous potentials, other conditions can be derived. In particular, Calogero[7] has obtained

$$\int \sqrt{V(r)}\, dr < \frac{\pi}{2} \ . \tag{6}$$

Conditions involving powers of $V < 1$ will be investigated in a forthcoming publication by one of us (H. Grosse).

2. *Presentation of the New Inequalities*

The ground state of the Schrödinger equation

$$-\Delta\psi + V(x)\psi(x) = E\psi(x) \tag{7}$$

if it exists, must have a negative energy E_0 given by the variational principle

$$E_0 = \inf_{\|\psi\|_2=1} H(\psi) \quad H(\psi) = \int (\nabla\psi)^2 d^3x + \int V(x)(\psi(x))^2 d^3x \tag{8}$$

$$= \|\nabla\psi\|_2^2 + (\psi, V\psi) \ .$$

Here it is assumed that the potential V is a (not necessarily centrally symmetric) real function which vanishes sufficiently rapidly at infinity. In (8), the infimum is taken over the Hilbert space $\mathcal{H}'_V$ of functions for which $\nabla\psi$, ψ, and $|V|^{\frac{1}{2}}\psi$ are square integrable, but it may also be taken over any space of smoother functions which is dense in $\mathcal{H}'_V$, e.g. the space $\mathfrak{D}$ of infinitely differentiable functions with compact support. Also there is no loss of generality if we assume these functions to be real-valued.

Let $V = V_+ - V_-$, $V_\pm \geq 0$, be the decomposition of V into its positive and negative parts, and let $r = |x-y|$, where y is an arbitrarily chosen fixed point in R_3. Then we have for an arbitrary real a the Hölder inequality

$$\int V_-\psi^2 d^3x \leq \left\{\int (r^a V_-)^p d^3x\right\}^{1/p} \left\{\int (r^{-a}\psi^2)^q d^3x\right\}^{1/q} \tag{9}$$

$$\equiv \|r^a V_-\|_p \, \|r^{-a/2}\psi\|_{2q}^2$$

for any $1 \leq p \leq \infty$, $\frac{1}{p} + \frac{1}{q} = 1$. We thus obtain a lower bound for the functional H

$$H(\psi) \geq \|\nabla\psi\|_2^2 - \|r^{\alpha}V_-\|_p \|r^{-\alpha/2}\psi\|_{2q}^2 . \tag{10}$$

(Note that for some α and p, the right-hand side of (9) can be $+\infty$.) We shall now choose α so that the quantity $\|r^{\alpha}V_-\|_p \equiv N_p(V_-, y)$ is dimensionless. Since V has dimension $(\text{length})^{-2}$, we get the relation

$$p(2-\alpha) = 3, \quad \text{i.e.} \quad q(1+\alpha) = 3 . \tag{11}$$

Under this condition, the norm $\|r^{-\alpha/2}\psi\|_{2q}$ has the same dimension as $\|\nabla\psi\|_2$, which means that the functional

$$F_q(\psi) = \frac{\|\nabla\psi\|_2^2}{\|r^{(q-3)/2q}\psi\|_{2q}^2} = \frac{\int |\nabla\psi|^2\, d^3x}{\left[\int r^{q-3}\psi^{2q}\, d^3x\right]^{1/q}} \tag{12}$$

remains unchanged under the scale transformations $\psi(x) \to \lambda\psi(\rho x)$ $(\lambda, \rho \neq 0)$. In (12) we have assumed $r = |x|$ without loss of generality. If we denote

$$\mu_q = \inf_{0 \neq \psi \in \mathfrak{D}} F_q(\psi) \geq 0 \tag{13}$$

then (10) can be written

$$H(\psi) \geq \{\mu_q - N_p(V_-, y)\} \|r^{(q-3)/2q}\psi\|_{2q}^2 \tag{14}$$

where

$$N_p^p(V_-, y) = \int |y-x|^{2p-3} V_-^p(x)\, d^3x . \tag{15}$$

The inequality (14) is the starting point of our paper. For suppose a) that for some $1 \leq q \leq \infty$, μ_q is strictly positive, and b) that for some $y \in \mathbf{R}^3$, we have the inequality

$$\mu_q - N_p(V_-, y) > 0 . \tag{16}$$

Then it follows from (14) that the potential V cannot give rise to a bound state. In the case $\mu_q > 0$, (14) can be written in the equivalent form

$$H(\psi) \geq \|\nabla\psi\|_2^2 \{1 - N_p/\mu_q\} \; . \tag{14'}$$

Our aim is to determine the numbers μ_q. Since the functional F_q is invariant under rotations around the origin, we might make the naive supposition that the infimum of F_q is to be sought among centrally symmetric functions $\psi = \psi(r)$. It turns out that the minimization of the functional

$$F_q^R = \text{restriction of } F_q \text{ to centrally-symmetric } \psi \tag{17}$$

is a relatively simple task: the numbers μ_q can be explicitly computed and turn out to be strictly positive for $1 \leq q \leq \infty$ (see Theorem 1 below).

This naive argument is, however, wrong in the case $q > 3$, i.e. $1 \leq p < \frac{3}{2}$: although $\mu_q^R = \inf F_q^R > 0$, we have

$$\mu_q = 0 \quad \text{for} \quad q > 3 \, , \quad \text{i.e.} \quad 2p - 3 < 0 \; . \tag{18}$$

For suppose μ_q were positive, take a potential $V = -V_-$ of compact support deep enough so that it can bind a particle (a spherical square-well will do). Then because of $2p - 3 < 0$, the integral (15) can be made as small as we like by taking $|y|$ big enough, in particular so small that inequality (16) is fulfilled. This contradicts the fact that there is a negative bound state and hence (18) follows.

The above argument does not work for $2p - 3 > 0$. In fact we have the

PROPOSITION 1. *For* $1 < q \leq 3$, *i.e.* $p \geq \frac{3}{2}$,

$$\inf F_q = \inf F_q^R \; . \tag{19}$$

For the proof we first remark that we can restrict ourselves to non-negative ψ's, because replacing ψ by $|\psi|$ will not change $F_q [\nabla|\psi| = \varepsilon(\psi)\nabla\psi]$. Then we use the following theorem:

THEOREM 1. *Given* $\psi(x) \geq 0$, *define* $\psi_R(|x|)$, *the spherical decreasing rearrangement of* ψ : ψ_R *is a decreasing function of* $|x| = r$ *such that for every non-negative constant* M, *the Lebesgue measure* $\mu[\psi_R(|x|)\geq M] = \mu[\psi(x)\geq M]$. *Then*

$$\text{a)} \quad \int |\nabla\psi|^2\, d^3x \geq \int |\nabla\psi_R|^2\, d^3x$$

and

$$\text{b)} \quad \int \chi\psi\, d^3x \leq \int \chi_R\psi_R\, d^3x$$

where χ *and* ψ *are any two positive functions.*

Part (b) of this theorem has been known for a long time, while part (a) is presumably new, so that its proof is given in Appendix A.

We take $\chi = r^{q-3}$. For $q \leq 3$, χ is decreasing and $\chi_R = \chi$. We have also evidently $(\psi^{2q})_R = \psi_R^{2q}$, so that $F_q(\psi) \geq F_q(\psi_R)$. This is just our statement (19).

For the spherically symmetric functional, we have

THEOREM 2. *For* $1 < q < \infty$, *the functional* F_q^R *has the strictly positive infimum*

$$\mu_q^R = \frac{p}{p-1}\left[4\pi\,\frac{(p-1)\Gamma^2(p)}{\Gamma(2p)}\right]^{1/p}, \quad \frac{1}{p}+\frac{1}{q}=1 \tag{20}$$

which is attained by the uniquely determined family of functions

$$\psi_q = \frac{a}{(1+br^{1/(p-1)})^{p-1}} \tag{21}$$

where the arbitrary constants a *and* b *reflect the scale invariance of the problem.*

We can prove this theorem by using an old result of Bliss.[(8)] However, we prefer to give a new straightforward proof. Even this proof is a bit delicate and it will be given in Appendix B. Let us give here only the formal calculation leading to (20) and (21).

By the change of variables

$$\phi = \sqrt{r}\,\psi \qquad x = \ln r \tag{22}$$

the functional F_q^R takes the form

$$F_q^R = (4\pi)^{1/p} G_{2q}(\phi),\ G_{2q}(\phi) = \frac{\int_{-\infty}^{+\infty}\left\{\left(\frac{-d\phi}{dx}\right)^2 + \frac{1}{4}\phi^2\right\}dx}{\left[\int_{-\infty}^{+\infty}\phi^{2q}dx\right]^{1/q}} = \frac{I}{J^{1/q}} \tag{23}$$

and the naive variation equation $\delta G(\phi) = 0$ gives us the differential equation

$$\phi'' = \frac{1}{4}\phi - \kappa\phi^{k-1},\quad \kappa = \frac{I}{J},\quad 2 < k = 2q < \infty \tag{24}$$

which we have to solve under the initial condition

$$\phi(\pm\infty) = \phi'(\pm\infty) = 0 \tag{25}$$

since the integrals I and J have to converge. The first integral of (24) is given by

$$\phi'^2 = \frac{1}{4}\phi^2 - \frac{2\kappa}{k}\phi^k . \tag{26}$$

The arbitrary additive constant in the right-hand side of (26) was set equal to zero in accordance with (25). By the change of scale $\phi \to (k\kappa/8)^{1/k2}\phi$, the equations (26) and (24) take the simpler form

$$\phi'^2 = \frac{1}{4}(\phi^2 - \phi^k),\qquad \phi'' = \frac{1}{4}\phi - \frac{k}{8}\phi^{k-1} . \tag{26'}$$

[Notice that G_k is invariant under the transformations $\phi(x) \to \lambda\phi(x-a)$.] Up to a translation, the solutions of (26′) are given by inversion of the integral

$$\int_\phi^1 \frac{2dt}{t\sqrt{1-t^{k-2}}} = |x| ,\qquad 2 < k < \infty . \tag{27}$$

The substitution $t^{k-2} = 1 - u^2$ leads to an elementary integral with the result

$$\phi_q(x) = \left[\cosh\,(q-1)\,\frac{x}{2}\right]^{-1/(q-1)} \tag{28}$$

which is precisely formula (21) in the old variables.

It remains to compute the minimal value $\nu_k = G_k(\phi_k)$. For the sake of comparison with the results of Appendix B, we shall give the details of this calculation. By multiplying the second equation (26′) with ϕ and integrating over $(-\infty, +\infty)$, we get $I(\phi_k) = (k/8)\,J(\phi_k)$ so that $\nu_k = I \cdot J^{-k/2} = (k/8)\,J^{1-2/k}$. The integral J is computed with the change of variables $dx = 2dt/\sqrt{t^2 - t^k}$ to

$$J = 4 \int_0^1 \frac{t^{k-1}\,dt}{\sqrt{1 - t^{k-2}}} \; . \tag{29}$$

This integral can be expressed in terms of Γ-functions, which leads to formula (20) of Theorem 2.

Note: It is easily verified on the expression (17) that

$$\lim_{p\to+1} \mu_q^R = 4\pi\,, \qquad \lim_{p\to\infty} \mu_q^R = \frac{1}{4} \tag{30}$$

so that μ is strictly positive for $1 \leq p \leq \infty$.

The case $p \to 1$ corresponds to the Bargmann inequality[2] for the absence of bound states $\int r|V(r)|\,dr < 1$. The case $p \to \infty$ corresponds to the condition $\sup r^2|V(r)| < \frac{1}{4}$ which can be found in Courant and Hilbert.[4] The only translation-invariant condition is the one obtained for $p = \frac{3}{2}$.

Let us end this section by pointing out another amusing fact which illustrates the necessity of a rigorous proof of Theorem 1. Let F_q^a be the restriction of the functional F_q^R to functions which vanish outside and on the boundary of a sphere of finite radius a. Then, as shown in Appendix B, the infimum of F^a is the same as that of F^R, but there is no function which saturates that minimum.

3. *Discussion of Inequality (14)*

Proposition 1 and Theorem 2 show that the criterion (16) for the absence of bound states is indeed valid as it stands for all $p \geq \frac{3}{2}$ with μ_q given by formula (20) for any potential. For spherically-symmetric potentials $V = V(r)$ however, Theorem 2 allows us to exploit the whole range $1 < p < \infty$ in the following way.

For a wave function of angular momentum ℓ of the form $\psi(r) P_\ell(\cos\theta)$, the expectation value of the total energy takes the form

$$H_\ell(\psi) = \int_0^\infty \left\{ \left(\frac{d\psi}{dr}\right)^2 + \frac{\ell(\ell+1)}{r^2}\psi^2 + V\psi^2 \right\} r^2\, dr \ . \tag{31}$$

If we subject this functional to the same manipulations, we are led to the minimization of the functional

$$(4\pi)^{-1/p} F_q^\ell(\psi) = \int_0^\infty \left\{ \psi'^2 + \frac{\ell(\ell+1)}{r^2}\psi^2 \right\} r^2\, dr \left[\int_0^\infty r^{q-3}\psi^{2q} r^2\, dr \right]^{-1/q} . \tag{32}$$

The change of variables (22) then leads to the functional G_{2q}^ℓ, which differs from the old one (23) only through the replacement $\frac{1}{4}\phi^2 \to [(2\ell + 1)^2/4]\phi^2$. By the scale transformation $\phi(x) = \phi_1[(2\ell+1)x]$, we get back to the old functional

$$(4\pi)^{-1/p} F_q^\ell(\psi) = G_{2q}^\ell(\phi) = (2\ell+1)^{1+1/q} G_{2q}(\phi_1) \tag{33}$$

for which we know the infimum. Therefore the criterion for the absence of a bound state in the ℓth partial wave reads

$$\frac{(p-1)^{p-1}\Gamma(2p)}{(2\ell+1)^{2p-1} p^p \Gamma^2(p)} \int_0^\infty r^{2p-1} V_-^p(r)\, dr < 1 \ . \tag{34}$$

Of course, when $p \geq \frac{3}{2}$ and $\ell = 0$, we may replace r^{2p-1} by $|x-y|^{2p-1}$ in this formula, where y is an arbitrary point in space. The corresponding minimizing functions are given by

$$\phi_{q,\ell}(x) = \left[\cosh(q-1)\frac{(2\ell+1)x}{2}\right]^{-\frac{1}{q-1}}, \quad \psi_{q,\ell} = \frac{ar^{\ell}}{\left[1+br^{\frac{2\ell+1}{p-1}}\right]^{p-1}}. \tag{35}$$

We want to show that, given a fixed p, the bound we have obtained is the best one in the following sense: the numerical constant μ_q appearing in the inequalities (16) and (34) cannot be replaced by any smaller number. The reason is the following: let $\psi_{q,\ell}$ be a function (35) minimizing the functional (34) and $u_{q,\ell} = r\psi_{q,\ell}$ the corresponding reduced wave function. Then we define a potential $v_{q,\ell}$ by the equation

$$\left\{-\frac{d^2}{dr^2} + \frac{\ell(\ell+1)}{r^2}\right\} u_{q,\ell} + v_{q,\ell}\, u_{q,\ell} = 0 \tag{36}$$

so that $u_{q,\ell}$ may be regarded as the zero energy solution of the Schrödinger equation (36) with the potential $v_{q,\ell}$. Now, according to (35), $u_{q,\ell} \approx a \cdot r^{\ell+1}$ for $r \to 0$ and $\approx$ const $r^{-\ell}$ for $r \to \infty$, which are precisely the conditions of a zero energy bound state. From (36) we find

$$v_{q,\ell} = -\kappa\, r^{q-3} \psi_{q,\ell}^{2q-2}. \tag{37}$$

Now if we choose $V = v$ in the Schrödinger equation (7), we find that in the Hölder inequality (9), actually the equality sign holds because the three integrands $v_{-}\psi^2$, $(r^{\alpha}v_{-})^p$, and $(r^{-\alpha}\psi^2)^q$ are all proportional to each other in view of (37). This also implies $\mu_q - N_p(v^{-}_{q,\ell}) = 0$, which proves our assertion.

Note: From (35) and (37), it follows that "the saturating potentials" have the following behavior:

$$-v_{q,\ell} \simeq \text{const. } r^{q-3+2(q-1)\ell}, \quad -v_{q,\ell} \simeq \text{const. } r^{-(q+1)-2\ell(q-1)}. \tag{38}$$
$$\text{for } r \to 0 \qquad\qquad \text{for } r \to \infty$$

Since they depend on several parameters, they are well suited as "comparison potential" for a given potential V.

We give a practical illustration of this fact in Table 1, where we give the minimum strength of some classical potentials (square well, exponential, Yukawa, Gaussian) necessary to produce a bound state. The "exact" result is taken from Blatt and Weisskopf[9] and comes from a numerical solution of the Schrödinger equation. We also give the Bargmann bound, the bound for $p = \frac{3}{2}$, and the optimal bound. Except for the square well, the $p = \frac{3}{2}$ bound is already excellent (within 2-3% of the exact results). Optimizing with respect to p reduces the discrepancy to less than 1% for the smooth potentials and 4% for the square well.

For the case of spherically-symmetric potentials, another generalization can be made to the case of more bound states. It is well known that if we have ν_ℓ bound states with strictly negative energy with angular momentum ℓ (not counting the $2\ell + 1$ degeneracy), the zero energy radial reduced wave function has ν_ℓ zeros, excluding the origin. Then, if r_p and r_{p+1} are successive nodes, we get

$$0 = \int_{r_p}^{r_{p+1}} \left\{ \left(\frac{du_\ell}{dr}\right)^2 + \frac{\ell(\ell+1)}{r^2} u_\ell^2 + V u_\ell^2 \right\} dr$$

and we can apply to this finite interval all the chain of inequalities previously derived because they are valid for continuous functions with compact support. In this way, adding up the inequalities, we get a bound on the number of bound states:

$$\nu_\ell (2\ell + 1)^{2p-1} < \frac{(p-1)^{p-1}\,\Gamma(2p)}{p^p\,\Gamma^2(p)} \int_0^\infty r^{2p-1}\, V_-^p(r)\, dr \ . \qquad (39)$$

It is known that, at least in the case $p = 1$, inequality (39) cannot be improved, even for $\nu_\ell > 1$. We believe that no substantial improvement can be achieved for different values of p. However, let us point out that if inequality (39) is saturated, it will be only for one given value of ℓ. If, for instance, we try to sum (39) over the various values of ℓ, we will get an overestimate of the number of bound states. For instance, if we take $p = \frac{3}{2}$, we get for the total number of bound states in a spherically-symmetric potential

$$N = \sum_{0}^{\ell_{max}} \nu_\ell (2\ell+1) < I \sum_{0}^{\ell_{max}} 1/(2\ell+1) \tag{40}$$

with $2\ell_{max} + 1 = \sqrt{I}$

$$I = \frac{16}{3\sqrt{3}\pi} \int_0^\infty r^2 \, |V\_(r)|^{3/2} \, dr \; . \tag{41}$$

We have therefore

$$N < I \left[1 + \frac{1}{4} \ln I\right] \qquad \text{for } I \geq 1 \; . \tag{42}$$

This is a strict bound valid for spherically-symmetric potentials. This can be compared with the asymptotic estimate

$$N \sim \frac{1}{6\pi^2} \int |V^-|^{3/2} \, d^2x = \frac{\sqrt{3}}{8} I \tag{43}$$

valid for $V^- = \lambda v^-$, $\lambda \to +\infty$, v^- having a fixed shape.

Though this asymptotic estimate may not be a strict bound, we believe that the logarithmic factor should not be present in (42). On the other hand, for the case without spherical symmetry, we know that the asymptotic estimate cannot be an upper bound, for, by taking N distant potential wells saturating separately the inequality with $I = 1 + \varepsilon$, it is possible to build a system with N bound states. The best one can hope to prove is therefore

$$N \leq I \; .$$

At present we only know that this inequality holds for $I = 1$ and also for $I = 2$. It holds for $I = 2$ for, if we have two bound states, the wave function of the higher state cannot have a constant sign since it is orthogonal to the ground state, which has a positive definite wave function. Therefore the space is divided at least into two regions where $\psi > 0$, $\psi < 0$, with ψ vanishing on the border. Our inequalities can be applied to these two regions separately. This gives the factor 2.

4. *Muonic Atoms*

As a simple illustration of the use of no-binding theorems, we shall derive bounds for the ground-state energy E_0 of an atom with a μ^- and N electrons. We shall take the nuclear charge Z sufficiently small so that relativistic and nuclear size effects can be neglected. Naively one would assume that the electrons just see $Z - 1$ and thus one should get (in the atomic unit):

$$E_0 = -\frac{Z^2\mu}{2} \quad \text{(i.e. the energy of the muon in its ground state)}$$

$$+ E(N, Z-1) \quad \text{(i.e. the energy of N electrons in the potential of a charge } Z-1\text{)}.$$

It is trivial to see[10] that this represents an upper bound for the exact energy. To prove a certain accuracy of this naive expectation, we shall derive a lower bound nearby. Designating the muon variables by (x, p) and those of the electrons by (x_i, p_i), $i = 1, \cdots, N$, we can write the total Hamiltonian

$$H = H_0 + \sum_{i=1}^{N} V_i$$

$$H_0 = \frac{p^2}{2\mu} + \sum_{i=1}^{N} \frac{p_i^2}{2} - Z\left(|x|^{-1} + \sum_{i=1}^{N} |x_i|^{-1}\right) + \sum_{i>j} |x_i - x_j|^{-1} \tag{44}$$

$$V_i = |x - x_i|^{-1} > 0 \ .$$

To obtain lower bounds we shall employ the projection method[10]

$$H \geq H_0 + \sum_i P(P\,V_i^{-1}P)\,P \tag{45}$$

where P is the projector onto the ground state of the muon (r, $r_i = |\vec{x}|$, $|\vec{x}_i|$), and V_i^{-1} is the inverse of V_i in configuration space:

$$P = |2(Z\mu)^{3/2}\,e^{-Z\mu r}><2(Z\mu)^{3/2}\,e^{-Z\mu r}| \tag{46}$$

$$P(P|x-x_i|P)^{-1}P = \frac{1}{r_i} - \frac{1}{r_i}\left[1 + \frac{Z^2\mu^2 r_i^2}{1 - e^{-2Z\mu r_i}\left(1 + \frac{Z\mu r_i}{2}\right)}\right]^{-1} \tag{47}$$

$$= \frac{1}{r_i} + V(r_i)\ .$$

Since

$$H_0 \geq \left[-\frac{Z^2\mu}{2} + \sum \frac{p_i^2}{2} - Z\sum r_i^{-1} + \sum_{i>j} |x_i - x_j|^{-1}\right] P$$
$$+ (1-P)\left[-\frac{Z^2\mu}{8} + E(N,Z)\right] \tag{48}$$

we have the operator inequality

$$H > \left[-\frac{Z^2\mu}{2} + \sum_{i=1}^{N} \frac{p_i^2}{2}\left(1 - \frac{1}{m_1}\right) - (Z-1)\sum_{i=1}^{N} r_i^{-1} + \sum_{i>j} |x_i - x_j|^{-1}\right] P$$
$$+ (1-P)\left[-\frac{Z^2\mu}{8} + E(N,Z)\right] + P\left\{\sum_{i=1}^{N}\left(\frac{p_i^2}{2m_1} + V(r_i)\right)\right\}.$$

Now

$$\{\quad\} \geq 0$$

if

$$\frac{2m_1}{Z\mu} < \frac{p}{p-1}\left[\frac{(p-1)\Gamma^2(p)}{\Gamma(2p)}\right]^{1/p}\left[\int U^p\, r^{p-1}\, dr\right]^{-1/p} \tag{50}$$

with

$$U = \left[1 + \frac{r^2}{1 - e^{-2r}\left(1 + \frac{r}{2}\right)}\right]^{-1}$$

for some value of p.

The maximum of the right-hand side is reached for

$$p = 1.8242\,, \quad \frac{2m_1}{Z\mu} = 1.2706\ . \tag{51}$$

Thus using scaling in the distances, we get the inequality,

$$-\frac{Z^2\mu}{2} + \left(1 - \frac{1.574}{Z\mu}\right)^{-1} E(N,Z-1) \le E_0 < -\frac{Z^2\mu}{2} + E(N,Z-1) \tag{52}$$

which holds as long as the left-hand side is smaller or equal to

$$-\frac{Z^2\mu}{8} + E(N,Z) \quad \text{provided} \quad \frac{Z\mu}{m_1} > 1.574\ .$$

However, from scaling, we easily get[10]

$$E(N,Z) > \left(\frac{Z}{Z-1}\right)^2 E(N,Z-1)\,c\,, \quad c \le 2\ . \tag{53}$$

Hence the inequality holds if

$$|E(N,Z-1)|\left[c\left(\frac{Z}{Z-1}\right)^2 - \left(1 - \frac{1.574}{Z\mu}\right)^{-1}\right] < \frac{3}{8}\, Z^2\mu\ .$$

Even the crudest upper bound on $|E(N,Z-1)|$ shows that for $\mu \ge 1$, this inequality is always satisfied.

Thus we see that the uncertainty in the electron energy is always less than 1%, and with increasing Z, soon becomes smaller than the relativistic corrections.

Another by-product is that we prove at the same time that a system composed of a proton, an electron, and a particle of negative charge is not bound if the mass of this particle is larger than 1.574 electron masses. Notice that if one solves the Schrödinger equation numerically for the potential v, the figure 1.574 is replaced by 1.570!

Acknowledgements

This work was done while we were studying a broader subject in collaboration with Elliott Lieb and Jack Barnes. We acknowledge many stimulating exchanges with them. We thank B. Bonnier and H. Epstein for discussions.

Table 1

Potential		Exact	Bargmann Bound	p = 3/2	Optimal p	Optimal V
Square well	$V\,\theta[1-r]$	V = 2.467	V = 2	V = 2.108	1.1764	2.359
Exponential	$V\,\exp(-2r)$	V = 5.783	V = 4	V = 5.669	1.4473	5.753
Yukawa	$V\,\frac{\exp-r}{r}$	V = 1.680	V = 1	V = 1.648	1.6875	1.664
Gaussian	$V\,\exp(-r^2)$	V = 2.684	V = 2	V = 2.615	1.336	2.660

Appendix A

We want to prove

$$\int |\nabla\psi_R|^2\, d^3x \le \int |\nabla\psi|^2\, d^3x \tag{A.1}$$

and in fact, more generally

$$\int_{\psi_R > M} |\nabla\psi_R|^2\, d^3x \le \int_{\psi > M} |\nabla\psi|^2\, d^3x \tag{A.2}$$

we define

$$I(M) = \int_{\psi > M} |\nabla\psi|^2\, d^3x \tag{A.3}$$

$$I_R(M) = \int_{\psi_R > M} |\nabla\psi_R|^2\, d^3x\ . \tag{A.4}$$

We repeat that, by definition,[(11),(12)]

(i) ψ_R is a decreasing function of $r = |x|$;

(ii)

$$\int_{\psi > M} d^3x = \int_{\psi_R > M} d^3x = V(M) \tag{A.5}$$

(the domain of integration of the left-hand side is not necessarily connected). We call $d\sigma_M$ the (scalar) surface element of the surface $\psi = M$ and $d\sigma_{MR}$ the surface element of the surface $\psi_R = M$. Notice that $\nabla\psi$ is normal to the surface $\psi = M$ and $\nabla\psi_R$ normal to the surface $\psi_R = M$. Then differentiating (A.5), we get

$$\left|\frac{dV}{dM}\right| = \int \frac{d\sigma_M}{|\nabla\psi|} = \int \frac{d\sigma_{MR}}{|\nabla\psi_R|} \tag{A.6}$$

and differentiating (A.3) and (A.4)

$$\left|\frac{dI}{dM}\right| = \int |\nabla\psi| \, d\sigma_M \tag{A.7}$$

$$\left|\frac{dI_R}{dM}\right| = \int |\nabla\psi_R| \, d\sigma_{MR} \tag{A.8}$$

multiplying (A.7) and (A.8), we get

$$\left|\frac{dV}{dM}\right| \left|\frac{dI}{dM}\right| = \int \frac{d\sigma_M}{|\nabla\psi|} \int |\nabla\psi| \, d\sigma_M \geq \left|\int d\sigma_M\right|^2 . \tag{A.9}$$

by Schwarz's inequality, while

$$\left|\frac{dV}{dM}\right| \left|\frac{dI_R}{dM}\right| = \int \frac{d\sigma_{MR}}{|\nabla\psi_R|} \int |\nabla\psi_R| \, d\sigma_{MR} = \left|\int d\sigma_{MR}\right|^2 . \tag{A.10}$$

Here the *equality* sign comes from the fact that $|\nabla\psi_R|$ is constant, by construction, along the surface $\psi_R = \text{const.}$ Now the surfaces $\psi = M$ and $\psi_R = M$ contain by (A.5) the same volume. By standard isoperimetric inequalities, we know that the minimal surface for a given volume is that of a sphere. Hence

$$\int d\sigma_M \geq \int d\sigma_{MR} \tag{A.11}$$

and

$$\left|\frac{dV}{dM}\right| \left|\frac{dI}{dM}\right| \geq \left|\frac{dV}{dM}\right| \left|\frac{dI_R}{dM}\right| \tag{A.12}$$

and

$$\left|\frac{dI}{dM}\right| \geq \left|\frac{dI_R}{dM}\right| . \tag{A.13}$$

Then (A.1) follows by integration of (A.13) from $M = 0$ to $M =$ maximum of ψ.

There are several subtleties that we have ignored in this proof. For instance, we have assumed that there are no three-dimensional regions where ψ is constant. We have assumed that the regions $\psi = \text{const.}$ are two-dimensional surfaces made of a finite number of pieces, sufficiently smooth, etc. We leave it to specialists to make this proof completely clean. Notice also that the proof works in any number of dimensions.

An alternative proof, avoiding the use of isoperimetric inequalities and using the Green's function of the diffusion equation, has been proposed by Elliott Lieb.[(13)] See also[(14)].

Appendix B

This Appendix is devoted to the proof of Theorem 2. What we have to show is that the formal calculations leading to the functions (28) and the numbers ν_k (29) indeed furnish the true "ground state" resp. the true minimum of the functional $G_k(\phi)$ (23).

To start with, let us remark that the functional (23) $G(\phi) = I/J^{1/q}$ (we work with a fixed $2 < 2q = k < \infty$, so we omit from now on any indices referring to this number) is *a priori* meaningful on a space of functions ϕ on which the integrals I and J converge. The largest such space is the Banach space $\mathcal{B}$ of functions with finite norm

$$N^2(\phi) = I(\phi) + J^{1/q}(\phi) \ . \tag{B.1}$$

$\mathcal{B}$ can be regarded as the completion of the space $\mathcal{D} = \{C^\infty$ functions with compact support$\}$ with respect to the norm N.

The trouble with the functional G is that it is translationally invariant, so that if a ground state exists, it is necessarily infinitely degenerate: if ϕ minimizes G, then so do all the functions $\phi_{\lambda,a}(x) = \lambda\phi(x-a)$ [ϕ cannot be translationally invariant since for $\phi = \text{const.}$ $N(\phi) = \infty$; this is therefore a case of "broken symmetry"]. The idea of the proof is the following: we shall break this invariance by first considering the case of a compact interval, say $K = [-R,+R]$, $0 < R < \infty$. For the corresponding functional $G_K(\phi)$, we shall then easily prove the

LEMMA 1.

a) *There exists a function* $\phi \in \mathcal{H}_K$ *which minimizes the functional, where* $\mathcal{H}_K$ *is the Hilbert space of (real-valued) functions on* $K = [-R,+R]$ *obtained by completion of* $\mathcal{D}(K)$ *with respect to the norm*

$$N_2^2(\phi) = \int_K \left(\phi'^2 + \frac{1}{4}\phi^2\right) dx \equiv \|\phi'\|_2^2 + \frac{1}{4}\|\phi\|_2^2 \ . \tag{B.2}$$

b) *The minimizing function is at least twice continuously differentiable on* K *and satisfies there the differential equation* (24) *and the boundary condition*

$$\phi(-R) = \phi(R) = 0 \tag{B.3}$$

which determine ϕ *uniquely up to a multiplicative constant. The corresponding minimal value* $\nu(R) = G_K(\phi)$ *is strictly positive.*

Theorem 2 will then follow from this lemma by taking the limit $R \to \infty$ if we notice that

$$G_K(\phi) = G(\phi) \quad \text{for} \quad \phi \in \mathcal{D}(K) \tag{B.4}$$

Because of the density of $\mathcal{D}(K)$ in $\mathcal{H}_K$ and of $\mathcal{D}(R)$ in $\mathcal{B}$, it follows from $\mathcal{D}(K_1) \subset \mathcal{D}(K_2) \subset \mathcal{D}(R)$ when $K_1 \subset K_2$ that the function

$$\nu(R) = \inf_{0 \neq \phi \in \mathcal{H}_K} G_K(\phi) = \inf_{\phi \in \mathcal{D}(K)} G_K(\phi) > 0$$

is a decreasing function of R and that

$$\lim_{R \to \infty} \nu(R) = \nu = \inf_{\phi \in \mathcal{D}(R)} G(\phi) = \inf_{\phi \in B} G(\phi) \ . \tag{B.5}$$

A direct computation of this limit will turn out to coincide with the value ν given in the text, where we have shown that this value is attained by the function (28). It is then easy to see that up to translations and multiplication by a constant, this solution is unique.

Proof of Lemma 1. Let us first remark that the space $\mathcal{H}_K$ consists of continuous functions on K vanishing at both ends of K, for which the norm N_2 is finite. This automatically ensures the existence of the integral $J(\phi)$, so that G_K is meaningful for all $0 \neq \phi \in \mathcal{H}_K$. Indeed for any $\phi \in \mathcal{D}(K)$, the Schwarz inequality gives

$$|\phi(x_1) - \phi(x_2)| = \left| \int_{x_1}^{x_2} \phi' \, dx \right| \leq |x_1 - x_2|^{\frac{1}{2}} N_2(\phi) \tag{B.6}$$

where x_1, x_2 are any two points in K. Also by setting $x_1 = x$ and $x_2 = \pm R$, we get from here in view of $\phi(\pm R) = 0$

$$|\phi(x)| \leq \min |x \pm R|^{\frac{1}{2}} N_2(\phi) \equiv \chi(x) N_2(\phi) \ . \tag{B.7}$$

If we take now a Cauchy sequence $\phi_n \in \mathfrak{D}(K)$ which converges to an arbitrary element $\phi \in \mathcal{H}_K$ in the norm N_2, it follows that (B.6) and (B.7) are valid also for any element of $\mathcal{H}_K$.

By the very definition of $\nu(R)$, there exists a sequence of $0 \neq \phi_n \in \mathfrak{D}(K)$ such that $G_K(\phi_n) \to \nu(R)$ for $n \to \infty$. Since G_K is scale-invariant, we may normalize the ϕ_n so that

$$J(\phi_n) = 1 \quad \text{and hence} \quad I(\phi_n) = N_2^2(\phi_n) \equiv \nu_n \searrow \nu(R) . \tag{B.8}$$

The inequalities (B.7) and (B.8) tell us that the family of functions $\{\phi_n\}_1^\infty$ is equicontinuous and uniformly bounded. The Arzela-Ascoli theorem tells us that this family is compact in the sup-norm, so that we may assume that the sequence ϕ_n converges uniformly on K to the $\frac{1}{2}$-Hölder continuous function $\phi(x)$ bounded by $\chi(x)\nu_R$. Because of uniform convergence, we have

$$1 = \lim_{n\to\infty} J(\phi_n) = J(\phi) , \tag{B.9}$$

so that $\phi \neq 0$. We have to show that $\phi \in \mathcal{H}_K$ and $G_K(\phi) = \nu(R)$.

Because of uniform convergence, we see that the sequence $\sigma_n = \frac{1}{4} \|\phi_n\|_2^2$ converges to a strictly positive limit:

$$\lim_{n\to\infty} \sigma_n = \frac{1}{4} \int_K \phi^2 \, dx = \sigma > 0 . \tag{B.10}$$

This implies also the convergence of the sequence

$$\rho_n = \int_K \phi_n'^2 \, dx = \nu_n - \sigma_n \to \rho \geq 0 . \tag{B.11}$$

If we denote by (u, v) the scalar product $\int_K uv \, dx$, then for any $v \in \mathfrak{D}(K)$, the limit

$$\lim_{n\to\infty} (\phi_n', v) = - \lim_{n\to\infty} (\phi_n, v') = -(\phi, v') = \langle \phi', v \rangle \tag{B.12}$$

exists again because of uniform convergence. Hence the derivative ϕ' of ϕ exists in the sense of distributions as indicated by the last expression in (B.12). On the other hand,

$$|(\phi_n', v)| \leq \|\phi_n'\|_2 (v, v)^{\frac{1}{2}} = \rho_n^{\frac{1}{2}} (v, v)^{\frac{1}{2}} .$$

By taking the limit $n \to \infty$, we find that the linear functional (B.12) is bounded by

$$|<\phi', v>| \leq \rho^{\frac{1}{2}} (v, v)^{\frac{1}{2}} \quad \text{for all} \quad v \in \mathfrak{D}(K) . \tag{B.13}$$

Because $\mathfrak{D}$ is dense in $L_2(K)$ with scalar product (u, v), the linear functional $<\phi', v>$ can be uniquely extended to the whole of L_2. By the Fischer-Riesz theorem, there exists a $w \in L_2$ such that $<\phi', v> = (w, v)$. Hence $\phi' = w \in L_2(K)$ and thus $\phi \in \mathcal{H}_K$. By putting $v = \phi'$ in (B.13), we further obtain $(\phi', \phi') < \rho$. Combining all these results, we get the inequality

$$G_K(\phi) \leq \rho + \sigma = \nu(R) .$$

But since $\nu(R)$ was supposed to be the infimum of G_K on $\mathcal{H}_K$, the equality sign must hold:

$$G_K(\phi) = \nu(R) > 0 . \tag{B.14}$$

It remains only to show that the ϕ we have obtained is twice continuously differentiable and satisfies the differential equation (24):

$$\phi'' = \frac{1}{4}\phi - \kappa |\phi|^{k-1} \varepsilon(\phi) \equiv L(\phi), \quad \kappa = I/J > 0 . \tag{B.15}$$

As is usual in the variational calculus, it follows immediately from the observation that the function $\lambda \to G_K(\phi + \lambda v)$ takes its absolute minimum at $\lambda = 0$ for all $v \in \mathfrak{D}(K)$. Here ϕ'' has to be understood in the sense of distributions, but since the right-hand side $L(\phi)$ has been proved to be continuous, the second derivative exists also in the ordinary sense (after having maybe changed the L_2-function ϕ' on a set of Lebesgue measure zero). Finally, the initial condition (B.3) holds for any element of $\mathcal{H}_K$ since $\chi(\pm R) = 0$ in the inequality (B.7). This completes the proof of our lemma.

We now have to compute the number $\nu(R)$. The first integral of Equation (B.15) is given by

$$\phi'^2 = \frac{1}{4}\phi^2 - \frac{2\kappa}{k}|\phi|^k + C \equiv P(\phi) . \tag{B.16}$$

The initial condition (B.3) implies that $c = \phi'^2(+R) = \phi'^2(-R)$, which is a strictly positive number, since $\phi(R) = \phi'(R) = 0$ would imply $\phi \equiv 0$. The solutions of (B.16) are obtained as the inverse functions of the multi-valued integral

$$x = \pm \int \frac{d\phi}{\sqrt{P(\phi)}} . \tag{B.17}$$

The strict positivity of the constants κ, C and $k > 2$ implies, as is easily seen, that the function $p(t) = \frac{1}{4}t - (2\kappa/k)t^k + C$ has exactly one simple zero $t_0 > 0$ for $t \geq 0$. Hence $P(\phi) > 0$ for $-\phi_0 < \phi < \phi_0$, $\phi_0 = t_0^{\frac{1}{2}}$, it has a simple zero at $\phi = \pm\phi_0$ and is negative elsewhere. As it is well known from the theory of the pendulum, this implies that the inverse functions of (B.17) (which differ from each other only by a translation) are periodic functions defined on the whole real line that oscillates between the extremal values $\pm\phi_0$; their simple zeros are half a period apart and they monotonically increase resp. decrease between two consecutive extremal values. The property $P(-\phi) = P(+\phi)$ furthermore implies that they are symmetric with respect to their extremal points and antisymmetric with respect to their zeros.

It is convenient to normalize $\phi_0 = 1$ by a change of scale $\phi \to \lambda\phi$. With this normalization

$$C = \frac{2\kappa}{k} - \frac{1}{4} > 0 \quad \text{which implies} \quad \kappa > \frac{k}{8} . \tag{B.18}$$

The quarter-period of the solutions of (B.16) is given by

$$T(\kappa) = \int_0^1 \frac{dt}{\sqrt{P(t,\kappa)}} . \tag{B.19}$$

$\phi(\pm R) = 0$ imposes the condition

$$T(\kappa) = R/n\,, \qquad n = 1, 2, 3, \cdots \tag{B.20}$$

which can always be solved for κ:

$$\kappa_n = T^{-1}(R/n)\ . \tag{B.21}$$

T is namely a strictly decreasing real-analytic function of κ for $\kappa > k/8$, and moreover

$$\lim_{\kappa\to\infty} T(\kappa) = 0 \qquad \lim_{\kappa\to k/8} T(\kappa) = \infty\ . \tag{B.22}$$

[The last statement comes from the logarithmic divergence of the integral (B.19) at $t = 0$ for $\kappa = k/8$.] Hence T^{-1} exists on $(0, \infty)$, is real-analytic and strictly decreasing, and satisfies:

$$\lim_{R\to 0} T^{-1}(R) = \infty \qquad \lim_{R\to\infty} T^{-1}(R) = k/8\ . \tag{B.23}$$

This discussion shows that the functional G_K has a denumerable infinity of "stationary states" $\phi_n\,(n = 1, 2, 3, \cdots,)$, the nth of which has exactly $n-1$ distinct zeros in the interior of K. It is expected that $\phi = \phi_1$ is the sought-for ground state. To show this, we compute $G_K(\phi_n)$. By multiplying the differential equation (B.15) for ϕ_n by ϕ_n and integrating, we obtain, because of $\phi_n(\pm R) = 0$: $I(\phi_n) = \kappa_n J(\phi_n)$, and hence

$$G_K(\phi_n) = \frac{I}{J}\, J^{1/p} = \kappa_n \left[\int_{-R}^{+R} |\phi_n|^k dx\right]^{1/p} = [n\,L(\kappa_n)]^{1/p} \quad \text{where}$$

$$L(\kappa) = \kappa^p \int_0^1 \frac{2t^k\,dt}{\sqrt{P(t,\kappa)}}\ . \tag{B.24}$$

In the last step we have used the fact that $(-R, +R)$ contains exactly 2n quarter-periods of ϕ_n, and we made the change of variables $dx = P^{-\frac{1}{2}}(t)\,dt$ on the quarter-period. Here $p^{-1} = 1 - 2k^{-1}$. It turns out that $L(\kappa)$ is a strictly monotonically increasing function of κ for $(k/8) < \kappa < \infty$, and since κ_n increases with n according to (B.21), the ground state indeed corresponds to the value $n = 1$. We now define, for $n \geq 2$, $\Psi_n = \phi_n$ on the last half period $[R - (2/n)R, R]$, $\Psi_n = 0$ elsewhere. Evidently $\Psi_n \in \mathcal{H}_K$ and $G_K(\Psi_n) = L^{1/p}(\kappa_n) < G_K(\phi_n)$, so that ϕ_n cannot be the ground state of G_K. Neither can ψ_n, since its derivative has a discontinuity at $x = R(1-2/n)$, which contradicts Lemma 1. Hence ϕ_1 is the minimizing function determined uniquely up to a multiplicative constant and

$$\nu(R) = L^{1/p}(\kappa) \quad \text{where} \quad \kappa = T^{-1}(R) \; . \tag{B.25}$$

From (B.23) it follows that $\lim R \to 0$ corresponds to $\lim \kappa = k/8$, and we readily see that formula (24) of Section 2 is obtained in this limit. This proves formula (20) of Theorem 2. That the functions (28) saturate this lower bound, we have shown in the text. Their uniqueness comes from the following argument. As in the proof of Lemma 1, it follows from inequality (B.6) (which holds also for all $\phi \in \mathcal{B}$ on which the functional G is defined), that any function ϕ minimizing G also satisfies the differential equation (B.15). Its first integral is given by (B.16) with $C = 0$, because there must exist a sequence of points $x_n \to \infty$ such that $\lim \phi(x_n) = \lim \phi'(x_n) = 0$ [otherwise the integral $I(\phi)$ would diverge]. But all solutions of the last equation are given by (28), as shown in the text. This completes the proof of Theorem 2.

Note: It is interesting to remark that the functional G_H of a half-infinite interval, say $H = (-\infty, 0)$, has the same infimum ν as the functional G of the whole real line. On the one hand we have namely $\inf G_H \geq \nu$, on the other hand, one easily sees that $\lim_{n\to\infty} G_H(\phi_n) = \nu$ for the sequence of functions $\phi_n(x) = \alpha(x)\,\phi_q(x+n)$, where ϕ_q is a minimizing function (28) of the functional G and α is any positive C^∞-function such that $\alpha(x) = 0$ for $x \geq 0$; $0 \leq \alpha \leq 1$ for $-1 \leq x < 0$ and $\alpha = 1$ for $x < 1$. The infimum is, however, *not* attained. For suppose some $\phi \in \mathcal{B}_H$ does minimize G_H, then it must be a solution of (B.16) with $C = 0$.

Since no solution of this equation vanishes at $x = 0$, $\phi \notin \mathcal{B}_H$. This proves the remark made at the end of Section 2.

V. GLASER and A. MARTIN
CERN – GENEVA

H. GROSSE and W. THIRRING
INSTITUT FÜR THEORETISCHE PHYSIK
DER UNIVERSITÄT WIEN

REFERENCES

(1) R. Jost and A. Pais, Phys. Rev. *82*, 840 (1951).

(2) V. Bargmann, Proc. Nat. Acad. Sci. (US) *39*, 961 (1952).

(3) M. Birman, Mat. Sb. *55*, 125 (1961),
J. Schwinger, Proc. Nat. Acad. Sci. (US) *47*, 122 (1961).

(4) R. Courant and D. Hilbert, Methods of mathematical physics (Interscience Publishers, New York, 1953), Vol. 1, p. 446.

(5) W. Faris, Battelle Mathematics Report 87 (Battelle Memorial Institute, Geneva, 1974).

(6) A. Martin, Helv. Phys. Acta *45*, 142 (1972).
See also B. Simon, J. Math. Phys. *10*, 1123 (1969).

(7) F. Calogero, Commun. Math. Phys. *1*, 80 (1965).
See also K. Chadan, Nuovo Cimento *58A*, 191 (1968).
W. M. Frank, J. Math. Phys. *8*, 466 (1967).

(8) G. A. Bliss, J. London Math. Soc. *5*, 40 (1930).
See also G. Rosen, SIAM J. Appl. Math. *21*, 30 (1971).

(9) J. M. Blatt and V. F. Weisskopf, Theoretical nuclear physics (John Wiley and Sons, New York, 1952), p. 55.

(10) W. Thirring, Vorlesungen über mathematische Physik T7: Quanten mechanik (Institut für Theoretische Physik der Universität, Wien, 1974), p. 133.

(11) G. H. Hardy, J. E. Littlewood and G. Polya, Inequalities (Cambridge University Press, 1973), p. 280

(12) H. J. Brascamp, J. M. Luttinger and E. H. Lieb, J. Functional Analysis *17*, 227 (1974).

(13) E. Lieb, private communication.

(14) J. M. Luttinger J.M.P. *14*, 586 (1973).

SPINNING TOPS IN EXTERNAL FIELDS*

Sergio Hojman† and Tullio Regge††

I. *Introduction*

The Bargmann Michel Telegdi (1959) equations solve completely the problem of determining the precession of the spin of a particle with magnetic moment from its translational motion in a homogeneous field. Therefore, any theory which endeavors to describe the motion of a like spinning particle in an arbitrary field has to reduce to the BMT equations in the appropriate limit. One such a theory in Lagrangian form has been proposed by one of us (T. R.) in collaboration with A. Hanson (1974).

We would like to discuss briefly here the connection, as well as the differences, between this theory and previous approaches. A similar discussion will be carried out for the well-known Papapetrou (1951) equation for the motion of a top in a gravitational field (Hojman, 1975).

A definite advantage of a canonical formalism is the possibility of defining exactly the role and the physical interpretation of the variables appearing in the equations. Furthermore, the variational procedure insures the consistency of the equations of motion which is not easily achieved in other theories.

Finally, the approach in I is not restricted to homogeneous fields and this leads to a definite generalization of the conventional BMT approach.

This does not imply that the kind of particle described in I is necessarily physically relevant since the particular electromagnetic interaction

---

*Publication supported in part by National Science Foundation Grant No. GP 30799X to Princeton University

†Absent on leave from Universidad de Chile.

††Research supported in part by National Science Foundation Grant No. GP 40768X.

proposed there need not be the only possible one, moreover, particles are not spherical tops since they do not exhibit the peculiar degeneracy of the mass spectrum associated to the extra symmetry of the Lagrangian. However, most of the qualitative features of the spherical top should be retained with minor modifications in the general case.

We are particularly interested here in evaluating the effect of the spin precession through the inertia of the spin on the translational motion of the top. This feedback effect is usually totally negligible and as such it has never been extensively looked into in previous attempts. We may not exclude, however, that in high enough and highly inhomogeneous fields, both electromagnetic and gravitational, this term may be observable and we have accordingly discussed some of the related effects.

It should be noted, finally, that the conventional Papapetrou equations are of third order in the coordinates. This is a most unpleasant feature which forces one to introduce ad hoc boundary conditions and/or limitations on the magnitude of the spin. This is not needed and in fact, some results of Papapetrou theory obtained through such an approximation can be seen to be exact in our formalism (for details, see II).

II. *Constraints*

The description of a relativistic top is usually done by means of four dimensional tensorial quantities (specifying its velocity, momentum, angular velocity and spin) for the sake of explicit relativistic invariance. However, the aforementioned tensors contain more degrees of freedom than the system one is dealing with and therefore constraints must be imposed on them in order to be able to identify such tensors with the appropriate physical variables of a relativistic spinning top.

The way in which one constraints momentum and velocity is very similar to the one one uses for spinless particles and it will not be discussed in this section (for details see Hanson and Regge (1974)). The constraint equations for spin and angular velocity are more interesting and have presented more difficulties and we will briefly discuss them in what follows.

Consider an antisymmetric spin tensor $S^{\mu\nu}$ (which will be precisely defined in the next section). $S^{\mu\nu}$ has six independent components, but only three of them are needed to describe the rotation of the top. To get rid of the extra degrees of freedom, two different kinds of constraints have been proposed

$$S^{\mu\nu} u_\nu = 0 \tag{II.1}$$

and

$$S^{\mu\nu} P_\nu = 0 \ . \tag{II.2}$$

We will refer to these as the Pirani (1956) and Tulczyjew (1959) constraints respectively. (See also Pryce (1948).) u_ν and P_ν are the velocity and momentum of the spinning particle, respectively.

It is easy to show that the Pirani constraint for a free top gives rise to a motion in an arbitrary large circle in the "rest" frame of the momentum ($P^i = 0$) (for details see Weyssenhoff and Raabe (1947) and Hojman (1975)). This fact means that the Pirani constraint is not restrictive enough and allows solutions with more than six degrees of freedom that is the right number to describe a top (three for translation and three for rotation). This circular motion is contradicted by experiment and clearly does not possess the right non-relativistic limit. Therefore, we will avoid the use of the Pirani constraint.

We are left, then, with the Tulczyjew constraint, and we will see in the next section that it is the appropriate constraint to adopt because it implies that a free top does not move in the rest frame of the momentum. The associated gauge (and invariant relations) for the angular velocity will be dealt with in the next section.

III. *Lagrangian Formalism for a Free Top*

We will describe a top by the variables $x^\mu(\tau)$, $e_{(a)}{}^\mu(\tau)$ where x^μ are four coordinates that give the position of the top in a Minkowski space-time, $e_{(a)}{}^\mu$ are four vectors attached to the top to describe its rotation and τ is an arbitrary parameter in the top worldline. (The index a has no tensorial character and it is the name of the vector, μ is a tensorial index: $e^2_{(0)}$ means the 2 component of the timelike vector of the tetrad) (see I and II).

The four vectors $e_{(a)}$ constitute an orthonormal complete frame

$$g_{\mu\nu}\, e^{\mu}_{(a)}\, e^{\nu}_{(\beta)} = \eta_{(a\beta)} \tag{III.1}$$

$$\eta^{(a\beta)}\, e^{\mu}_{(a)}\, e^{\nu}_{(\beta)} = g^{\mu\nu} \tag{III.2}$$

where $g_{\mu\nu}$ is the metric of the Minkowski spacetime and $\eta_{(a\beta)} = \eta^{(a\beta)} = \mathrm{diag}(1,-1,-1,-1)$ is an invariant matrix. Therefore, only six of the sixteen components of the tetrad vectors are independent. Define now the velocity u^{μ} and angular velocity $\sigma^{\mu\nu}$ by

$$u^{\mu} \equiv \dot{x}^{\mu} \tag{III.3}$$

$$\sigma^{\mu\nu} = \eta^{(a\beta)}\, e^{\mu}_{(a)}\, \dot{e}^{\nu}_{(\beta)} = -\sigma^{\nu\mu} \tag{III.4}$$

where the dot means differentiation with respect to τ. The antisymmetry of $\sigma^{\mu\nu}$ follows from the completeness relation and implies that the angular velocity has only six independent components. It is useful to define, for variational purposes, the antisymmetric tensor $\delta\theta^{\mu\nu}$

$$\delta\theta^{\mu\nu} = \eta^{(a\beta)}\, e^{\mu}_{(a)}\, \delta e^{\nu}_{(\beta)} = -\delta\theta^{\nu\mu}\ . \tag{III.5}$$

One can easily prove that

$$\delta\sigma^{\mu\nu} = \delta\theta^{\mu\nu} + \sigma^{\mu\lambda}\,\delta\theta_{\lambda}^{\ \nu} - \delta\theta^{\mu\lambda}\,\sigma_{\lambda}^{\ \nu} \tag{III.6}$$

which will be needed when devising the equations of motion.

Define the four invariants

$$\left.\begin{aligned} a_1 &= u_{\mu}\, u^{\mu} \equiv u^2 \\ a_2 &= \sigma^{\mu\nu}\,\sigma_{\mu\nu} \equiv \sigma\cdot\sigma \\ a_3 &= u_{\mu}\,\sigma^{\mu\nu}\,\sigma_{\nu\lambda}\, u^{\lambda} \equiv u\sigma\sigma u \\ a_4 &= \det\sigma = (1/16)(\sigma^{\mu\nu}\sigma^{*}_{\mu\nu})^2 \equiv (1/16)(\sigma\cdot\sigma^{*})^2 \end{aligned}\right\} \tag{III.7}$$

where

$$\sigma^{*\mu\nu} = (1/2)\,\epsilon^{\mu\nu\lambda\rho}\,\sigma_{\lambda\rho}$$

with $\epsilon^{0123} = +1$.

The free relativistic spherical top Lagrangian is of the form

$$L_0 = \sqrt{a_1}\,\mathcal{L}(a_2/a_1, a_3/a_1^2, a_4/a_1^2) \tag{III.8}$$

which insures reparametrization invariance under the change of time variable $\tau' = \tau'(\tau)$ (see I).

The linear momentum P^μ and spin $S^{\mu\nu}$ are defined by

$$P^\mu = -\partial L_0/\partial u_\mu = -2u^\mu L_1 - 2\sigma^{\mu\nu}\sigma_{\nu\lambda}u^\lambda L_3$$

$$S^{\mu\nu} = -\partial L_0/\partial\sigma_{\mu\nu} = -4\sigma^{\mu\nu}L_2 - 2(u^\mu\sigma^{\nu\lambda}u_\lambda - u^\nu\sigma^{\mu\lambda}u_\lambda)L_3$$
$$-\frac{1}{2}\sigma^{*\mu\nu}(\sigma\cdot\sigma^*)L_4 \tag{III.9}$$

where

$$L_i(a_1, a_2, a_3, a_4) = (\partial/\partial a_i)L_0(a_1, a_2, a_3, a_4)\ . \tag{III.10}$$

The equations of motion for arbitrary variations δx_μ, $\delta\theta_{\mu\nu}$ are obtained with the help of (III.6) and (III.9)

$$\dot{P}^\mu = 0 \tag{III.11}$$

$$\dot{S}^{\mu\nu} = S^{\mu\lambda}\sigma_\lambda{}^\nu - \sigma^{\mu\lambda}S_\lambda{}^\nu$$
$$= P^\mu u^\nu - u^\mu P^\nu\ . \tag{III.12}$$

From these equations and Tulczyjew constraint, one has

$$0 = (S^{\mu\nu}P_\nu)^{\boldsymbol{\cdot}} = \dot{S}^{\mu\nu}P_\nu + S^{\mu\nu}\dot{P}_\nu = \dot{S}^{\mu\nu}P_\nu = (P^\mu u^\nu - P^\nu u^\mu)P_\nu \tag{III.13}$$

so that u is parallel to P and $u^i = 0$ in the frame where $P^i = 0$. The top is at rest in the rest frame of the momentum when one chooses Tulczyjew constraint.

This theory has two first class constraints (in the sense of Dirac (1964), see also I)

$$P_\mu P^\mu = f\left(\frac{1}{2} S_{\mu\nu} S^{\mu\nu}\right) \qquad \text{(III.14)}$$

$$\det S = (1/16)(S^{\mu\nu} S^*_{\mu\nu})^2 = 0 \, . \qquad \text{(III.15)}$$

The first one can be used to fix the time gauge and the second one is used to choose

$$e_{(0)}{}^\mu = \frac{P^\mu}{M} \qquad \text{(III.16)}$$

where

$$P_\mu P^\mu = M^2 \qquad \text{(III.17)}$$

(M is the rest mass of the top). There are also second class constraints. For a detailed treatment of the canonical formalism, see I.

Relation (III.14) is the trajectory equation and contains a function of a single variable. The trajectory function can be found from the Lagrangian as the envelope of a family of straight lines in the M–J plane parametrized by ξ given by

$$\mathcal{L}(\xi) = -M + (\xi/2)^{\frac{1}{2}} J \qquad \text{(III.18)}$$

where

$$J^2 = \frac{1}{2} S_{\mu\nu} S^{\mu\nu} \qquad \text{(III.19)}$$

and

$$\mathcal{L}(\xi) = \mathcal{L}(\xi, 0.0) \qquad \text{(III.20)}$$

($\mathcal{L}$ defined in (III.8)).

The Tulczyjew constraint follows and implies

$$2L_1 L_2 = L_3(a_1 L_1 + a_2 L_2 + a_3 L_3 + a_4 L_4) \qquad \text{(III.21)}$$

$$2L_2 L_3 = L_1 L_4 \qquad \text{(III.22)}$$

and these differential equations can be used to construct the Lagrangian uniquely from the knowledge of the trajectory function (see I and II).

The system defined in this way has exactly the same number of degrees of freedom as the nonrelativistic theory of a rotator.

IV. *Electromagnetic Interactions of a Top*

We can treat charged tops by simply adding $-e u_\mu A^\mu$ to the free Lagrangian, where e is the charge and A^μ is the electromagnetic potential. A magnetic moment requires a much more complicated treatment and we know of no *a priori* reason why the choice made in I should be unique or even the simplest one. The device is to define $\tilde{\sigma}^{\mu\nu}$ from the implicit nonlinear equation

$$\tilde{\sigma}^{\mu\nu} = \sigma^{\mu\nu} + g L_0(u, \tilde{\sigma}) F^{\mu\nu} \qquad \text{(IV.1)}$$

and the new Lagrangian is then obtained by replacing σ with $\tilde{\sigma}$ everywhere.

One obtains the equations of motion

$$\pi^\mu = - e F^{\mu\nu} u_\nu + (1/2) g L_0 S_{\alpha\beta} F^{\alpha\beta,\mu} \qquad \text{(IV.2)}$$

$$\begin{aligned} \dot{S}^{\mu\nu} &= S^{\mu\lambda} \sigma^\nu_\lambda - \sigma^{\mu\lambda} S^\nu_\lambda \\ &= \pi^\mu u^\nu - u^\mu \pi^\nu + g L_0 (F^{\mu\lambda} S^\nu_\lambda - S^{\mu\lambda} F^\nu_\lambda) \end{aligned} \qquad \text{(IV.3)}$$

where

$$\left.\begin{aligned} \pi^\mu &= - \frac{\partial L_0(u, \tilde{\sigma})}{\partial u_\mu} \\ S^{\mu\nu} &= - \frac{\partial L_0(u, \tilde{\sigma})}{\partial \sigma_{\mu\nu}} \end{aligned}\right\} \qquad \text{(IV.4)}$$

From these equations, one finds the equation of motion for $W^\mu = S^{*\mu\nu} \pi_\nu$

$$\pi^2 \dot{W}^\mu = - \pi^\mu W_\alpha \dot{\pi}^\alpha - g L_0 (\pi^2 F^{\mu\nu} W_\nu + \pi^\mu W_\alpha F^{\alpha\beta} \pi_\beta) \,. \qquad \text{(IV.5)}$$

This equation reduces to the Bargmann-Michel-Telegdi equation for weak homogeneous fields and we must identify

$$g\mathcal{L}\big|_I = -\frac{ge}{2m}\Big|_{BMT}$$

$$\pi^\mu\big|_I = mu^\mu\big|_{BMT} \,.$$

The resulting condition is that the magnetic interaction should be small compared to the rest mass of the top $(M^2 \equiv \pi^\mu \pi_\mu)$

$$M^2 >> \frac{e}{2} F_{\mu\nu} S^{\mu\nu} \,.$$

One can solve (IV.2) and (IV.3) in a constant homogeneous magnetic field B along the 3 direction. The particle moves in a circular cylinder of constant radius R and the cyclotron frequency Ω is now spin orientation dependent

$$\Omega = -\frac{\pi_0\, eB\left(M^2 - eBS^{12} - \left(\frac{\tilde{g}}{2} - 1\right) eBS^{12}\right)}{\pi_0^2(M^2 - eBS^{12}) - e^2B^2R^2\, eBS^{12}\left(\frac{\tilde{g}}{2} - 1\right)}$$

and it reduces to the usual value of the cyclotron frequency for $\tilde{g} = 2$ or $S^{12} = 0$. The energy π_0 is a constant

$$\pi_0 = (M^2 + e^2B^2R^2)^{\frac{1}{2}}$$

and S^{12} is also constant.

The time evolution of the components of the spin orthogonal to the field in the laboratory frame are given by

$$g^{23} = S\left[\cos(\omega_2 t + \psi) + \frac{\pi_0 + M}{\pi_0 - M} \cos(\omega_1 t - \psi)\right]$$

where S and ψ are constants of integration and ω_1 and ω_2 are

$$\omega_{1,2} = \Omega\left[1 \pm \frac{\pi_0 M\left(\frac{\hat{g}}{2} - 1\right)}{M^2 - \frac{\tilde{g}}{2}\, eBS^{12}}\right].$$

(These results agree with the standard BMT theory to lowest order in the field B.) We see then that the amplitudes of S^{23} and S^{31} in the laboratory frame are modulated because $\omega_1 \neq \omega_2$.

π^3 is a constant of motion and in our example $\pi^3 = 0$, however, u^3 is not constant of motion and oscillates with a small amplitude around $u^3 = 0$ as a result of the influence of the rotational inertia on the motion of the top. The equations (IV.2) and (IV.3) (for $g = 0$) can also be solved exactly for the plane motion of a top in a central potential (for details and more examples, see II).

V. *Do Classical Top Trajectories Bend Backwards in Time?*

It was found in I that, for $g = 0$, the four velocity u^μ can be written as

$$u^\mu = -2v B^\mu \tag{V.1}$$

where

$$B^\mu = G^{\mu\nu} \pi_\nu \tag{V.2}$$

$$G^{\mu\nu} = g^{\mu\nu} - \frac{e}{NM} S^{\mu\alpha} F_\alpha^{\ \nu} \tag{V.3}$$

$$N = M - \frac{e}{2M} S_{\alpha\beta} F^{\alpha\beta} \tag{V.4}$$

and the authors suggested that very intense fields could make particle trajectories bend backwards in time. We will briefly discuss that possibility here.

Let us first consider $N > 0$. $B^\mu \neq 0$ because $B^0 = \pi^0 = M > 0$ in the frame where $\pi^i = 0$ (and $S^{0i} = 0$ due to the constraint). We choose τ such that

$$v = -\frac{1}{2}\,\mathrm{sgn}(N) \left[(B^0)^2 + (B^1)^2 + (B^2)^2 + (B^3)^2\right]^{-\frac{1}{2}} \tag{V.5}$$

(so that $u_0^2 + u_1^2 + u_2^2 + u_3^2 = 1$) and v does not vanish or go to infinity because of the definition of B^μ and $N > 0$. Furthermore, x^μ, π^μ, $S^{\mu\nu}$ and $F^{\mu\nu}(x^\rho)$ and therefore N are differentiable and therefore continuous.

Hence, B^μ and also u^μ are continuous and u^0 has the same sign of B^0 and is positive in the frame where $\pi^i = 0$. Therefore, for $N > 0$, u^0 cannot be negative in every frame as is needed for particle creation (annihilation). For free particles $u^\mu = -2v\pi^\mu$, therefore $u^0 > 0$, $u_\mu u^\mu = 4v^2M^2 > 0$.

Now, consider $N \leq 0$. Equation (V.5) implies that u^μ remains continuous through the point where $N = 0$. In the neighborhood of $N = 0$, the second term in (V.3) is predominant and

$$u^\mu \approx \frac{(SF\pi)^\mu}{\left[((SF\pi)^0)^2 + ((SF\pi)^1)^2 + ((SF\pi)^2)^2 + ((SF\pi)^3)^2\right]^{\frac{1}{2}}} + \text{order }(N). \tag{V.7}$$

If the particle is finally in a field free region, we have again $N > 0$, $u^\mu = -2v\pi^\mu$, $u^0 > 0$, $u_\mu u^\mu = 4v^2 M^2 > 0$.

That is, if the particle is initially and finally free, we have that $\mathrm{sgn}(u^0)_{\mathrm{initial}} = \mathrm{sgn}(u^0)_{\mathrm{final}}$ and the trajectory does not bend backwards in time, no matter how strong the field may be.

This does not exclude the possibility of u^μ being spacelike at some points of the trajectory where the electromagnetic is intense enough. In fact, this is true wherever $N = 0$ because of (V.7). In a negative N region, u^μ could also be timelike, but with $u_0 < 0$ indicating a temporary pair creation (annihilation). However, as soon as a weak field region is entered, u_0 is again positive.

VI. *Gravitational Interactions of a Top*

To study the gravitational interactions of a top we will, first of all, obtain the equations of motion of the spinning particle using General Relativity. According to the Equivalence Principle. One may follow very closely the development carried out in Section III interpreting the equations as valid in a locally inertial frame. Most of the equations will be formally the same, but now $g_{\mu\nu}$ is the metric in a Riemannian spacetime. Instead of rewriting what we have done in Section III, we will point out where the main differences with flat spacetime arise. Equation (III.6) that has been obtained by interchanging time derivatives and variations (that commute in flat spacetime but not in a Riemannian spacetime), looks now different and can be written as

$$D\sigma^{\mu\nu} = \frac{D(\delta\theta^{\mu\nu})}{D\tau} + \sigma^{\mu\lambda}\,\delta\theta_{\lambda}^{\ \nu} - \delta\theta^{\mu\lambda}\,\sigma_{\lambda}^{\ \nu} - g^{\mu\rho}\,R^{\nu}_{\ \rho\tau\lambda}\,u^{\tau}\,\delta x^{\lambda} \tag{VI.1}$$

where $(D/D\tau)$ and D mean covariant time derivative and covariant variation along the worldline of the top respectively and $R^{\alpha}_{\ \beta\gamma\delta}$ is the Riemann tensor. (For details see II.)

Equation (VI.1) has an important role to play in the variation of the Lagrangian to get the equations of motion, so it is clear that the equations of motion will reflect the fact that (VI.1) differs from (III.6). We construct the Lagrangian in the same way as it is done in Section III and after some work one finds the equations of motion

$$\frac{DP^{\mu}}{D\tau} = -\frac{1}{2}\,R^{\mu}_{\ \nu\alpha\beta}\,u^{\nu}\,S^{\alpha\beta} \tag{VI.2}$$

$$\frac{DS^{\alpha\beta}}{D\tau} = S^{\alpha\lambda}\,\sigma_{\lambda}^{\ \beta} - \sigma^{\alpha\lambda}\,S_{\lambda}^{\ \beta} \tag{VI.3}$$

or

$$\frac{DS^{\alpha\beta}}{D\tau} = P^{\alpha}\,u^{\beta} - u^{\alpha}\,P^{\beta} \tag{VI.4}$$

when P^{μ} and $S^{\mu\nu}$ are defined by (III.9).

From (III.6), one immediately sees that the spin $J^2 = \frac{1}{2}\,S_{\alpha\beta}\,S^{\alpha\beta}$ is conserved even before imposing any constraint on the spin tensor. Equations (VI.2)-(VI.4) are very similar to those found by Papapetrou (1951), but our equations are second order while his are third order differential equations and ours are more suitable for identifying the proper physical variables (for a more complete discussion, see II).

We will adopt here the same constraints as the ones used in I for flat spacetime, that is, we will have

$$\left.\begin{aligned} S^{\mu\nu}P_\nu &= 0 \\ e^\mu_{(0)} &= \frac{P^\mu}{M} \\ P^\mu P_\mu - f\left(\tfrac{1}{2} S_{\alpha\beta} S^{\alpha\beta}\right) &= 0 \\ x^0 - \tau &= 0 \end{aligned}\right\} \qquad \text{(VI.5)}$$

to supplement (VI.2) - (VI.4).

One can show that $P^\mu P_\mu = M^2$ is constant of motion (see II).

The equations of plane motion for a top moving in a Schwarzschild field can be exactly solved. The top energy and total angular momentum are conserved and can be interpreted as the sum of the energy of a point particle plus a dipole interaction due to the spin and the total angular momentum as the sum of orbital angular momentum plus spin, respectively.

More generally, one can show that if the geometry has a Killing vector ξ^μ, there is a conserved quantity C_ξ associated to it defined by

$$C_\xi = P^\mu \xi_\mu - \frac{1}{2} S^{\mu\nu} \xi_{\mu;\nu}$$

where ; denotes covariant derivative, (for details, see II).

The effective potential from the (exact) solution of the plane motion agrees with the one found by Rasband (1973), using an approximate version of Papapetrou equations.

VII. *Angular Velocity and Dynamical Fermi-Walker Transport*

Instead of constructing the Lagrangian as one has done it in Section III, one can use the knowledge of the constraints equations and the fact that the action is reparametization invariant and build the Lagrangian using Hamilton's principle (see II). One can show in that process that the angular velocity $\sigma^{\mu\nu}$ can be written in the following way

$$\sigma^{\mu\nu} = \frac{1}{M^2} (P^\mu \dot{P}^\nu - P^\nu \dot{P}^\mu) - \lambda f' S^{\mu\nu} \qquad \text{(VII.1)}$$

where $\lambda = (u_\mu P^\mu)/M^2$ is a Lagrange multiplier and $f' = df/dJ^2$ where f is the trajectory function.

One can see that (VII.1) is the counterpart of Equation (13.61) of Misner Thorne and Wheeler (1973)

$$\sigma^{\mu\nu} = u^\mu a^\nu - u^\nu a^\mu + \varepsilon^{\mu\nu\alpha\beta}\omega_\alpha u_\beta \qquad \text{(VII.2)}$$

when one has chosen $e^\mu_{(0)} = P^\mu/M$ instead of the usual choice $e^\mu_{(0)} = u^\mu$. a^μ and ω^μ are orthogonal to u^μ, and they are the acceleration and the rotational velocity with respect to Fermi-Walker (F – W) transported (non-rotating) vectors, respectively. One then sees that $\sigma^{\mu\nu}$ can be different from zero if $\dot{u}^\mu \neq 0$ $(\dot{P}^\mu \neq 0)$, even if the particle is not rotating and the information about rotation is contained in ω^μ (the rotational velocity). Comparing to (VII.1), we see that the information about rotation is given by $S^{\mu\nu}$ (ω^μ is proportional to $W^\mu = S^{*\mu\nu} P_\nu$) and a particle is rotating iff its spin is different from zero. The transport defined in (VII.1) (that we propose to call dynamical F–W transport as opposed to the usual geometrical F–W transport) is such that $e_{(0)}$ is preserved by it along the worldline of the top and it does not coincide with the usual F–W transport except when u^μ is parallel to P^μ. The role of the velocity in the usual treatments is played here by the momentum. This is also true for the BMT equation where the formalism presented here shows that what is normally assumed to be the velocity is actually the momentum. This points out to the fact that P^μ is physically more relevant than u^μ, after all, P^μ is a dynamical variable while u^μ is not. In practice, however, the difference between P^μ/M and the normalized velocity u^μ is very slight and for weak fields well below detectability. The time behavior of P^μ is also simpler (for example, in the uniform magnetic field B, π^3 is constant of the motion while u^3 is not. For details, see II). Also π^μ is always timelike while u^μ may become spacelike or even reversed in time in limited spacetime regions.

One can easily see from the equations of motion that $S^{\mu\nu}$ and W^μ are (dynamically) F–W transported.

ACKNOWLEDGMENTS

It is a pleasure to thank Professor A. Hanson for interesting comments. One of us (S. Hojman) would also like to thank Professor J. A. Wheeler for constant encouragement.

SERGIO HOJMAN
PRINCETON UNIVERSITY, PRINCETON, N. J.

TULLIO REGGE
INSTITUTE FOR ADVANCED STUDY, PRINCETON, N. J.

REFERENCES

V. Bargmann, L. Michel and V. L. Telegdi (1959) Phys. Rev. Lett. *2*, 435. (Also referred to as BMT.)

P. A. M. Dirac (1964) "Lectures on Quantum Mechanics," Belfer Graduate School of Science, Yeshiva University, New York.

A. J. Hanson and T. Regge (1974) Ann. Phys. (N. Y.) *87* No. 2, 498. (Also referred to as I.)

S. Hojman (1975) Ph.D. Thesis, Princeton University (unpublished). (Also referred to as II.)

C. W. Misner, K. S. Thorne, and J. A. Wheeler (1973) "Gravitation," W. H. Freeman and Co. San Francisco. (Also referred to as MTW.)

A. Papapetrou (1951) Proc. Roy. Soc. *209A*, 248.

F. A. E. Pirani (1956) Acta Phys. Pol. *15*, 389.

M. H. L. Pryce (1948) Proc. Roy. Soc. *195A*, 62.

S. N. Rasband (1973) Phys. Rev. Lett. *30*, 111.

W. Tulczyjew (1959) Acta Phys. Pol. *18*, 393.

J. Weyssenhoff and A. Raabe (1947) Acta Phys. Pol. *9*, 7.

MEASURES ON THE FINITE DIMENSIONAL SUBSPACES OF A HILBERT SPACE: REMARKS TO A THEOREM BY A. M. GLEASON*

Res Jost

The results of this paper are in Section 6. They generalize the well-known theorem by A. M. Gleason on Measures on the Closed Subspaces of a Hilbert Space [2].

§0. *Introduction*

The theme of this note has its origin in George Mackey's [4] foundations of quantum mechanics. He introduces the notion of a (positive, normed) σ-additive measure on the lattice of closed subspaces of a separable Hilbert space. A. M. Gleason [2] succeeded in a famous paper to prove that Mackey's measures are in fact exactly the statistical operators of Johnny von Neumann. In spite of my admiration for Gleason's proof, I could not help feeling that somehow the true essence of the problem had not yet been uncovered. This paper is the result of an attempt to get rid of this uneasiness. I have little doubt that it will not satisfy any possible reader. But let him judge.

§1. *Reduction to 3 Dimensions*

Let H be a complex Hilbert space. If S_1 and S_2 are two *orthogonal finite dimensional subspaces* of H, then $S_1 \perp S_2$ denotes their orthogonal sum. The family of finite dimensional subspaces of H is $L_f(H)$. A function $\mu : L_f(H) \to R_+$ is a *measure* if

*Dedication: An Valja als bescheidenes Zeugnis meines Dankes für eine lebenslange, nie getrübte, Freundschaft.

$$\mu(S_1 \perp S_2) = \mu(S_1) + \mu(S_2) , \tag{1.1}$$

whenever S_1 and S_2 are orthogonal.

The one-dimensional subspaces (rays) of H form the *projective Hilbert space* PH. We shall use the notation ξ, η, ζ for points of H; $x, y, z, a, b, \cdots$ for points in PH. $x : H \setminus \{0\} \to PH$ maps $\xi \neq 0$ onto the ray $x(\xi)$, which contains ξ. $x(\xi_1)$ and $x(\xi_2)$ are orthogonal, if $(\xi_1, \xi_2) = 0$. $x^\perp := \{y | y$ orthogonal to $x\}$ is the *polar* to the *pole* x. The restriction of a measure μ to the one-dimensional subspaces is a *frame function* $f_\mu : PH \to R_+$. The frame function f determines the measure μ completely. A full characterization of frame functions is given by the

CRITERION. $f : PH \to R_+$ *is a frame function if*

$$x_1 \perp x_2 \perp \cdots \perp x_k = x_1' \perp x_2' \perp \cdots \perp x_k' \tag{1.2}$$

always implies

$$f(x_1) + f(x_2) + \cdots + f(x_k) = f(x_1') + f(x_2') + \cdots + f(x_k') . \tag{1.3}$$

The following proposition seems to me a weak generalization of the famous theorem by A. M. Gleason.

PROPOSITION 1. *Let* $\operatorname{Dim} H \geq 3$, *and* f *be a frame-function, and* $F : H \to R_+$ *be defined by*

$$\begin{aligned} F(0) &= 0 \\ F(\xi) &= \|\xi\|^2 f(x(\xi)), \quad \xi \in H \setminus \{0\} . \end{aligned} \tag{1.4}$$

There exists a positive, symmetric, sesquilinear form

$$G : H \times H \to C$$

with the property

$$F(\xi) = G(\xi, \xi) . \tag{1.5}$$

At first sight I was struck by the fact that the main difficulty in the extraordinarily subtle and economical proof of Gleason appears for Dim H = 3. It is of lesser importance that his proof starts with a real Hilbert space (i.e. on the Euclidean real vector space), and that the transition to an arbitrary complex Hilbert space (of dimension $\geqq 3$) is somewhat tedious. In the above formulation of Proposition 1 it seems, however, evident that the proposition itself deals with 3-dimensional subspaces only.

In order to see this, it suffices to recall the defining properties of a symmetric sesquilinear form G:

$$\left.\begin{aligned} G(\xi,\eta_1) &= \overline{G(\eta_1,\xi)} \\ G(\xi,\lambda\eta_1) &= \lambda G(\xi,\eta_1), \quad \lambda \in C \\ G(\xi,\eta_1+\eta_2) &= G(\xi,\eta_1)+G(\xi,\eta_2) \,. \end{aligned}\right\} \tag{1.6}$$

For fixed ξ, η_1, η_2, one stays in a complex Hilbert space of at most 3 dimensions. In addition, by polarization, F determines G uniquely. Using (1.4), (1.6) can be expressed as a statement about f. Proposition 1 is therefore equivalent to

PROPOSITION 1_3. (*Proposition 1 restricted to* Dim H = 3).

In what follows we deal, unless otherwise stated, exclusively with Hilbert spaces of dimensions 1, 2, and 3 which we denote respectively by H_1, H_2, H_3. H stands for H_3 unless it is defined differently.

§2. *Reduction of Proposition 1_3 to a Proposition on Continuity*

It is a consequence of Proposition 1_3 that $f: PH \to R_+$ is continuous. On the other hand, the continuity of f implies Proposition 1_3.

THEOREM 1. *If the frame-function* $f: PH \to R_+$ *is continuous, it satisfies* (1.4) *and* (1.5) *of Proposition* 1_3.

PROOF. Since PH is compact, f assumes its maximum at a point $e_1 = x(\varepsilon_1)$. The restriction f_1 of f to $e_1^\perp$ assumes its maximum at a point $e_2 = x(\varepsilon_2)$. The restriction of f to $e_1^\perp \cap e_2^\perp$ is the restriction to one point $e_3 = x(\varepsilon_3)$. The vectors $\varepsilon_1, \varepsilon_2, \varepsilon_3$ are not unique. They can and will be chosen to form an orthonormal basis of H. Let $\xi_\nu \in \mathbf{R}$, $\nu \in \{1,2,3\}$, $\sum_\nu \xi_\nu^2 = 1$.

$$\phi(\xi) = f\left(x\left(\sum_\nu \xi_\nu \varepsilon_\nu\right)\right) \tag{2.1}$$

defines a continuous spherical function on S^2: the unit-sphere in $\mathbf{R}^3$. ϕ has the property that

$$\eta_1 \perp \eta_2 = \eta_1' \perp \eta_2' , \tag{2.2}$$

$\eta_\nu \in S^2$, $\eta_\nu' \in S^2$ always implies

$$\phi(\eta_1) + \phi(\eta_2) = \phi(\eta_1') + \phi(\eta_2') \tag{2.3}$$

(and, according to an argument which underlies the parametrization of $SO(3,\mathbf{R})$ by Euler-angles, for any orthonormal triplet $\varepsilon_1', \varepsilon_2', \varepsilon_3'$ on S^2

$$\phi(\varepsilon_1') + \phi(\varepsilon_2') + \phi(\varepsilon_3') = W , \tag{2.4}$$

$W \in \mathbf{R}_+$ a constant). From (2.2) and (2.3), one easily obtains the result(1) that ϕ is the restriction of a positive definite quadratic form in $\mathbf{R}^3$ to the unit sphere S^2, and further

$$\begin{aligned}\phi(\xi) &= \xi_1^2 f(e_1) + \xi_2^2 f(e_2) + \xi_3^2 f(e_3) \\ &= f\left(x\left(\sum_\nu \xi_\nu e_\nu\right)\right) .\end{aligned} \tag{2.5}$$

If $\theta_\nu \in \mathbf{C}$ are of modulus 1, we still have

$$f\left(x\left(\sum_\nu \xi_\nu \theta_\nu \varepsilon_\nu\right)\right) = \sum_\nu \xi_\nu^2 f(e_\nu) , \tag{2.6}$$

(1)[2], p. 887, Theorem 2.3.

due to the arbitrariness of the choice of the orthonormal basis $\varepsilon_1, \varepsilon_2, \varepsilon_3$. But (2.6) implies, for $\zeta_\nu \in C$, $\sum_\nu |\zeta_\nu|^2 = 1$,

$$f\left(x\left(\sum_\nu \zeta_\nu \varepsilon_\nu\right)\right) = \sum_\nu |\zeta_\nu|^2 f(e_\nu) . \tag{2.7}$$

(2.7) is a strengthening of (1.4), (1.5).

What remains to be proved is the

PROPOSITION 2_3. *Let* $f: PH_3 \to R_+$ *be a frame function, then* f *is continuous.*

§3. *The Excluded Case* H_2

Although H_2 has been carefully excluded till now, a further analysis of this case is still interesting and in fact gives us a clue for the proof of Proposition 2_3.

It is useful to introduce an (arbitrary) orthonormal basis $\varepsilon_1, \varepsilon_2$ in H_2 and to define points in H_2 by their coordinates $\zeta_\nu = (\varepsilon_\nu, \xi)$. The anti-unitary involution I_ε (it depends on the basis chosen)

$$\left.\begin{aligned} I_\varepsilon : \{\zeta_1, \zeta_2\} &\mapsto \{\zeta_1', \zeta_2'\} \\ \zeta_1' &= -\overline{\zeta_2} \\ \zeta_2' &= \overline{\zeta_1} \end{aligned}\right\} \tag{3.1}$$

represents orthogonality in PH_2:

$$(\xi, I_\varepsilon \xi) = 0 \tag{3.2}$$

for all $\xi \in H_2$.

If one introduces an inhomogeneous coordinate in PH_2 by $z = \zeta_2/\zeta_1$, then I_ε takes the form

$$I_\varepsilon : z \mapsto z' = -(\overline{z})^{-1} . \tag{3.3}$$

From (3.2) and (3.3), $z^{\perp} = -(\bar{z})^{-1}$ and, in addition, the functional equation

$$f + f \circ I_{\varepsilon} = W \quad (W \in R_{+}\text{, constant}) \tag{3.4}$$

characterizes a frame function in PH_2. It is clear that the values of f e.g. in $\{z \mid |z| < 1\}$ are completely arbitrary. It is obvious that a proposition of the kind of Proposition 1 must in general not hold in H_2. For H_1 such a proposition is true and void.

§4. *Elementary Geometry*

In this section H stands throughout for H_3. Let $<x_1, x_2>$ be the straight line joining x_1 and x_2 $(x_1 \neq x_2)$. The following formula is trivial

$$<x_1, x_2> = (x_1^{\perp} \cap x_2^{\perp})^{\perp} . \tag{4.1}$$

For a given p, let $N_p^0 = PH \setminus \{p\} \cup p^{\perp}$. We define an involution $\theta_p : N_p^0 \to N_p^0$ by

$$\theta_p(x) = <x, p> \cap x^{\perp} = x^{\perp} \cap (x^{\perp} \cap p^{\perp})^{\perp} . \tag{4.2}$$

By definition

$$\left.\begin{aligned} &\theta_p(x) \in <x, p> \\ &\theta_p(x) \in x^{\perp} \\ &\theta_x(p) \in <x, p> \\ &\theta_x(p) \in p^{\perp} . \end{aligned}\right\} \tag{4.3}$$

We have, therefore, for any frame function

$$f(x) + f(\theta_p(x)) = f(p) + f(\theta_x(p)) . \tag{4.4}$$

In this connection, we need the following simple

LEMMA 1. *Let* $x \neq y \in \theta_p(x)^{\perp}$. *Then*

$$\theta_y(x) = p^{\perp} \cap x^{\perp} . \tag{4.5}$$

PROOF.

$$\begin{aligned}\theta_y(x) &= \langle x, y\rangle \cap x^{\perp} = \theta_p(x)^{\perp} \cap x^{\perp} \\ &= ((x^{\perp} \cap p^{\perp})^{\perp} \cap x^{\perp})^{\perp} \cap x^{\perp} = \langle x^{\perp} \cap p^{\perp}, x\rangle \cap x^{\perp} \\ &= x^{\perp} \cap p^{\perp} .\end{aligned}$$

For the following definitions and arguments, it is again useful to introduce an orthonormal basis in H. Let $p = x(\varepsilon_3)$ and $\varepsilon_1, \varepsilon_2, \varepsilon_3$ be orthonormal. For $\xi \in H$, the coordinates are $\zeta_k = (\varepsilon_k, \xi)$. In $N_p = PH \setminus p^{\perp}$ we can introduce inhomogeneous coordinates by $z_1 = \zeta_1/\zeta_3$, $z_2 = \zeta_2/\zeta_3$. The scalar product

$$(x, x')_p := \bar{z}_1 z_1' + \bar{z}_2 z_2' \tag{4.6}$$

depends on p only. The involution θ_p takes the form $(x \in N_p^0)$

$$\theta_p(x) = -(x, x)_p^{-1} x . \tag{4.7}$$

Gleason's proof, which we will follow closely, is based on the discussion of two sets, which both depend on a pair of points $p;\ y \in N_p^0$. These sets are defined as

$${}_pK_y := \{x \mid y \in \theta_p(x)^{\perp}\} = \{x \mid (x, x-y)_p = 0\} \qquad (4.8)^{(1)}$$

and

$${}_pK_y^2 := \{x \mid \exists z : (x, x-z)_p = 0 \ \& \ (z, z-y)_p = 0\} . \qquad (4.9)^{(1)}$$

The set (4.8) corresponds to the circle of Thales in elementary geometry.

We need

THEOREM 2. *For* $y \in N_p^0$, *the set* ${}_pK_y^2$ *has a non-void open kernel.*

---

(1) The notations ρx and $x+y$ refer to the usual operations of multiplication and addition of vectors, defined in terms of the inhomogeneous coordinates z_1, z_2 in the affine plane N_p.

PROOF. 1. Let $(u,y)_p = 0$ and $4(u,u)_p = (y,y)_p$. Then $x_0 = \frac{1}{2}y + u \in {}_pK_y^2$. In order to see this, take $z = x_0$.

2. We have to show now that we can vary x arbitrarily around x_0. For an infinitesimal variation $x = x_0 + \delta x$, $z = x_0 + \delta z$, the conditions require

$$(x_0, \delta z)_p = (x_0, \delta x)_p \quad \text{and} \quad (\delta z, x_0 - y)_p + (x_0, \delta z)_p = 0 , \tag{4.10}$$

or

$$\left.\begin{aligned} (x_0, \delta z)_p &= (x_0, \delta x)_p \\ (y - x_0, \delta z)_p &= (\delta x, x_0)_p \end{aligned}\right\} . \tag{4.11}$$

Equations (4.11) have a unique solution.

REMARK. It is not difficult to give a more accurate discussion of ${}_pK_y^2$.

§5. *The proof of continuity*

From now on we follow quite slavishly Gleason. As before, $H = H_3$ and f is a frame function. For any set $U \subset PH$

$$\operatorname{osc}(f, U) = \sup_{x \in U} f(x) - \inf_{x \in U} f(x) . \tag{5.1}$$

THEOREM 3. *If, for* U *open and nonempty,*

$$\operatorname{osc}(f, U) = \rho \tag{5.2}$$

then for any $x \in PH$ *there exists an open neighborhood* $V_x \ni x$ *such that*

$$\operatorname{osc}(f, V_x) \leqq 4\rho . \tag{5.3}$$

PROOF.

1. Let $a \in U$ and $x \not\perp a$. $y = x^\perp \cap a^\perp$ is unique. (5.3) follows, therefore, from the

PROPOSITION. $a \in U$, $x \in a^\perp$, *and* (5.2) *imply the existence of an open set* $W_x \ni x$ *for which* $\operatorname{osc}(f, W_x) \leqq 2\rho$.

2. *Proof of the proposition.* Let $b \neq x$, $b \in \langle a, x \rangle$, and $\theta_x(b) \in U$. Then $\theta_b(x) = a \in U$ and

$$W_x = \{y \mid \theta_b(y) \in U \;\&\; \theta_y(b) \in U\} \tag{5.4}$$

is open and contains x as an element. Let $y_\nu \in W_x$ and define

$$y'_\nu = \theta_b(y_\nu) \quad \text{and} \quad b'_\nu = \theta_{y_\nu}(b) \ . \tag{5.5}$$

Now from (4.4)

$$f(y_\nu) + f(y'_\nu) = f(b) + f(b'_\nu) \tag{5.6}$$

and by subtraction

$$|f(y_1) - f(y_2)| \leqq |f(y'_1) - f(y'_2)| + |f(b'_1) - f(b'_2)| \ , \tag{5.7}$$

since $y'_\nu \in U$, $b'_\nu \in U$, the result follows.

The two following lemmas form part of the proof of continuity of frame functions (Proposition 2_3). Isolated, they are of no interest.

LEMMA 2. *Let* $g : PH \to R_+$ *be a frame function that restricted to* $p^\perp (p \in PH)$, *is constant and equal to* W:

$$g|_{p^\perp} = W \ . \tag{5.8}$$

Let in addition

$$g(p) \leqq \rho \ , \qquad \rho > 0 \tag{5.9}$$

then, for any $x \in N_p^0$,

$$g(x) \leqq W + \rho \ , \tag{5.10}$$

and, for any $y \in \theta_p(x)^\perp$,

$$g(x) \leqq g(y) + \rho \ . \tag{5.11}$$

PROOF.

1. Let $x' = \theta_p(x)$, then from (4.4)

$$g(x) + g(x') = g(p) + g(\theta_x(p)) \leqq \rho + W \tag{5.12}$$

and, from $g: PH \to R_+$, the result (5.10).

2. Let $x \neq y \in \theta_p(x)^\perp$ and $y' = \theta_x(y)$. Since by Lemma 1

$$\theta_y(x) = p^\perp \cap x^\perp \tag{5.13}$$

we find from (4.4)

$$g(y) + g(y') = g(x) + W \tag{5.14}$$

or

$$g(x) = g(y) + g(y') - W \tag{5.15}$$

and with (5.10)

$$g(x) \leqq g(y) + \rho \ . \tag{5.16}$$

LEMMA 3. *Under the same assumptions as in Lemma 2, every* $x \in N_p^0$ *has an open neighborhood* $V_x \ni x$ *such that*

$$\operatorname{osc}(g, V_x) \leqq 12\rho \ . \tag{5.17}$$

PROOF. Let

$$\inf_{x \in N_p^0} g(x) = \sigma \tag{5.18}$$

and $a \in N_p^0$ such that

$$\sigma \leqq g(a) < \sigma + \rho \ . \tag{5.19}$$

If ${}_pK_a^2$ is defined by (4.9), then for $x \in {}_pK_a^2$, one can find y such that

$$y \in \theta_p(x)^\perp \quad \text{and} \quad a \in \theta_p(y)^\perp \ . \tag{5.20}$$

From Lemma 2,

$$\sigma \leqq g(x) \leqq g(y) + \rho \leqq g(a) + 2\rho \leqq \sigma + 3\rho$$

or

$$\operatorname{osc}(g, {}_pK_a^2) \leqq 3\rho \ . \tag{5.21}$$

The result follows from Theorem 2 and Theorem 3.

THEOREM 4. *If* $f : PH_3 \to R_+$ *is a frame function, then* f *is continuous.*

PROOF. Since the positive constants are frame functions, we can assume without loss of generality $\inf f = 0$. Let $\rho > 0$, $\rho' > 0$ be given and let $f(p) = \rho'$ and $104\rho' < \rho$. Choose an orthonormal basis $\varepsilon_1, \varepsilon_2, \varepsilon_3$ of H such that $p = x(\varepsilon_3)$. The coordinates of a point $\xi \in H$ are defined by $\zeta_k(\xi) = (\varepsilon_k, \xi)$. With respect to these coordinates, the *antiunitary* involution I is determined by

$$\left. \begin{aligned} I &: \xi \mapsto \xi' \\ \zeta_1(\xi') &= -\overline{\zeta_2(\xi)} \\ \zeta_2(\xi') &= \overline{\zeta_1(\xi)} \\ \zeta_3(\xi') &= \overline{\zeta_3(\xi)} \ . \end{aligned} \right\} \tag{5.22}$$

Since I is antiunitary, $f \circ I$ is a frame function too, and so is

$$g = f + f \circ I \ . \tag{5.23}$$

We know from Section 3 that g is constant on $p^{\perp}$. Since $I : p \to p$, we also know the value

$$g(p) = 2f(p) \leqq 2\rho' \ . \tag{5.24}$$

According to Lemma 3, there exists an open neighborhood $V \ni p$ in which

$$0 \leqq g(x) \leqq 24\rho' + g(p) \leqq 26\rho' \ . \tag{5.25}$$

But $f \leqq g$ and therefore

$$\operatorname{osc}(f, V) \leqq 26\rho' \; . \tag{5.26}$$

Now Theorem 3 yields

$$\operatorname{osc}(f, U_x) \leqq 104\rho' < \rho \tag{5.27}$$

for any point x and a suitable open neighborhood $U_x \ni x$. This implies continuity of f.

§6. *The Main Theorem, Discussion*

We know now that Proposition 1 holds. The following theorem states slightly more.

THEOREM. *Let* H *be a complex Hilbert space,* $\operatorname{Dim} H \geqq 3$. *Let* $f: PH \to R_+$ *satisfy*

$$f(x_1) + f(x_2) = f(y_1) + f(y_2) \tag{6.1}$$

whenever

$$x_1 \perp x_2 = y_1 \perp y_2 \; . \tag{6.2}$$

Define $F: H \to R_+$ *by*

$$\left.\begin{aligned} F(\xi) &= \|\xi\|^2 f(x(\xi)), \qquad \xi \in H \quad \{0\} \\ F(0) &= 0 \end{aligned}\right\} \tag{6.3}$$

and $G: H \times H \to C$ *by*

$$G(\xi, \eta) = \frac{1}{4}\{F(\xi+\eta) - F(\xi-\eta) - iF(\xi+i\eta) + iF(\xi-i\eta)\} \; . \tag{6.4}$$

G *is a positive, symmetric, sesquilinear form.*

REMARK. Checking our proofs, one realizes that (1.2) and (1.3) have been used only in the case $k = 2$.

COROLLARY 1.(1) *The following statements are equivalent*

(i) F *strongly continuous at* $\xi = 0$

(ii) $G(\xi, \eta) = (\xi, P\eta)$, $P^* = P \geqq 0$ *bounded.*

(iii) f *bounded.*

COROLLARY 2.(2) *The following statements are equivalent*

(i) $G(\xi, \eta) = (\xi, P\eta)$, $P^* = P \geq 0$ P *of trace class,* $\mathrm{Tr}\, P = W$,

(ii) *for any orthonormal basis* $\{\varepsilon_\nu | \nu \in J\}$,

$$\sum_{\nu \in J} f(x(\varepsilon_\nu)) = W$$

(iii) *for some orthonormal basis* $\{\varepsilon_\nu | \nu \in J\}$,

$$\sum_{\nu \in J} f(x(\varepsilon_\nu)) = W < \infty .$$

The theorem together with Corollary 2 imply Gleason's Theorem.

The following proposition generalizes our main theorem slightly.

THEOREM. *Let* H *be a complex Hilbert-space,* $\mathrm{Dim}\, H \geqq 3$. *Let* $\Sigma_\varepsilon = \{z | z \in C,\ |\arg z| < \frac{\pi}{2} - \varepsilon\}$, $\varepsilon > 0$. *Let* $f : PH \to \Sigma_\varepsilon$ *satisfy* (6.1), (6.2). *Define* $F : H \to \Sigma_\varepsilon$ *by* (6.3) *and* G *by* (6.4). G *is a sesquilinear form with a positive symmetric part.*

---

(1)[3], p. 315 f., Theorems 1.17 and 1.20.

(2)From (iii) the boundedness of f follows from Schwarz's inequality

$$|G(\xi, \eta)|^2 \leqq F(\xi) F(\eta) ,$$

which leads to

$$F\Big(\sum_\nu \zeta_\nu \varepsilon_\nu\Big) \leqq (\zeta, \zeta) \sum_\nu F(\varepsilon_\nu) = W(\zeta, \zeta) .$$

For the rest, see [1], p. 38 Hilfssatz 1, where one substitutes $P^{\frac{1}{2}}$ for A.

In our modification of Gleason's result, we were mainly motivated by the desire to eliminate the requirement of σ-additivity of the measure and separability of H. These hypotheses have, in the original paper, exactly two purposes. One is to ensure the existence of a nontrivial frame function, the other, to guarantee that the measure is completely determined by this frame function. We, on the other hand, discuss exclusively measures which do not vanish identically on PH. Corollary 2 then gives the necessary and sufficient restrictions which allow the unique extension to σ-additive measures. These extensions are always carried by a subspace of at most $\aleph_0$ dimensions.

The following question arises. What are the (not necessarily σ-additive) measures μ on the lattice L(H) of the closed subspaces of H? This question only makes sense for $\operatorname{Dim} H \geqq \aleph_0$ and offers new aspects if the restriction of μ to $L_f(H)$ vanishes. I have no ambition to solve this problem. It is, however, of interest, to exclude the existence of $\{0,1\}$ measures (i.e. measures $\mu : L(H) \to \{0,1\}$ with $\mu(H) = 1$), for the following reason. If one interprets such a measure – and this corresponds to the motivation of the whole enterprise – as a (generalized) quantum-mechanical state of a system which has as questions the elements of L(H), then such a state would answer every question with certainty by "yes" (1) or "no" (0). Such a state could rightly be called a classical answer to the questions L(H). Such answers are excluded even if one disregards σ-additivity and separability.

THEOREM. *Let* $\operatorname{Dim} H \geqq \aleph_0$ *and* $\mu : L(H) \to [0,1]$ *be a normed measure. The range of* μ *contains all rational numbers in* $[0,1]$.

PROOF. For any $n \in \mathbf{N}$, we decompose H into a tensor product $H = H' \otimes H_n$, $\operatorname{Dim} H_n = n$. For any $S \in L(H_n)$, define

$$\mu_n(S) = \mu(H' \otimes S) \; .$$

$\mu_n : L(H_n) \to [0,1]$ is a normed measure on $L(H_n)$. It is easy to see that the range of μ_n covers $\{k/n \mid k \in \{0, 1, 2, \cdots, n\}\}$.

This almost trivial remark is a very weak form of a deep and beautiful result of S. Kochen and E. P. Specker [5]. They state and prove essentially the following

PROPOSITION. *Let* $\operatorname{Dim} H \geqq \aleph_0$. *There is a set* $Q \subset L(H)$ *with* $\operatorname{Card} Q = 109$ *and with the following property. Let* $\mu : Q \to [0,1]$ *be extendable to satisfy*

$$\mu(S_1 \perp S_2) = \mu(S_1) + \mu(S_2), \qquad S_1, S_2 \in Q \tag{6.5}$$

$$\mu(S_1 \perp S_2 \perp S_3) = 1, \qquad S_1, S_2, S_3 \in Q . \tag{6.6}$$

Then

$$\mu(Q) \neq \{0,1\} .$$

In the remaining part I would like to elucidate this proposition from the point of view of this note. Let me again write $H = H' \otimes H_3$ ($\operatorname{Dim} H_3 = 3$). The subspaces in Q are of the form $H' \otimes x$, where $x \in PH_3$. Q is therefore represented by a set Q' of 109 points of the projective plane. We choose the elements of Q' in such a way that they span a real subspace of PH_3, which we denote by $(PH_3)_r$. We shall eventually give a description of Q' in the real projective plane.

The relevant relation between points of $(PH_3)_r$ is their orthogonality. A configuration of points together with this relation can be represented by a graph. The vertices represent the points, the lines indicate orthogonality of the vertices which they join. The following two graphs Γ_1 and Γ_2 of Kochen and Specker are the building-blocks for the graph of Q' (Figure 1).

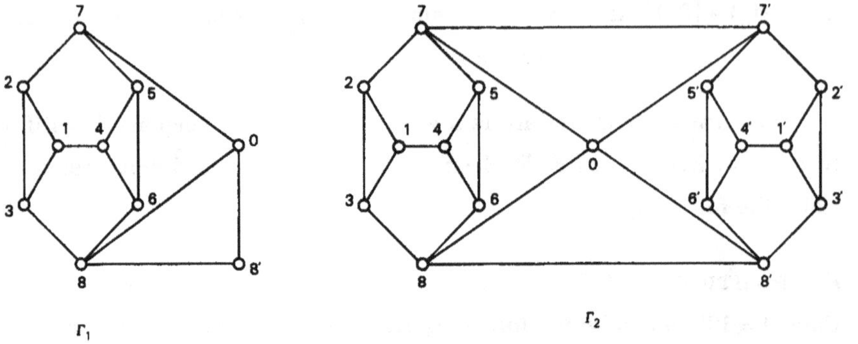

Figure 1. The elementary Kochen-Specker graphs Γ_1 and Γ_2 of 10 and 17 vertices.

Note that Γ_1 is a subgraph of Γ_2. These graphs are realizable in $(PH_3)_r$, as the following list of homogeneous "coordinates" for the corresponding points indicates

$$\alpha > 0, \qquad \alpha\beta = 2$$

1. $(1, 0, 1)$		1′. $(0, -1, 1)$
2. $(-1, \alpha, 1)$		2′. $(\alpha, 1, 1)$
3. $(-1, -\beta, 1)$		3′. $(-\beta, 1, 1)$
4. $(-1, 0, 1)$	0. $(0, 0, 1)$	4′. $(0, 1, 1)$
5. $(1, -\alpha, 1)$		5′. $(-\alpha, -1, 1)$
6. $(1, \beta, 1)$		6′. $(\beta, -1, 1)$
7. $(\alpha, 1, 0)$		7′. $(1, -\alpha, 0)$
8. $(-\beta, 1, 0)$		8′. $(1, \beta, 0)$.

These points (and a few others) are plotted in Figure 2 in the affine plane N_0. Points at infinity $7^{\#}$ and $8^{\#}$ are indicated by the straight lines through 0 which intersect the straight line at infinity in the corresponding points. Let δ be the angle between the lines through 7 and 8′. One finds

$$\sin\delta = [1 + (\alpha+\beta)^2]^{-\frac{1}{2}} .$$

The values

$$\alpha + \beta = (2\sqrt{5} + 5)^{\frac{1}{2}}$$

$$\alpha - \beta = (2\sqrt{5} - 3)^{\frac{1}{2}}$$

lead to $\delta = \pi/10$. This value is chosen for Figure 2. By rotation by integer multiples of $\pi/10$ around 0, we obtain sets of 41 (Γ_1) or 71 (Γ_2) points. These points are also drawn in Figure 2.

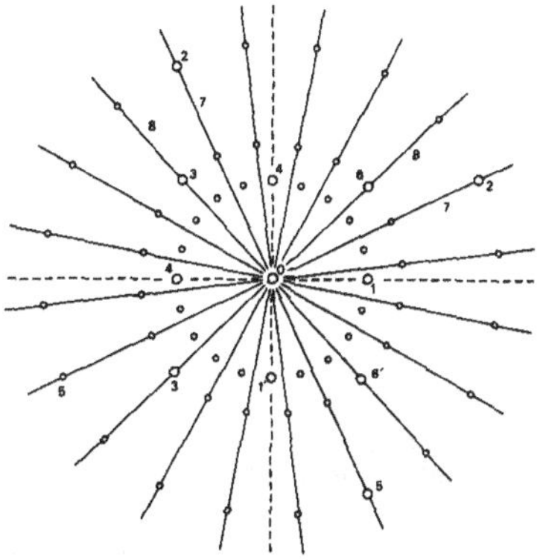

Figure 2. Representation of the points $Q'(\Gamma_1)$, $Q'(\Gamma_2)$, $Q'(\Gamma_1')$, $Q'(\Gamma_2')$ in the projective plane $(PH_3)_r$. The numbers correspond to the numbers in Figure 1.

We indicate the respective graphs by Γ_1', Γ_2' (Figure 3).

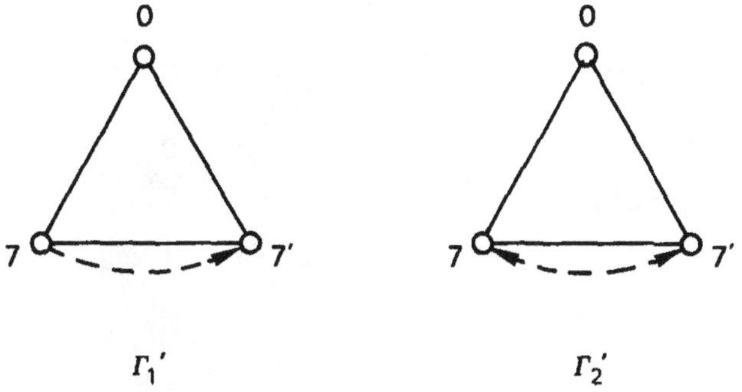

Figure 3. Symbols for the graphs Γ_1' of 41 vertices and Γ_2' of 71 vertices.

By combining Γ_1' and Γ_2' we obtain a graph of $109 = 71 + 38$ vertices, indicated by Γ_3 (Figure 4).

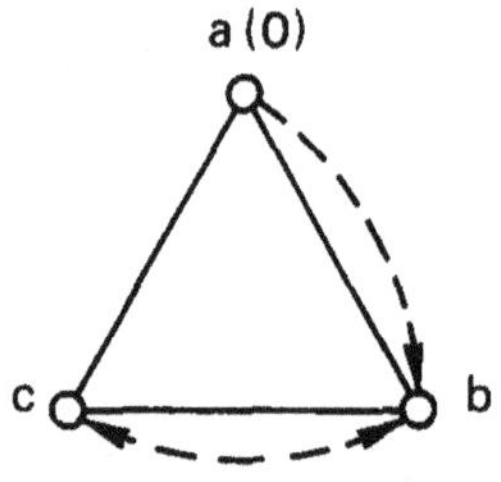

Figure 4. Symbol for the graph Γ_3 of 109 vertices.

Assume for the moment the existence of a nontrivial $\{0, 1\}$ measure $\mu : Q(\Gamma_3) \to \{0, 1\}$ which can be extended uniquely by (6.5) (6.6). μ can be restricted to any subgraph of the type $\Gamma_1^{\#}$ or $\Gamma_2^{\#}$. Let μ_{Γ_ν} be such a restriction. μ_{Γ_ν} attributes numbers 0 or 1 to the vertices of Γ_ν in such a way that in each triangle, 1 appears exactly once. The product of the values in vertices joined by a line is 0. This *excludes*

for Γ_1 the case $\mu_{\Gamma_1}(7) = 1, \quad \mu_{\Gamma_1}(8') = 0$

for Γ_2 the cases $\mu_{\Gamma_2}(7) = 1, \quad \mu_{\Gamma_2}(8') = 0$

and

$\mu_{\Gamma_2}(7) = 0, \quad \mu_{\Gamma_2}(8') = 1$

and by construction

for Γ_1' the case $\mu_{\Gamma_1'}(7) = 1, \quad \mu_{\Gamma_1'}(7') = 0$

for Γ_2' the cases $\quad \mu_{\Gamma_2'}(7) = 1, \quad \mu_{\Gamma_2'}(7') = 0$

and

$$\mu_{\Gamma_2'}(7) = 0, \quad \mu_{\Gamma_2'}(7') = 1$$

and it *leaves* for $\mu_{\Gamma_2'}$ the case

$$\mu_{\Gamma_2'}(0) = 1, \quad \mu_{\Gamma_2'}(7') = 0, \quad \mu_{\Gamma_2'}(7) = 0 .$$

This case, however, is finally excluded by Γ_3. This contradiction demonstrates the proposition.

The original graph of Kochen and Specker has 117 vertices and can be represented by Γ_4 (Figure 5).

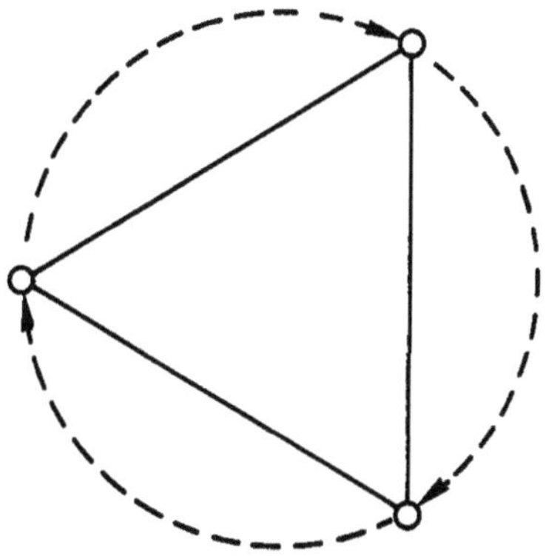

Figure 5. Symbol for the original Kochen-Specker graph of 117 vertices.

The reduction from 117 to 109 points in Q' is, of course, immaterial as is the fact that all the points of Q' can be constructed by ruler and compass. It is, perhaps, slightly more interesting that all questions in Q can be represented as (quadratic) functions of 39 observables.

Our motivation was the elimination of operations which require infinitely many arguments (like σ-additivity). This motivation has necessarily led us to the profound investigations of Simon Kochen and Ernst Specker.

R. JOST
SEMINAR FÜR THEORETISCHE PHYSIK DER ETH, ZÜRICH

[1] *I. M. Gelfand und N. J. Wilenkin,* Verallgemeinerte Funktionen Bd. IV, DVW Berlin 1964.

[2] *A. M. Gleason,* Measures on the Closed Subspaces of a Hilbert Space. Journal of Math. and Mech. *6* (1957), 885-893.

[3] *Tosio Kato,* Perturbation Theory for Linear Operators. Springer, Berlin. Heidelberg. New York 1966.

[4] *G. W. Mackey,* Mathematical Foundations of Quantum Mechanics. Benjamin, New York. Amsterdam 1963.

[5] *S. Kochen and E. P. Specker,* The Problem of Hidden Variables in Quantum Mechanics. Journal of Math. and Mech. *17* (1967), 59-88.

THE FROISSART BOUND AND CROSSING SYMMETRY*

N. N. Khuri

There are several logical chains that connect Axiomatic Quantum Field Theory to physical and measurable quantities. One of the best known and best established is the chain that leads from the axioms to the Froissart bound.[1] This gives, among other bounds, an upper bound for the total cross section for the collision of two particles which is of the form:

$$\sigma_{tot}(s) \leq C\,(\ln s)^2 \ , \tag{1}$$

where s is the square of the total center of mass energy, and $C = (\pi/\mu^2)$ in the case of pion-pion scattering with μ being the mass of the pion.

The most important links in the chain that lead to (1) were provided by the work of Martin[2] and the work of Bros, Epstein, and Glaser.[3] The chain starts with the Wightman axioms, which lead us to the L.S.Z. formalism,[4] which gives us analyticity properties for the two body elastic scattering amplitude.[5] These analyticity properties when coupled with unitarity lead to the bound of Equation (1).

In deriving Equation (1), not all the information that one can get from axiomatic field theory is used. The most important missing feature is crossing symmetry. From the mathematical as well as the physical point of view, the question whether the bound (1) can be improved is very much an open question. This must be affirmed even though it has been conjectured that the bound of Equation (1) is actually saturated in some field theories.[6] Furthermore, the derivation of the Froissart bound plays a

*Work supported in part by the U. S. Energy Research and Development Administration under Contract Number E(11-1)-2232B.

central role in the whole effort to find general physical properties of quantum field theories that do not depend on dynamical details. Because of this, any improvement of the power of $\ln s$ in (1) would be of major significance, as it would certainly open the way to an improvement of many other bounds and asymptotic theorems.

In this brief note, we shall review the problem of whether the Froissart bound is the best bound, with special attention to the question of how one can include the information coming from full crossing symmetry. We shall also discuss the explicit examples of functions that have been constructed which satisfy, full crossing, analyticity, temperedness, and the unitarity inequalities for the partial waves.

Let us first state the results of axiomatic field theory which lead to the inequality (1). For simplicity we only consider equal mass, spin zero, neutral pseudoscalar particles, and set the mass $\mu = 1$. The scattering amplitude, $F(s,t)$, for the process $\pi^0\pi^0 \to \pi^0\pi^0$ is a function of s defined above, and the momentum transfer t. In this case, $t = -2k^2(1-\cos\theta)$, where k and θ are respectively the c.m. momentum and the c.m. scattering angle. The elastic amplitude can be expanded in the usual partial wave expansion,

$$F(s,t) = \frac{\sqrt{s}}{k} \sum_{\ell=0}^{\infty} (2\ell+1) f_\ell(s) P_\ell(\cos\theta) \ . \tag{2}$$

The s-channel absorptive part of $F(s,t)$ is defined by $A(s,t) = \frac{1}{2i}[F(s+i\varepsilon,t) - F(s-i\varepsilon,t)]$, for $s > 4$ and $t_0 \leq t \leq 0$. A similar expansion holds for $A(s,t)$ with $\mathrm{Im} f_\ell$ replacing f_ℓ in Equation (2).

The bound (1) follows from three properties all consequences of axiomatic field theory. These are,

a) *Analyticity*: For real $s > 4$, $A(s,t)$ is an analytic function of t regular in the Lehmann-Martin Ellipse,[2] $E(s)$, which has foci at $t = 0$ and $t = -4k^2$ and right extremity at $t = 4$.

b) *Unitarity*: The partial wave amplitudes, $f_\ell(s)$, satisfy the unitarity inequalities:

$$0 \leq |f_\ell(s)|^2 \leq \mathrm{Im} f_\ell(s) \leq 1\,; \qquad 4 \leq s < \infty\ . \tag{3}$$

Actually all that one needs to derive (1) is positivity and boundedness, $0 \leq \mathrm{Im} f_\ell(s) \leq 1$, and the optical theorem.

c) *Polynomial Boundedness*: For any fixed $t \in E(s)$, the absorptive part $A(s,t)$ has for large real s the bound,[2]

$$|A(s,t)| \leq \text{Const. } s^2\ . \tag{4}$$

A similar polynomial bound holds for $F(s,t)$.

The Froissart bound also holds in more general axiomatic schemes.[7] However, in this note we shall restrict ourselves to the Wightman axioms and tempered distributions.

Starting with just the three properties a), b), and c), the problem is a mathematically completed one. Namely, given only these three conditions, one knows that the bound (1) is the best that can be obtained. Kinoshita, Loeffel, and Martin[8] have constructed an explicit example which satisfies a), b), and c) and which saturates the $(\ln s)^2$ behavior of the bound.

However, unfortunately, as was stressed in Reference 8, this does not settle the problem. In the derivation of the bound, one of the most important features of field theory, namely, crossing symmetry has not been used as an input. Froissart's original intuititive motivation for the bound is based on a potential scattering picture with an energy dependent coupling constant. The proof of the bound by Martin,[1] and the arguments of Cheng and Wu[6] are all s-channel constrained.

For the $\pi^0\pi^0 \to \pi^0\pi^0$ case crossing symmetry can be very simply expressed by introducing the third Mandelstam variable $u = -2k^2(1+\cos\theta)$; $s+t+u = 4$. Then we have the additional restriction on F:

d) The amplitude $F = F(s,t,u)$ is a completely symmetric function of the variables s, t, and u.

The question is whether including condition d) in our input will improve the Froissart bound. The first thing to do in answering this question is to try to construct explicit examples that satisfy a), b), c), d). There exist, mainly due to the work of Atkinson[9] and Kupsch,[10] many such examples. The remarkable fact, however, is that in all these examples, the one that gives the fastest growth in $\sigma_{tot}(s)$ behaves for large s as,

$$\sigma_{tot}(s) \sim \ln s \ . \tag{5}$$

This example was obtained by Kupsch.[10] He constructed explicitly a fully crossing symmetric function, $F(s,t,u)$, that has the following properties: i) It satisfies the Mandelstam representation with a finite number of subtractions. ii) The partial waves satisfy the inelastic inequality, $\mathrm{Im}\, f_\ell(s) \geq |f_\ell(s)|^2$, $s > 4$. iii) $F(s,t=0) \to C s \ln s$ as $s \to \infty$. We shall not write down this example here, but refer the interested reader to Kupsch's paper.

Practically everyone working in this field has tried to do better than Kupsch. Namely, one tries to construct a function $F(s,t,u)$ that satisfies a), b), c), and d) for which σ_{tot} grows faster than $\ln s$ preferably as fast as $(\ln s)^2$. No one has succeeded so far. Whether this is an indication that Equation (5) gives the bound one should aim for, or an indication that we have not been clever enough is a open question. The absence of an example with the growth $\sigma_{tot}(s) \sim (\ln s)^2$ certainly leaves the question up in the air.

Suppose for the moment we take the point of view that the absence of a counter-example is an indication that crossing symmetry might improve the bound. How can one then proceed to include d) in the derivation of a new bound? The property d) is very simple to state in terms of $F(s,t,u)$, but in that form it is almost useless for the problem of getting asymptotic bounds. One can translate d) to a set of conditions on the physical partial waves $f_\ell(s)$, $s > 4$. A set of necessary and sufficient conditions on $\mathrm{Im}\, f_\ell(s)$ for all integer ℓ and $s > 4$ which will guarantee full crossing symmetry has been obtained by Roskies[11] and by Auberson and Khuri.[12] These conditions are quite unwieldy. They take the form of the vanishing of integrals over $4 \leq s \leq \infty$ of certain infinite linear combinations of

$\mathrm{Im}\, f_\ell(s)$. There are an infinite number of them, and there does not seem to be an easy way to extract restrictions on the individual $\mathrm{Im}\, f_\ell(s)$ from them.

In our opinion, a better and simpler way to get a potentially useful necessary and sufficient condition to guarantee full crossing is by dealing with the absorptive part $A(s,t)$. This was done in Reference 12 and we review it here.

The problem is the following: Supposing we are given an s-channel absorptive part $A_1(s,t)$ with the necessary analyticity properties in t, how can we make sure it is the absorptive part of a fully symmetric amplitude? Of course, we can easily choose the u-channel absorptive part $A_3(u,t)$ such that $A_1 \equiv A_3$. This guarantees $s \leftrightarrow u$ crossing. However, it also effectively determines $F(s,t)$ modulo subtractions, via fixed-t dispersion relations. The difficult part to guarantee now is that the F determined by $A_1 = A_3 = A$ should also have a t-channel absorptive part $A_2(t,s)$ which is identical with A_1 and A_3, i.e. $A_1(x,y) = A_2(x,y) = A_3(x,y)$.

In Reference 12, this problem was solved leading to a simple necessary and sufficient condition on $A(s,t)$ which guarantees full crossing. Although there probably exist other equivalent conditions, we choose for simplicity to stick to the one derived in Reference 12 and discuss it below.

To state this condition, we need to define two new variables a, and ξ to replace s and t. These are given by

$$a = \frac{\bar{s}\,\bar{t}\,\bar{u}}{\bar{s}\bar{t} + \bar{u}\bar{t} + \bar{s}\bar{u}} + 4/3 \,, \tag{6}$$

and

$$\xi = \frac{\bar{s}^3}{(\bar{s} - \bar{a})} \,, \tag{7}$$

where $\bar{s} = s - 4/3$, $\bar{a} = a - 4/3$, etc. We also consider the curve given by

$$\bar{t}(\xi, a) = \frac{1}{2}\,[-\bar{s}(\xi, a) + (4\xi - 3\bar{s}^2(\xi, a))^{\frac{1}{2}}] \,, \tag{8}$$

where $\overline{s}(\xi, a)$ is obtained by inverting (7). Note that as $\xi \to \infty$, $t \to a$. We define a new function $\mathfrak{a}(\xi, a)$ from $A(\overline{s}, \overline{t})$ by:

$$\mathfrak{a}(\xi, a) = A(\overline{s}(\xi, a); \overline{t}(\xi, a)); \qquad 0 \leq a \leq 4 . \tag{9}$$

One then considers a simple Mellin transform of $a(\xi, a)$,

$$\int_{\xi_0(a)}^{\infty} \mathfrak{a}(\xi, a) \xi^{-\nu-1} d\xi = G(\nu, a) , \tag{10}$$

with

$$\xi_0(a) = \frac{512}{27(4-a)} . \tag{11}$$

The function $G(\nu, a)$, for $0 \leq a < 4$, is analytic in ν in the half-plane $\mathrm{Re}\, \nu > \left(\frac{1}{2} + \frac{1}{4}\sqrt{a}\right)$. This follows from the fact that for large s, and $0 < t < 4$, the absorptive part $A(s, t)$ has the upper bound[13]

$$A(s, t) \leq C s^{1+\sqrt{\frac{t}{4}}} (\ln s)^2 . \tag{12}$$

One can also show that for $\mathrm{Re}\, \nu > 1$, $G(\nu, a)$ is also analytic in a in some neighborhood of $a = 0$.

The necessary and sufficient condition that insures full crossing symmetry can now be stated as follows:

$$G(\nu = n, a) = P_n(a) , \qquad n = 1, 2, \cdots, \infty , \tag{13}$$

where $P_n(a)$ is any polynomial in a of degree n. Indeed the coefficients of $P_n(a)$ for all $n \geq 1$ are enough to uniquely determine a fully symmetric $F(s, t, u)$. Thus all we need to do to make sure a certain $A(s, t)$ is the absorptive part of a fully symmetric amplitude, is first to form $\mathfrak{a}(\xi, a)$ through Equation (9), and then check that the Mellin transform in Equation (10) gives an answer which is a polynomial of degree n when $\nu = n$.

The condition (13) looks very simple – but it is hard to see intuitively how strong it is when coupled with a), b), and c). The next step one needs is to solve the following mathematical problem: to describe fully the class of functions of two variables, $W(x, y)$, which have the property

$$\int_0^{\infty} W(x, y) x^{-n-1} dx = P_n(y) , \qquad (14)$$

where $P_n(y)$ is a polynomial of degree n. Given a full set of properties of this class, one can proceed further by checking whether these properties when coupled with a), b), and c) can lead to an improvement of Equation (1). We do not know if a useful classification of the functions $W(x, y)$ exists or can be derived.

Until one has a simple listing of the full properties of the W-class of functions, no progress can be made towards an improvement of Equation (1). There might be other ways to guarantee crossing different from Equation (13), but we doubt that they can be simpler.

At this stage, one might ask, given the fact that Equation (13) is so simple, why not try and use it to construct a counterexample. Namely, one has several examples that satisfy a), b), and c) and saturate the Froissart bound. One can try to add a function to one of these examples which asymptotically grows slower than Froissart, but such that the sum satisfies Equation (13) and unitarity is not violated. For example, if one takes the Cheng-Wu[6] ansatz for $A(s, t)$, then its behavior for large s will necessarily lead to a singularity in the ν-plane for the function $G(\nu, a)$, $a > 0$ and small. This singularity will be of the form $\sim \left[(2\nu - 1)^2 - \frac{a}{4}\right]^{-3/2}$. For $a = 0$, this singularity reduces to the third order Regge pole that must exist if Equation (1) is saturated. One has to find a correction term $A_E(s, t)$ such that in the Mellin transform (10) of $A = A_{C-W} + A_E$, the above singularity is cancelled for $\nu = n$, $n = 1, 2, \cdots$; and in addition to getting Equation (13), one has to guarantee anew that A_E does not make the sum violate s-channel unitarity. Again this is not an easy task.

We summarize the rather unfortunate but nevertheless challenging situation. None of the published models that satisfy analyticity, s-channel unitarity, and saturate the Froissart bound have been shown to satisfy t-channel unitarity, much less full crossing symmetry. This statement certainly applies to the Cheng-Wu model.[6] None of the explicit examples that have been constructed so far which satisfy analyticity, full crossing, unitarity inequalities, and temperedness, do saturate the Froissart bound. In fact, the best that one can do in this latter case is to obtain the asymptotic behavior $\sigma_{tot}(s) \sim \ln s$. This makes one hopeful that crossing-symmetry when included will improve the bound. We have given what in our opinion is the simplest *necessary and sufficient* condition that guarantees crossing – but its relation to the bounds so far seems obscure.

It is our hope that in the near future this problem will be resolved. Either one finds an example satisfying a), b), c), and d) which saturates the inequality (1); or a new bound is derived. The latter possibility, the breaking of the Froissart barrier, will, we are certain, lead to a whole series of new and significant bounds in addition to settling some still unsolved problems like the proof of the Pomeranchuk theorem. Finally, one should add that in this note we have concentrated on only one additional property that could improve the Froissart bound, namely crossing. There are others, such as insuring full elastic unitarity,[9] or trying to make use of some minimal information on particle production.[14] However, crossing-symmetry is the problem that must be settled first.

THE ROCKEFELLER UNIVERSITY
NEW YORK, NEW YORK

REFERENCES

1. M. Froissart, Phys. Rev. *123*, 1053 (1961); a simpler proof was later given by A. Martin, Phys. Rev. *129*, 1432 (1963).

2. A. Martin, Nuovo Cimento *42*, 930 (1966); *ibid.*, *44*, 1219 (1966).

3. J. Bros, H. Epstein and V. Glaser, Nuovo Cimento *31*, 1265 (1964); and Comm. Math. Phys. *1*, 240 (1965).

4. K. Hepp, Helv. Phys. Acta *37*, 639 (1964).

5. H. Lehmann, Nuovo Cimento *10*, 578 (1958), and also Reference 2.

6. H. Cheng and T. T. Wu, Phys. Rev. Lett. *24*, 1456 (1970).

7. H. Epstein, V. Glaser, and A. Martin, Comm. Math. Phys. *13*, 257 (1969).

8. T. Kinoshita, J. Loeffel and A. Martin, Phys. Rev. *135*, B 1464 (1964).

9. D. Atkinson, Nuclear Physics *B7*, 375 (1968); *B8*, 377 (1968); *B13*, 415 (1969).

10. J. Kupsch, Fort. der Physik *19*, 783 (1971). This is a review article and contains references to the earlier literature.

11. R. Roskies, Phys. Rev. *D2*, 247 (1970); *2*, 1649 (1970).

12. G. Auberson and N. N. Khuri, Phys. Rev. *D6*, 2953 (1972).

13. A. Martin, reference 1.

14. N. N. Khuri, Phys. Rev. *D8*, 2702 (1973).

INTERTWINING OPERATORS FOR SL(n, R)

A. W. Knapp* and E. M. Stein*

In 1947 Bargmann [1] derived the list of the irreducible unitary representations of the group $G = SL(2, \mathbf{R})$ of real two-by-two matrices of determinant one. For the most part, he grouped these into three series – the principal series,[1] the discrete series,[2] and the complementary series.[1] He showed that only the first two series are needed for the analysis of $L^2(G)$ and gave a number of formulas that can be interpreted [7] as the Plancherel formula for G. The principal series representations were realized in L^2 of the circle, and he observed that one representation of that form was reducible, with the others irreducible. The inner product for the complementary series was more subtle, and exhibiting it amounted to the proof of the existence of complementary series.

Similar results – the Plancherel formula, an irreducibility criterion for principal series, and conditions for existence of complementary series – are now known for a wide class of groups. We shall give in this paper a survey of these results in the context of $G = SL(n, \mathbf{R})$, the group of real n-by-n matrices of determinant one. Our survey is intended as an illustration of the use of an analytical tool, the theory of intertwining operators. The intertwining operators, developed in [10, 12-15, 18] give complete information about reducibility of principal series and provide an inner product

*Preparation of this paper was supported by grants from the National Science Foundation and the Institute for Advanced Study.

[1]Bargmann used the term "continuous series outside the exceptional interval" for the principal series. "The exceptional interval" refers to the complementary series.

[2]Discrete series now refers to irreducible representations whose matrix coefficients are square integrable. Bargmann allowed as well two other representations in his "discrete series."

giving existence of some complementary series. It is not known whether the list of complementary series they produce is complete. For the Plancherel formula, which was proved by Harish-Chandra in a long series of papers ending with [8, 9], the intertwining operators make the Plancherel measure more explicit.

Our program will be to develop the intertwining operators for $SL(2, \mathbf{R})$, combine these operators suitably to handle a special case in $SL(n, \mathbf{R})$, and then use the special case to handle the general case. Finally, we shall turn our attention to the three problems we have mentioned, obtaining explicit results for each. A last section of the paper contains comments about a wider class of groups than $SL(n, \mathbf{R})$.

We should caution that a number of results about $SL(n, \mathbf{R})$ can be obtained by various methods that have a different scope. Typical illustrations of these approaches are the attacks on the irreducibility question by Gelfand and Graev [3], Zelobenko [22], and Wallach [20]. For the most part, we shall avoid these other methods.

1. *Operators for* $SL(2, \mathbf{R})$

For $G = SL(2, \mathbf{R})$, the principal series is indexed by a two-element set and a real parameter, so by pairs $(\pm, it)$ with t in $\mathbf{R}$. We give three realizations.

In the noncompact picture, the Hilbert space is $L^2(\mathbf{R})$. If $g = \begin{pmatrix} a & b \\ c & d \end{pmatrix}$ and if f is in $L^2(\mathbf{R})$, the representations $U(+, it, \cdot)$ and $U(-, it, \cdot)$ are given by

$$U(+, it, g) f(x) = |-bx+d|^{-1-it} f\left(\frac{ax-c}{-bx+d}\right)$$

$$U(-, it, g) f(x) = \operatorname{sgn}(-bx+d)\, |-bx+d|^{-1-it} f\left(\frac{ax-c}{-bx+d}\right).$$

All of these are unitary, and all are irreducible except $U(-, 0, \cdot)$, which is reducible and splits into two inequivalent irreducible pieces.

For the induced picture, we introduce the subgroups

$$M = \begin{pmatrix} \varepsilon & 0 \\ 0 & \varepsilon \end{pmatrix}, \ A = \begin{pmatrix} r & 0 \\ 0 & r^{-1} \end{pmatrix}, \ N = \begin{pmatrix} 1 & y \\ 0 & 1 \end{pmatrix}, \ V = \begin{pmatrix} 1 & 0 \\ x & 1 \end{pmatrix}, \ K = \begin{pmatrix} \cos\theta & \sin\theta \\ -\sin\theta & \cos\theta \end{pmatrix},$$

where $\varepsilon = \pm 1$ and $r > 0$. Fix $(\pm, it)$. Let the two characters of the two-element group M be defined by $\sigma_+(m) = 1$ and $\sigma_-(m) = \varepsilon$ if $m = \begin{pmatrix} \varepsilon & 0 \\ 0 & \varepsilon \end{pmatrix}$. The space of the induced representation is

$$\{f : G \to \mathbb{C} \mid f(x\,man) = r^{-1-it}\sigma_\pm(m)^{-1} f(x)\} \ .$$

Since every g in G decomposes as $g = kan$, according to the Iwasawa decomposition,[3] f is determined by its restriction to K. The norm on f is taken as

$$\|f\| = \|f|_K\|_2 \ ,$$

and the representation is $U(g)f(x) = f(g^{-1}x)$. Restriction from G to V provides a mapping that shows the induced picture and the noncompact picture are equivalent.[4]

In the compact picture, the induced picture is merely restricted to K. The space is the space of L^2 functions on K with $f(km) = \sigma(m)^{-1}f(k)$ and with the L^2 norm. The space is independent of t, and the formula for the group action makes sense with it replaced by a complex parameter z, except that the representations $U(\pm, z, \cdot)$ are not necessarily unitary. As z varies, the operators $U(\pm, z, g)$ act in the same space and vary analytically in z. We speak of the *nonunitary principal series*.

The other irreducible unitary representations occurring in the Plancherel formula are those of the discrete series, denoted D_n^+ and D_n^- with $n \geq 2$. We shall not write down their exact form, but give certain facts about them.

(1) $D_n^+ \oplus D_n^-$ is a subrepresentation of $U(+, n-1, \cdot)$ if n is even and of $U(-, n-1, \cdot)$ if n is odd. (This imbedding is done rigorously

---

[3]Gram-Schmidt decomposition in this group.

[4]This equivalence is a scalar multiple of a unitary operator if the measure on V is taken as Lebesgue measure dx.

in the compact picture and is understood as just an imbedding of the infinitesimal action of G on those Hilbert space elements that transform under K within a finite-dimensional subspace. It is not a unitary imbedding.)

(2) When we try to extend representations from $SL(2,R)$ to $SL^{\pm}(2,R)$[5] without enlarging the space, say be defining $U\begin{pmatrix}1 & 0\\ 0 & -1\end{pmatrix}$, we can extend the principal series but not individual discrete series. However, we can extend $D_n = D_n^+ \oplus D_n^-$. The representations D_n of $SL^{\pm}(2,R)$ comprise the discrete series of $SL^{\pm}(2,R)$.

Now we can state the Plancherel formula for G. If $U(g)$ is an irreducible representation and F is a sufficiently nice function on G, let $U(F) = \int_G F(g)U(g)\,dg$. Then

$$\|F\|_2^2 = \sum_{n=2}^{\infty} d_n(\|D_n^+(F)\|_{HS}^2 + \|D_n^-(F)\|_{HS}^2)$$

$$+ \int_{-\infty}^{\infty} \|U(+,it,F)\|_{HS}^2\, p_+(it)\,dt + \int_{-\infty}^{\infty} \|U(-,it,F)\|_{HS}^2\, p_-(it)\,dt\ ,$$

where HS denotes Hilbert-Schmidt norm and $\{d_n, p_+(it)\,dt, p_-(it)\,dt\}$ is the Plancherel measure. To give the Plancherel measure, we fix a normalization of Haar measure on G. Namely write

$$g = \begin{pmatrix}\cos\theta_1 & \sin\theta_1\\ -\sin\theta_1 & \cos\theta_1\end{pmatrix}\begin{pmatrix}e^s & 0\\ 0 & e^{-s}\end{pmatrix}\begin{pmatrix}\cos\theta_2 & \sin\theta_2\\ -\sin\theta_2 & \cos\theta_2\end{pmatrix},$$

with $0 \le \theta_1 \le 2\pi$, $0 \le \theta_2 \le 2\pi$, $0 \le s < \infty$. Then

$$dg = \frac{1}{4\pi^2}(e^{2s} - e^{-2s})\,d\theta_1\,ds\,d\theta_2$$

[5]The group of 2-by-2 real matrices of determinant ± 1.

is a Haar measure. The number d_n can be computed from the formula [6]

$$d_n^{-1} = \|f\|^{-4} \int_G |(D_n^{\pm}(g)f, f)|^2 \, dg$$

with the aid of the explicit form of $D_n^{\pm}$, and the result is that $d_n = \frac{1}{2}(n-1)$. We shall return to p_+ and p_- presently.

We come to the intertwining operators. It follows from Bargmann's classification that $U(\pm, it, \cdot)$ is equivalent with $U(\pm, -it, \cdot)$. Formally the operator given in the induced picture by

$$A(w, \pm, it) f(x) = \int_V f(xwv) \, dv \ ,$$

with $w = \begin{pmatrix} 0 & 1 \\ -1 & 0 \end{pmatrix}$ and f in the space of the induced representation,[6] implements this equivalence. However, this integral is divergent, and it is necessary to proceed with care. To see the problem, one can compute $A(w, \pm, it)$ in the noncompact picture. After a change of variables, the formula is

$$A(w, +, it) f(x) = \int_{-\infty}^{\infty} \frac{f(x-y)\, dy}{|y|^{1-it}}$$

$$A(w, -, it) f(x) = \int_{-\infty}^{\infty} \frac{f(x-y) \operatorname{sgn} y \, dy}{|y|^{1-it}} \ .$$

These integrals are convergent if it is replaced by z and if $\operatorname{Re} z > 0$. Thus the idea is to work with the nonunitary principal series and do an analytic continuation to get the intertwining operator. In order to avoid technical problems, it is helpful to work in the compact picture, carrying over the formal operator from the induced picture.

[6] We use the normalization that dv is Lebesgue measure. See footnote 4.

THEOREM 1. *In the compact picture,* $A(w, \pm, z)f$ *is convergent for* f *in* C^∞ *if* $\operatorname{Re} z > 0$ *and extends to a meromorphic function in the* z*-plane whose only singularities are at most simple poles at the nonpositive integers. Moreover,*

$$U(\pm, -z, \cdot)\, A(w, \pm, z) = A(w, \pm, z)\, U(\pm, z, \cdot)$$

as an identity of meromorphic functions. On trigonometric polynomials

$$A(w, \pm, z)^* = A(w^{-1}, \pm, \bar{z}) . \tag{1.1}$$

Formally the operator $A(w, +, it)$ is just fractional integration of order it. In terms of Fourier transforms[7] on the line, it is well known that

$$(A(w, +, it) f)\hat{}(\xi) = \gamma_+(it) |\xi|^{-it} \hat{f}(\xi) ,$$

where

$$\gamma_+(z) = \pi^{\frac{1}{2} - z}\, \Gamma\left(\tfrac{z}{2}\right) / \Gamma\left(\tfrac{1-z}{2}\right) .$$

See [19, p. 73]. Since $A(w^{-1}, +, it) = A(w, +, it)$, we expect that $A(w^{-1}, +, -it)\, A(w, +, it)$ is the multiple $\gamma_+(-it)\gamma_+(it)$ of the identity operator. Similarly we expect that $A(w^{-1}, -, -it)\, A(w, -, it)$ is the multiple $\gamma_-(-it)\gamma_-(it)$ of the identity, where

$$\gamma_-(z) = \pi^{\frac{1}{2} - z}\, \Gamma\left(\tfrac{1+z}{2}\right) / \Gamma\left(\tfrac{2-z}{2}\right) .$$

Simplifying the products of gamma functions and justifying matters by using the compact picture, we arrive at the following result.

---

[7] Here the Fourier transform is given by $\hat{f}(\xi) = \int_{-\infty}^{\infty} e^{2\pi i x \xi} f(x)\, dx$.

THEOREM 2. *In the compact picture*

$$A(w^{-1}, \pm, -z)\, A(w, \pm, z) = \eta_{\pm}(z) I$$

as identities of meromorphic functions, where

$$\eta_+(z) = 2\pi i z^{-1} \coth(-\pi i z/2)$$

and

$$\eta_-(z) = 2\pi i z^{-1} \tanh(-\pi i z/2) .$$

For future reference we note also that

$$\gamma_+(z)\gamma_-(-z) = \frac{2\pi}{z} . \qquad (1.2)$$

The connection between intertwining operators and the Plancherel measure is given by the following theorem, which is of a general nature and will have an analog in SL(n, R).

THEOREM 3. $p_{\pm}(it) = \frac{\pi}{4}\eta_{\pm}(it)^{-1}$ *for real* t. *Consequently*

$$p_+(it) = \frac{1}{8} t \tanh(\pi t/2)$$

$$p_-(it) = \frac{1}{8} t \coth(\pi t/2) .$$

This theorem can be proved in two steps, first by relating $\eta_{\pm}(z)$ with the asymptotic behavior of certain entry functions of $U(\pm, z, \cdot)$ and second by relating the asymptotic behavior with the Plancherel measure.

These intertwining operators will be combined in various ways in our discussion of SL(n, R). We introduce normalized operators, partly as a bookkeeping device, defining

$$\mathcal{A}(w, \pm, z) = \gamma_{\pm}(z)^{-1} A(w, \pm, z) .$$

The identity (1.1), in combination with Theorem 2 and the fact that $\gamma_{\pm}(\bar{z}) = \overline{\gamma_{\pm}(z)}$, implies that

(i) $\mathcal{A}(w^{-1}, \pm, -z)\mathcal{A}(w, \pm, z) = I$

(ii) $\mathcal{A}(w, \pm, z)^* = \mathcal{A}(w^{-1}, \pm, \bar{z})$

(iii) $\mathcal{A}(w, \pm, z)$ is unitary for z imaginary.

We have mentioned that only $U(-, 0, \cdot)$ is reducible among the $U(\pm, it, \cdot)$. For $z = 0$, $\mathcal{A}(w, \pm, 0)$ is unitary and intertwines $U(\pm, 0, \cdot)$ with itself. In the case of σ_+, $\mathcal{A}(w, +, 0)$ is scalar, whereas in the case of σ_-, $\mathcal{A}(w, -, 0)$ is exactly the Hilbert transform and is not scalar. In short, the intertwining operators we have constructed account for the only reducibility that occurs. A similar fact will hold for $SL(n, \mathbf{R})$.

2. *Some parabolic subgroups in* $SL(n, \mathbf{R})$

Following Gelfand and Graev [3], we introduce $[n/2] + 1$ series of representations in $G = SL(n, \mathbf{R})$, with several realizations for each. The intertwining operators will exhibit the equivalence of the several realizations. Each series will consist of representations induced from a generalized upper triangular subgroup, and we begin by defining the appropriate subgroups.

The parameter that points to the appropriate series will be called k, with $0 \leq k \leq [n/2]$. Choose ℓ so that $2k + \ell = n$. Let

$$M = \left(\begin{array}{c|c} \begin{matrix} \boxed{\begin{smallmatrix}\cdot&\cdot\\ \cdot&\cdot\end{smallmatrix}} & & & \\ & \boxed{\begin{smallmatrix}\cdot&\cdot\\ \cdot&\cdot\end{smallmatrix}} & & \\ 0 & & \ddots & \\ & & & \boxed{\begin{smallmatrix}\cdot&\cdot\\ \cdot&\cdot\end{smallmatrix}} \end{matrix} & 0 \\ \hline 0 & \begin{matrix} \varepsilon_{2k+1} & & 0 \\ & \ddots & \\ 0 & & \varepsilon_n \end{matrix} \end{array}\right), \quad N = \left(\begin{array}{c|c} \begin{matrix} \boxed{\begin{smallmatrix}1&0\\ 0&1\end{smallmatrix}} & & & \\ & \boxed{\begin{smallmatrix}1&0\\ 0&1\end{smallmatrix}} & * & \\ 0 & & \ddots & \\ & & & \boxed{\begin{smallmatrix}1&0\\ 0&1\end{smallmatrix}} \end{matrix} & * \\ \hline 0 & \begin{matrix} 1 & & * \\ & 1 & \\ & & \ddots & \\ 0 & & & 1 \end{matrix} \end{array}\right).$$

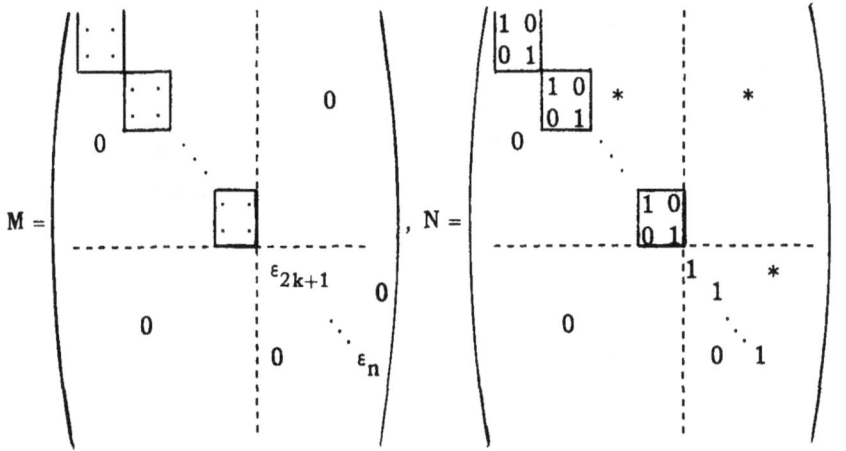

Members of M have k two-by-two blocks, each with determinant ± 1, down the diagonal, followed by ℓ diagonal entries $\varepsilon_{2k+1}, \cdots, \varepsilon_n$ equal to ± 1. Members of N are the sum of the identity and matrices strictly upper triangular relative to M. Let $V = N^{tr}$ and let K be $SO(n)$, the rotation subgroup.

Let $\mathfrak{a}_p$ be the vector space of diagonal matrices of trace 0, and let e_j be the linear functional on $\mathfrak{a}_p$ that picks out the jth diagonal entry. For $1 \leq j \leq k$, let

$$f_j = \frac{1}{2}(e_{2j-1} + e_{2j}) .$$

Let $\mathfrak{a}_M$ be the subspace of $\mathfrak{a}_p$ on which all the f_j are 0, and let $\mathfrak{a}$ be the subspace for which $e_{2j-1}(H) = e_{2j}(H)$ for $1 \leq j \leq k$. Then $\mathfrak{a}_p = \mathfrak{a}_M \oplus \mathfrak{a}$. Define A_p, A_M, and A to be the groups of exponentials of matrices in $\mathfrak{a}_p$, $\mathfrak{a}_M$, and $\mathfrak{a}$, respectively. For example,

$$A = \begin{pmatrix} r_1 & & & & & & & \\ & r_1 & & & & 0 & & \\ & & \ddots & & & & & \\ & & & r_k & & & & \\ & & & & r_k & & & \\ & 0 & & & & r_{2k+1} & & \\ & & & & & & \ddots & \\ & & & & & & & r_n \end{pmatrix} .$$

Then M and A normalize N and members of M and A commute with each other. Hence MAN is a group. The group MA is the group of all elements that commute with each element of A.

The groups that arise in the case $k = 0$, we shall single out by attaching subscripts p.[8] This convention defines M_p, N_p, and V_p, and it redefines A_p consistently. Put $N_M = N_p \cap M$. Note that $N_p = NN_M$ and that M is the product of M_p and the identity component M_0 of M.

---

[8] The subscript p is a vestige of the German $\mathfrak{p}$ in the Cartan decomposition $\mathfrak{g} = \mathfrak{k} \oplus \mathfrak{p}$.

The groups MAN are the basic groups we are concerned with, but we also want some variants of them, and we shall use the linear functionals f_i, $1 \leq i \leq k$, and e_j, $2k+1 \leq j \leq n$, to define these variants. To begin with, each nonzero difference of two of these linear functionals is called a *root*, or root of $\mathfrak{a}$, and we associate a subgroup of N or V to each. Let E_{ij} be the matrix that is 1 in the i-jth entry and 0 elsewhere. Define subgroups by

$$N_{f_i-f_j} = I + RE_{2i-1,2j-1} + RE_{2i,2j-1} + RE_{2i-1,2j} + RE_{2i,2j}$$

$$N_{f_i-e_j} = I + RE_{2i-1,j} + RE_{2i,j}$$

$$N_{e_i-f_j} = I + RE_{i,2j-1} + RE_{i,2j}$$

$$N_{e_i-e_j} = I + RE_{ij} .$$

If L is a root, the dimension of N_L is called the *multiplicity* of L. We associate a variant of N to each of the $(k+\ell)!$ enumerations of the $k+\ell$ linear functionals f_i, $1 \leq i \leq k$, and e_j, $2k+1 \leq j \leq n$. Fix such an enumeration, and adopt the convention that a functional minus a functional farther along in the list is a *positive root*. The remaining roots are the negative roots. With N_L defined above when L is a root, let

$$N_0 = \prod_{\substack{L=\text{root} \\ L>0}} N_L .$$

(If the functionals are enumerated in the original order, N_0 is N. If they are in reverse order, N_0 is V. There are $(k+\ell)! - 2$ other possibilities.) We shall be concerned with all the groups MAN_0 constructed this way.[9] Write V_0 for N_0^{tr}.

[9]These are some of the subgroups of G that are called parabolic in the literature. In fact, these are exactly all the parabolic subgroups with reductive part MA.

3. *Formal Intertwining Operators*

Fix one of the groups $P_0 = MAN_0$ constructed in the previous section. The series of representations of G that goes with P_0 is parametrized by (ξ, λ), where

ξ = irreducible discrete series representation of M on Hilbert space H^ξ

$\lambda = e^\Lambda$ = character of A, not necessarily unitary.

Here we can regard Λ as a complex-valued real-linear functional on $\mathfrak{a}$. To describe ξ more explicitly, let m in M be written as

$$m = (m_1, \cdots, m_k, \varepsilon_{2k+1}, \cdots, \varepsilon_n)$$

where the m_j's are the $SL^{\pm}(2, R)$ blocks comprising the top part of M. Then[10]

$$\xi(m) = \Big(\prod_{\text{certain } j} \varepsilon_j\Big) D_{N_1}(m_1) \oplus \cdots \oplus D_{N_k}(m_k) . \tag{3.1}$$

Let $\mu(a)$ be the positive number by which Lebesgue measure on N_0 is multiplied when N_0 is conjugated by A, so that μ is a certain positive character of A. (If $\{L_j\}$ is the set of positive roots defining N_0 and n_j is the multiplicity (1, 2 or 4) of L_j, then $\mu = \exp(\Sigma n_j L_j)$.)

In the induced picture the space for the representation $U_{P_0}(\xi, \lambda, \cdot)$ is

$$\{f : G \to H^\xi \,|\, f(x\, man_0) = \mu(a)^{-\frac{1}{2}} \lambda(a)^{-1} \xi(m)^{-1} f(x)\}$$

with norm

$$\|f\| = \|f|_K\|_2 ,$$

where Haar measure for K has total mass one. The group action is

$$U_{P_0}(\xi, \lambda, g) f(x) = f(g^{-1} x) .$$

The representation is unitary if λ is unitary.

[10] Equation (3.1) gives an irreducible ξ if $n > 2k$. However, if $n = 2k$, ξ is the sum of two inequivalent discrete series. An exact parametrization of the irreducible ξ's will be given in Section 6.

The compact picture is the restriction of the induced picture to K, and the noncompact picture is the restriction of the induced picture to V_0. Gelfand and Graev [3] describe the noncompact picture more explicitly.

Let $P_1 = MAN_1$ and $P_2 = MAN_2$ with the same MA. For f in the space of $U_{P_1}(\xi,\lambda,\cdot)$ in the induced picture, we consider

$$A(P_2 : P_1 : \xi : \lambda) f(x) = \int_{V_1 \cap N_2} f(xv)\,dv\ ,$$

where dv is Lebesgue measure (Haar measure) in the coordinates on $V_1 \cap N_2$. As with $SL(2,R)$, there are convergence problems, but at least on a formal level we have

$$U_{P_2}(\xi,\lambda,\cdot) A(P_2 : P_1 : \xi : \lambda) = A(P_2 : P_1 : \xi : \lambda) U_{P_1}(\xi,\lambda,\cdot)\ .$$

4. *Special case*, $k = 0$

In this section we assume that $k = 0$, and we drop the subscripts p to simplify notation. M is now finite abelian of order 2^{n-1}. The irreducible representations of M are one-dimensional, and we use σ for a typical one (instead of ξ).

One special feature of this situation is that the $n!$ possible choices for N_0 are all conjugate within G. In fact, let M' be the normalizer of A in K. Members of M' have one ± 1 in each row and column, and 0's elsewhere. A member of M' conjugates an N_0 to an N_1, leaving N_0 stable if and only if it is in M. The group M'/M is the full symmetric group on n letters and permutes the N_0's simply transitively. It acts also by permuting the e_i's, and the permutation that maps N_0 to N is exactly the one that restores the ordering of the e_i's to the natural one.

Thus let $P = MAN$ and let $P_0 = MAN_0 = w^{-1}Pw$ with w in M'. We investigate $A(P_0 : P : \sigma : \lambda)$. This operator is not given by a convolution integral in the noncompact picture as it stands. However, let us introduce

$R(w)f(x) = f(xw)$; $R(w)$ intertwines $U_{P_0}(\sigma,\lambda,\cdot)$ with $U_P(w\sigma, w\lambda,\cdot)$, where $w\sigma(m) = \sigma(w^{-1}mw)$, $w\lambda(a) = \lambda(w^{-1}aw)$. Thus we expect

$$A_P(w,\sigma,\lambda) = R(w)A(P_0 : P : \sigma : \lambda)$$

to intertwine $U_P(\sigma,\lambda,\cdot)$ with $U_P(w\sigma, w\lambda,\cdot)$, and this operator is given by a convolution integral in the noncompact picture.

For example, in $SL(3,R)$ with σ trivial and $w = \begin{pmatrix} 0 & 0 & -1 \\ 0 & -1 & 0 \\ -1 & 0 & 0 \end{pmatrix}$, we pass to the noncompact picture and make a change of variables to find f on V maps to the function of $v_0 \in V$ given by

$$\int_{-\infty}^{\infty}\int_{-\infty}^{\infty}\int_{-\infty}^{\infty} |z|^{-1+s_1}|xy-z|^{-1+s_2} f\left(v_0\begin{pmatrix} 1 & 0 & 0 \\ x & 1 & 0 \\ z & y & 1 \end{pmatrix}\right) dx\, dy\, dz\,.$$

This is essentially a convolution, but the singularity of the kernel is not limited to a single point. On the face of it, the problem of analytic continuation of this integral would seem to be much harder than the problem in $SL(2,R)$.

However, even in $SL(n,R)$, the problem reduces to the case of $SL(2,R)$. First, if w as a permutation is a consecutive transposition (i i+1), the operator $A_P(w,\sigma,\lambda)$ is an $SL(2,R)$ operator in disguise. To understand matters, we write out $A_P(w,\sigma,\lambda)f(v_0)$ in the noncompact picture,[11] decomposing v_0 as a product $v_0 = v_0''v_0'$, where v_0'' is 0 in the $(i+1,i)^{th}$ entry and v_0' is 0 in all off-diagonal entries but the $(i+1,i)^{th}$. If we regard v_0'' as fixed, then the operator is an $SL(2,R)$ operator for the imbedded subgroup in the ith and $(i+1)^{st}$ rows and columns. The σ and λ for the subgroup are obtained by restriction. Consequently the operators corresponding to consecutive transpositions admit analytic continuations. For general w, decompose the permutation

[11] To pursue matters rigorously, one uses also the compact picture.

as a product of consecutive transpositions in as short a fashion as possible and take the composition of the corresponding operators.[12] For normalizing factors γ we can use the corresponding product of $SL(2, R)$ factors. Put $\mathfrak{A}_P = \gamma^{-1} A_P$. The result is that A is given unambiguously by a convergent integral for certain $\lambda = e^{\Lambda}$'s and extends to a meromorphic function for Λ in C^{n-1}. Also $\mathfrak{A}$ is unambiguously defined, is meromorphic, and has the following properties.

THEOREM 4.

(i) $U_P(w\sigma, w\lambda, \cdot)\mathfrak{A}_P(w, \sigma, \lambda) = \mathfrak{A}_P(w, \sigma, \lambda) U_P(\sigma, \lambda, \cdot)$

(ii) $\mathfrak{A}_P(w_1 w_2, \sigma, \lambda) = \mathfrak{A}_P(w_1, w_2\sigma, w_2\lambda)\mathfrak{A}_P(w_2, \sigma, \lambda)$

(iii) $\mathfrak{A}_P(w, \sigma, \lambda)^* = \mathfrak{A}_P(w^{-1}, w\sigma, w\bar{\lambda}^{-1})$

(iv) $\mathfrak{A}_P(w, \sigma, \lambda)$ *is unitary if* λ *is unitary.*

5. *General Case,* k *Arbitrary*

Return to the case of general k and to the notation of Sections 2-3. Fix k and MA. It can happen that two choices of N_1 are not conjugate within G, and consequently the intertwining operator $A(P_2 : P_1 : \xi : \lambda)$ cannot be transformed into a convolution operator (in the noncompact picture) in any evident way. A typical example occurs with $SL(3, R)$, $k = 1$, when $P_1 = MAN$ and $P_2 = MAV$. The H^{ξ}-valued function f on $V \cong R^2$ is transformed as follows: If the image is evaluated at $\begin{pmatrix} I & x \\ 0 & 1 \end{pmatrix}$ with x in R^2, the value is

$$\int_{v \in R^2} |1+x\cdot v|^{-\frac{3}{2}+z} \xi\begin{pmatrix} |1+x\cdot v|^{-\frac{1}{2}}(I+xv^{tr}) & 0 \\ 0 & \operatorname{sgn}(1+x\cdot v) \end{pmatrix} f\begin{pmatrix} 1 & 0 \\ v^{tr} & 1 \end{pmatrix} dv .$$

This is not a convolution; in addition, the singularity of the kernel is one-dimensional, and for any $v \neq 0$, the behavior of the integrand in x depends on the asymptotic behavior of $\xi(m)$ as $m \to \infty$.

[12] This scheme for reducing the problem has a long history, beginning with Gelfand and Neumark [4, Chapter III]. It was developed further by Kunze and Stein [15] and completed by Schiffmann [18].

Despite these complications, there is a simple trick by which we can handle such integrals. The two key facts are (1) the imbedding of discrete series of $SL^{\pm}(2,R)$ in the nonunitary principal series and (2) the double induction formula for representations (induction in stages).

Recall from (3.1) that $\xi(m)$ is built from various discrete series $D_{N_j}(m_j)$ of $SL^{\pm}(2,R)$. Now M is essentially a direct sum of copies of $SL^{\pm}(2,R)$ and it follows from the imbedding of discrete series for $SL(2,R)$ and $SL^{\pm}(2,R)$ that ξ imbeds in

$$\omega = \operatorname*{ind}_{M_pA_MN_M \uparrow M} (\sigma \otimes \lambda_M \otimes 1) ,$$

where $A_M = A_p \cap M$, $N_M = N_p \cap M$, and

$$\lambda_M = \frac{1}{2}\sum_{j=1}^{k} (N_j - 1)(e_{2j-1} - e_{2j}) .$$

Moreover, if m_{ij} denotes the diagonal matrix with -1 in the ith and j^{th} diagonal entries and $+1$ in the other diagonal entries, then

$$\sigma(m_{2j-1,2j}) = (-1)^{N_j}, \qquad 1 \le j \le k$$

$$\sigma(m_{ij})I = \xi(m_{ij}), \qquad 2k < i < j \le n .$$

Since $\xi \subseteq \omega$, the double induction formula says that

$$\operatorname*{ind}_{MAN_0 \uparrow G} (\xi \otimes \lambda \otimes 1) \subseteq \operatorname*{ind}_{MAN_0 \uparrow G} \left(\operatorname*{ind}_{M_pA_MN_M \uparrow M} (\sigma \otimes \lambda_M \otimes 1) \otimes \lambda \otimes 1 \right)$$

$$= \operatorname*{ind}_{MAN_0 \uparrow G} \left(\operatorname*{ind}_{M_p(A_MA)(N_MN_0) \uparrow MAN_0} (\sigma \otimes (\lambda_M \otimes \lambda) \otimes 1) \right)$$

$$= \operatorname*{ind}_{M_pA_p(N_0)_p \uparrow G} (\sigma \otimes (\lambda_M \otimes \lambda) \otimes 1) ,$$

where $(N_0)_p = N_M N_0$. Therefore $U_{P_0}(\xi, \lambda, \cdot)$ is imbedded in the representation $U_{M_p A_p (N_0)_p}(\sigma, \lambda_M \otimes \lambda, \cdot)$, which is one of the representations of the special case $k = 0$ considered in Section 4.

Correspondingly the intertwining operator $A(P_2 : P_1 : \xi : \lambda)$, which is given formally by an integral over $V_1 \cap N_2$, can be identified with a restriction of the special case intertwining operator

$$A(M_p A_p (N_2)_p : M_p A_p (N_1)_p : \sigma : \lambda_M \otimes \lambda) ,$$

which is given formally by an integral over

$$(N_1)_p^{tr} \cap (N_2)_p = V_M V_1 \cap N_M N_2 = V_1 \cap N_2 .$$

Thus the convergence and analytic continuation of the formal intertwining operator in the case of general k is reduced to the case $k = 0$.[13]

It is a consequence of formula (1.2) that we can normalize $A(P_2 : P_1 : \xi : \lambda)$ by the same factor as the corresponding operator with $k = 0$ and arrive at the following conclusion, in which $\mathfrak{A} = \gamma^{-1} A$.

THEOREM 5.

(i) $U_{P_2}(\xi, \lambda, \cdot)\mathfrak{A}(P_2 : P_1 : \xi : \lambda) = \mathfrak{A}(P_2 : P_1 : \xi : \lambda) U_{P_1}(\xi, \lambda, \cdot)$

(ii) $\mathfrak{A}(P_3 : P_1 : \xi : \lambda) = \mathfrak{A}(P_3 : P_2 : \xi : \lambda)\mathfrak{A}(P_2 : P_1 : \xi : \lambda)$

(iii) $\mathfrak{A}(P_2 : P_1 : \xi : \lambda)^* = \mathfrak{A}(P_1 : P_2 : \xi : \bar{\lambda}^{-1})$.

Sometimes it happens that N_1 and N_2 are conjugate. Let M' be M times the normalizer of A in K, and let w be in $M' \cap K$. Suppose $P = MAN$ and $P_0 = MAN_0 = w^{-1} P w$. Then we can define

---

[13]This reduction works if f transforms within a finite-dimensional space under K. To make sense of part (i) of Theorem 5, we should multiply U_{P_1} and U_{P_2} on each side by the projection on such a subspace.

$$A_P(w,\xi,\lambda) = R(w)\,A(P_0 : P : \xi : \lambda)$$

$$\mathcal{A}_P(w,\xi,\lambda) = R(w)\,\mathcal{A}(P_0 : P : \xi : \lambda)$$

as in Section 4. Then $\mathcal{A}_P(w,\xi,\lambda)$ intertwines $U_P(\xi,\lambda,\cdot)$ and $U_P(w\xi, w\lambda, \cdot)$, and we obtain a result similar in form to Theorem 4, but with σ replaced by ξ.

6. *Plancherel Formula*

From the work of Romm [17] and Harish-Chandra [9, Theorem 11], the Plancherel formula for $G = SL(n,R)$ takes the following general form:

$$\|F\|_2^2 = \sum_{k=0}^{[n/2]} \sum_{\xi \text{ of } M} d(\xi) \int_{\mathfrak{a}'} \|U(\xi, i\Lambda, F)\|_{HS}^2 \, \mu_\xi(i\Lambda)\, d\Lambda\,, \qquad (6.1)$$

where the outside sum index is the number of two-by-two blocks in M, ξ is an irreducible discrete series representation of M with formal degree $d(\xi)$, μ_ξ is a function on the complexification of the dual $\mathfrak{a}'$ of $\mathfrak{a}$, and Lebesgue measure $d\Lambda$ on $\mathfrak{a}'$ is suitably normalized. We deal with the problem of making this formula totally explicit. The formulas of [9] would reduce this problem for $SL(n,R)$ to the cases $n = 2$, 3, and 4. The theory of intertwining operators, coupled with the results of [9], reduces the problem for $SL(n,R)$ immediately to the case $n = 2$.

Fix k. The first step is to parametrize the discrete series of M. Let M_0 be the identity component of M, and let $M^\#$ be the product of M_0 and the center of M. If ξ is an irreducible discrete series of M, then the restriction of ξ to $M^\#$ is a sum $\xi_1 \oplus \cdots \oplus \xi_j$ of irreducible representations, and $M/M^\#$ permutes the classes of the ξ_i's without fixed points. Therefore ξ is an induced representation, in fact is induced from any of the ξ_i's. Now ξ_i is still irreducible on M_0, which is the direct sum of k copies of $SL(2,R)$, and so is determined by a tuple $(N_1, \cdots, N_k)$ of

SL(2, R) parameters[14] (with each $|N_j| \geq 2$), together with the restriction of ξ_i to the center of M. Thus ξ_i (and hence ξ) is determined by the data

$$(N_1, \cdots, N_k) \quad \text{and} \quad \xi(m_{2k+1,j}) \quad \text{for} \quad 2k+2 \leq j \leq n .$$

For $2k+2 \leq j \leq n$, define

$$s_j = \begin{cases} +1 & \text{if } \xi(m_{2k+1,j}) = I \\ -1 & \text{if } \xi(m_{2k+1,j}) = -I . \end{cases}$$

The quotient $M/M^{\#}$ determines whether two sets of data lead to equivalent ξ's, and we arrive at the following criteria: If $k < n/2$, two sets of data lead to equivalent ξ's if and only if the tuples of s_j's are identical and the tuples of N_j's differ only by sign changes. If $k = n/2$, two sets of data lead to equivalent ξ's if and only if the tuples of N_j's differ by an even number of sign changes.

Apart from a normalization that we shall consider later, the numbers $d(\xi)$ are the products of the corresponding numbers for the SL(2, R)'s. If ξ has data $(N_1, \cdots, N_k, s_{2k+2}, \cdots, s_n)$, then

$$d(\xi) = \text{Const} \times \prod_{j=1}^{k} (|N_j| - 1) .$$

To get at μ_ξ, we combine (1) Harish-Chandra's theory that relates asymptotics with the Plancherel measure and (2) identities that relate the intertwining operators with asymptotics. The result is Theorem 6 below. By Theorem 5

$$A(P : P^{tr} : \xi : e^{\Lambda}) \, A(P^{tr} : P : \xi : e^{\Lambda}) = \eta_{P,\xi}(\Lambda) I$$

for a complex-valued meromorphic function $\eta_{P,\xi}$.

---

[14] Let us agree to associate the parameter N_j to $D^+_{N_j}$ and the parameter $-N_j$ to $D^-_{N_j}$.

THEOREM 6. *If* M *has* k *two-by-two blocks, then*

$$\mu_\xi(\Lambda) = \mathrm{Const}(k)\, \eta_{P,\xi}(\Lambda)^{-1}$$

for an explicitly given constant depending on k *and the normalization of Haar measure.*

The bookkeeping necessary to compute $\eta_{P,\xi}$ has been done by the normalizing factors for the intertwining operators. Let ξ be imbedded in

$$\operatorname*{ind}_{M_pA_MN_M \uparrow M} (\sigma \otimes e^{\Lambda_M} \otimes 1)\ .$$

The normalizing factors for the operators in the definition of $\eta_{P,\xi}$ are the same as for suitable operators when $k = 0$, and these in turn are products of $SL(2,R)$ factors. The product expansion for the normalizing factors yields a product expansion for $\eta_{P,\xi}$ in terms of η_+ and η_-, and the result is

$$\eta_{P,\xi}(\Lambda) = \prod_{i<j}{}' \eta_{\sigma(m_{ij})}\left(\frac{2(\Lambda + \Lambda_M, e_i - e_j)}{|e_i - e_j|^2}\right), \tag{6.2}$$

where $\prod'$ means that the factors corresponding to $(i,j) = (2i'-1, 2i')$ with $i' \leq k$ are omitted. The factors on the right side are of the form $\eta_+(z)$ and $\eta_-(z)$ and are given in Theorem 2.

We can be more explicit. We know that

$$\Lambda_M = \frac{1}{2}\sum_{j=1}^{k} (|N_j| - 1)(e_{2j-1} - e_{2j})\ .$$

Let

$$N_j' = |N_j| - 1$$

and

$$\Lambda = \sum_{i=1}^{k} \Lambda_i(e_{2i-1} + e_{2i}) + \sum_{j=2k+1}^{n} \Lambda_j e_j\ .$$

We do not know σ completely but know that $\xi(m) = \sigma(m)I$ for m in the center of M. Direct computation, even with this incomplete knowledge, shows that all the tanh and coth factors arising from the η_+'s and η_-'s cancel unless $2k < i < j$.[15]

We can now combine these computations with Theorems 6 and 2, obtaining the Plancherel measure except for a factor depending on k. Let $s_{2k+1} = +1$. Then

$$d(\xi)\mu_\xi(i\Lambda) = c_k\left[\prod_{j=1}^{k} N_j'\right]\left[\prod_{i<j\le k}\left\{\left(\tfrac{1}{4}(N_i'-N_j')^2 + (\Lambda_i-\Lambda_j)^2\right)\left(\tfrac{1}{4}(N_i'+N_j')^2 + (\Lambda_i-\Lambda_j)^2\right)\right\}\right]$$
$$\times\left[\prod_{\substack{i\le k\\ j>2k}}\left(\tfrac{1}{4}N_i'^2 + (\Lambda_i-\Lambda_j)^2\right)\right]\left[\prod_{2k<i<j}(\Lambda_i-\Lambda_j)\,{\tanh \atop \coth}(\pi(\Lambda_i-\Lambda_j)/2)\right] \tag{6.3}$$

with tanh if $s_is_j = +1$ and coth if $s_is_j = -1$.

To write down c_k, we have to specify normalizations for Haar measures. On $G = K(\exp \mathfrak{a}_p)N_p$, we use as a Haar measure

$$dg = e^{2\rho_p(H)}\,dk\,dH\,dn\ ,$$

where

- dk = on K has total mass one
- dH on $\mathfrak{a}_p$ is Lebesgue measure when $\mathfrak{a}_p$ has norm the square root of the sum of the squares of the entries
- dn on N_p is Lebesgue measure in the coordinates
- g in G is decomposed as $g = k(\exp H)n$ in KA_pN_p
- ρ_p on $\mathfrak{a}_p$ is half the sum of the positive roots defining N_p.

[15] The cancellation occurs root-by-root. That is, it occurs in lots of 4 when $i < j \le 2k$ and in lots of 2 when $i \le 2k < j$.

To fix $d\Lambda$ in the Plancherel formula, we normalize dH on $\mathfrak{a}$ to be Lebesgue measure when $\mathfrak{a}$ is equipped with norm the square root of the sum of the squares of the entries. Then we normalize $d\Lambda$ on the dual $\mathfrak{a}'$ so that

$$f(0) = \int_{\mathfrak{a}'} \left(\int_{\mathfrak{a}} e^{i\Lambda(H)} f(H)\, dH \right) d\Lambda \quad \text{for} \quad f \quad \text{in} \quad C^{\infty}_{com}(\mathfrak{a}) .$$

The constants contributing to c_k are the constant of Theorem 6, the constant contributing to $d(\xi)$, and the coefficients 2π that were dropped each time the factor η_+ or η_- appears in $\eta_{P,\xi}$. From Theorem 11 of [9] relating the Plancherel measure to "c-functions" and from Theorem 3 of [13] relating "c-functions" to intertwining operators, we find that the constant of Theorem 6 is

$$\text{Const}(k) = (k!\,\ell!)^{-1} \int_V e^{-2\rho H(v)}\, dv ,$$

where ρ is half the sum of the positive roots defining N. (This is different from ρ_p if $k \neq 0$.) This integral is computed in [2] and [5]. If

$$c(z) = \pi^{-\frac{1}{2}} \Gamma\left(\frac{z}{2}\right) \Big/ \Gamma\left(\frac{z+1}{2}\right) ,$$

then

$$\int_V e^{-2\rho H(v)}\, dv = \pi^{\frac{1}{2}(n^2-n)-k} \prod_{i<j}{}' c\left(\frac{2\langle 2\rho-\rho_p, e_i-e_j \rangle}{|e_i - e_j|^2} \right)$$

with $\prod'$ as in formula (6.2).

In the expression for $d(\xi)$, Haar measure for $SL^{\pm}(2,R)$ is to be normalized in the same fashion as Haar measure for $SL(n,R)$, and this is different from the normalization in Section 1. The representation ξ of M is induced from a representation ξ_0 of the subgroup $M^{\#}$, and

$$d(\xi_0) = (2\pi\sqrt{2})^{-k} \prod_{j=1}^{k} N_j' .$$

To pass to M, we write $M = M^{\#}F$ as a semidirect product with $M^{\#}$ normal. Each element of F has order 2. The normalized Haar measure of M, restricted to $M^{\#}$, is $|F|^{-1}$ times the normalized Haar measure of $M^{\#}$. We find that

$$d(\xi) = |F|\, d(\xi_0) .$$

Here

$$|F| = \begin{cases} 2^k & \text{if } k < n/2 \\ 2^{k-1} & \text{if } k = n/2 . \end{cases}$$

Thus the contribution to c_k from $d(\xi)$ is

$$|F|(2\pi\sqrt{2})^{-k} = \begin{cases} (\pi\sqrt{2})^{-k} & \text{if } k < n/2 \\ \frac{1}{2}(\pi\sqrt{2})^{-k} & \text{if } k = n/2 . \end{cases}$$

Finally, Theorem 2 says that a factor $(2\pi)^{-1}$ must be included in c_k for each η_+ or η_- that appears in $\eta_{P,\xi}$, and there are $\frac{1}{2}(n^2-n)-k$ such factors.

We conclude that

$$c_k = 2^{-(n^2-n)/2} \times \begin{Bmatrix} 1 & \text{if } k < n/2 \\ \frac{1}{2} & \text{if } k = n/2 \end{Bmatrix} \times (2\pi^{-2})^{k/2}(k!(n-2k)!)^{-1}$$

$$\times \prod_{i<j}{}' c\left(\frac{2\langle 2\rho-\rho_p, e_i-e_j\rangle}{|e_i-e_j|^2}\right). \tag{6.4}$$

7. *Reducibility Criterion*

For each k with $0 \leq k \leq [n/2]$, we shall use intertwining operators to decide which of the representations $U_P(\xi, \lambda, \cdot)$ of $G = SL(n, R)$ are reducible when λ is unitary. From the work of Gelfand and Graev [3], it follows that $U_P(\xi, \lambda, \cdot)$ splits into at most two irreducible pieces, necessarily inequivalent. With a little extra work, one can decide when this splitting actually occurs. However, we shall not follow this approach, but shall use the general framework of intertwining operators.

Fix k. As in Section 5, let M' be M times the normalizer of A in K. The quotient $W(\mathfrak{a}) = M'/M$ is called the Weyl group of $\mathfrak{a}$. It operates on the class of ξ and on λ. Namely if w is in $M' \cap K$, put

$$w\xi(m) = \xi(w^{-1}mw) \quad \text{and} \quad w\lambda(a) = \lambda(w^{-1}aw) .$$

The $w\lambda$ and the class $[w\xi]$ of $w\xi$ depend only on the coset of w in M'/M. The group $W(\mathfrak{a})$ is easily computed and can be regarded as all permutations of the f_i, $1 \leq i \leq k$, times all permutations of the e_j, $2k < j \leq n$.

Suppose $[w\xi] = [\xi]$. Then it is possible to extend ξ to a representation of the smallest group containing M and w, with the extended representation acting on the same space. That is, we can define $\xi(w)$. This definition is unique up to a scalar equal to a jth root of unity if j is the least positive integer such that w^j is in M.

In this case the operator $\xi(w)\mathfrak{A}_P(w, \xi, \lambda)$ satisfies

$$U_P(\xi, w\lambda, \cdot)\xi(w)\mathfrak{A}_P(w, \xi, \lambda) = \xi(w)\mathfrak{A}_P(w, \xi, \lambda)U_P(\xi, \lambda, \cdot) .$$

If also $w\lambda = \lambda$, then the operator $\xi(w)\mathfrak{A}_P(w, \xi, \lambda)$ commutes with $U_P(\xi, \lambda, \cdot)$ and will exhibit $U_P(\xi, \lambda, \cdot)$ as reducible if the operator is not scalar and if λ is unitary. The operator $\xi(w)\mathfrak{A}_P(w, \xi, \lambda)$ depends only on the coset $[w]$ of w in M'/M.

It is easy to determine whether $[w\xi] = [\xi]$. Let us enlarge the parameter set of $[\xi]$ to include $s_{2k+1} = 1$, writing it as

$$(N_1,\cdots,N_k,\ s_{2k+1},\cdots,s_n)\ ,$$

and let us admit

$$(N_1,\cdots,N_k,\ -s_{2k+1},\cdots,-s_n)$$

as a further equivalence. Then w acts as a permutation of the N_i's and of the s_j's, and $[w\xi] = [\xi]$ if and only if the final parameter set is equivalent with the original one.

For each $[w]$ in $W(\mathfrak{a})$, one can show that the normalizing factor for $A_P(w,\xi,\lambda)$ is nowhere vanishing for λ unitary. Let W_{ξ,λ_0} be the subgroup of $W(\mathfrak{a})$ of elements $[w]$ such that $[w\xi] = [\xi]$ and $w\lambda_0 = \lambda_0$. Then we have the following result [14].

THEOREM 7. *Let* λ_0 *be unitary. The operators* $\xi(w)\mathcal{A}_P(w,\xi,\lambda_0)$ *with* w *in* W_{ξ,λ_0} *such that the normalizing factor for* $A_P(w,\xi,\lambda)$ *is regular at* $\lambda = \lambda_0$ *form a basis for the vector space of bounded linear operators commuting with* $U_P(\xi,\lambda_0,\cdot)$.

Theorem 7 indicates a computation that will decide the reducibility question, since the normalizing factors ultimately are products of $SL(2,R)$ normalizing factors. If $[w]$ is W_{ξ,λ_0} and w does not act as the identity permutation on the indices $1,\cdots,k$, it is not hard to see that the normalizing factor fails to be regular. Thus we may assume w acts only on the s_j's. For simplicity, assume $\lambda = 1$. If $[w]$ is written as a product of consecutive transpositions in as short a fashion as possible, each factor in the corresponding decomposition of $\mathcal{A}_P(w,\xi,1)$ will be the identity or a Hilbert transform, and the condition of Theorem 7 is that all the factors be Hilbert transforms. One can then work out that w must map the parameter set

$$(N_1,\cdots,N_k,\ s_{2k+1},\cdots,s_n)$$

into

$$(N_1,\cdots,N_k,\ -s_{2k+1},\cdots,-s_n)\ .$$

THEOREM 8. *Let* ξ *have parameter set*

$$(N_1, \cdots, N_k, s_{2k+1}, \cdots, s_n)$$

and let λ_0 *be unitary. If there exists a permutation* $p \neq 1$ *of indices* $2k+1, \cdots, n$ *such that* $p(s_{2k+1}, \cdots, s_n) = (-s_{2k+1}, \cdots, -s_n)$ *and* $p\lambda_0 = \lambda_0$, *then* $U_p(\xi, \lambda_0, \cdot)$ *is reducible and splits into two inequivalent irreducible pieces. Otherwise* $U_p(\xi, \lambda_0, \cdot)$ *is irreducible. In particular, reducibility can occur only if* $\ell = n - 2k$ *is even and does occur when* ℓ *is even and positive if* ξ *and* λ_0 *are suitably chosen.*

8. *Complementary series*

The K-finite vectors for $U_p(\xi, \lambda, \cdot)$ are the members f of the representation space such that the span of $U_p(\xi, \lambda, K)f$ is finite-dimensional. Such vectors are dense in the representation space.

Informally $U_p(\xi, e^{\Lambda}, \cdot)$ is in the complementary series if Λ is not purely imaginary and if there exists an inner product on the space of K-finite vectors that makes $U_p(\xi, e^{\Lambda}, \cdot)$ unitary. The difficulty with this definition is that $U_p(\xi, e^{\Lambda}, x)$ need not leave stable the space of K-finite vectors. We can repair the difficulty by using the infinitesimal representation of $U_p(\xi, e^{\Lambda}, \cdot)$, i.e., the corresponding representation of the Lie algebra. Thus we say $U_p(\xi, e^{\Lambda}, \cdot)$ is in the *complementary series* if there exists an inner product $\langle \cdot, \cdot \rangle$ on the K-finite vectors with respect to which the infinitesimal representation is skew-Hermitian, i.e.,

$$\langle U_p(\xi, e^{\Lambda}, X)f, g \rangle = -\langle f, U_p(\xi, e^{\Lambda}, X)g \rangle$$

for X in the Lie algebra.

If we assume that $\langle f, g \rangle = (Lf, g)$ for an operator L and the usual L^2 inner product $(\cdot, \cdot)$ and if we take into account the identity $U_p(\xi, e^{\Lambda}, -X)^* = U_p(\xi, e^{-\bar{\Lambda}}, X)$, then we find the condition is that L be positive definite Hermitian and satisfy

$$U_P(\xi, e^{-\bar{\Lambda}}, X)L = LU_P(\xi, e^{\Lambda}, X) .$$

This equation is satisfied if L is a suitable intertwining operator and Λ is related suitably to $-\bar{\Lambda}$. Namely, if w is in $K \cap M'$, if $[w\xi] = [\xi]$, and if $w\Lambda = -\bar{\Lambda}$, this equation holds with

$$L = \xi(w)\mathfrak{A}_P(w, \xi, e^{\Lambda}) .$$

Moreover, this L is Hermitian if w, as a permutation, has order two. So a sufficient condition for complementary series is that this L be positive definite.

A technique for showing that L is positive definite is described in detail in [11] and [12]. The idea is that a continuous family of nonsingular Hermitian operators on a finite-dimensional space is everywhere positive definite if it is somewhere positive definite. We shall introduce assumptions that make L equal to the identity (which is positive definite) at $\Lambda = 0$. Because of the relations that intertwining operators satisfy, nonsingularity must persist until $\mathfrak{A}_P$ has a pole at Λ or $-\bar{\Lambda}$. Information about the poles of $\mathfrak{A}_P$ ultimately is largely a question about $SL(2, \mathbb{R})$.

In order to make maximum use of this technique, we must find all permutations $[w]$ of order two such that $[w\xi] = [\xi]$ and $\xi(w)\mathfrak{A}_P(w, \xi, 1)$ is scalar. From [14] one knows each such permutation is a product of transpositions with the same property. For such a transposition $[w]$, one can prove that $\xi(w)\mathfrak{A}_P(w, \xi, 1)$ is scalar if and only if $s_{w(j)} = s_j$ for $2k < j \leq n$ if ξ has parameters $(N_1, \cdots, N_k, s_{2k+1}, \cdots, s_n)$. Putting these facts together and making the necessary computations, we arrive at the following result.

THEOREM 9. *Let* ξ *have parameter set* $(N_1, \cdots, N_k, s_{2k+1}, \cdots, s_n)$. *Suppose* $[w]$ *is a permutation of order two such that* $[w\xi] = [\xi]$ *and* $s_{w(j)} = s_j$ *for* $2k < j \leq n$. *Then every complex* Λ *that is not purely imaginary and satisfies*

(i) $w\Lambda = -\bar{\Lambda}$

(ii) $\left|\operatorname{Re}\frac{\langle\Lambda,\alpha\rangle}{\langle\alpha,\alpha\rangle}\right| < 1$ *for every root* α *of* $\mathfrak{a}$ *of the form* $\alpha = f_i - f_j$

(iii) $\left|\operatorname{Re}\frac{\langle\Lambda,\alpha\rangle}{\langle\alpha,\alpha\rangle}\right| < \frac{1}{2}$ *for every root* α *of* $\mathfrak{a}$ *of the form* $\alpha = e_i - e_j$

is such that $U_P(\xi, e^{\Lambda}, \cdot)$ *is in the complementary series.*

9. *Wider Class of Groups*

Most of the results mentioned in this paper for SL(n, R) have generalizations to connected real semisimple Lie groups of matrices. The role of SL(2, R) in Section 1 is played by groups of real-rank one. Convergence for the intertwining integrals was handled by [15], analytic continuation was obtained independently in [18] and [12], and the normalization was done in [12]. The group that generalizes V is not always abelian, and the Fourier transform is not an appropriate tool; instead, the normalizing factors are constructed by means of Weierstrass canonical products.

For the general group, the representations that appear in the Plancherel formula are induced from parabolic subgroups MAN. The representations of MAN are assumed to be discrete series on M and unitary characters on A. In particular, MAN plays no role unless M has discrete series representations. The Plancherel formula is announced in [8] and [9].

The case $k = 0$ in SL(n, R) corresponds to the case of a minimal parabolic subgroup, in which M is compact. Schiffmann [18] realized that the intertwining operators in this case satisfied some relations even before normalization and exhibited them as compositions of real-rank-one operators. The normalization is done in [12].

The theory for the nonminimal parabolics is in [13] and [14]. The reduction to the case of minimal parabolics is in [13]. See also [21]. In the general case, theorems about reducibility appear in [14] and [10], and a theorem about complementary series appears in [14]. In the general case, the problem of deciding which intertwining operators are scalar and which are linearly independent is less transparent, but is solved by a detailed study of the group W_{ξ,λ_0}.

A. W. KNAPP
CORNELL UNIVERSITY

E. M. STEIN
PRINCETON UNIVERSITY

REFERENCES

[1] Bargmann, V., Irreducible unitary representations of the Lorentz group, *Ann. of Math.* (2) 48(1947), 568-640.

[2] Bhanu Murti, T. S., Plancherel's measure for the factor space SL(n, R)/SO(n, R), *Soviet Math. Dokl.* 1(1960), 860-862.

[3] Gelfand, I. M., and M. I. Graev, Unitary representations of the real unimodular group, *Amer. Math. Soc. Transl.* (2) 2(1956), 147-205.

[4] Gelfand, I. M., and M. A. Neumark, *Unitäre Darstellungen der Klassischen Gruppen*, Akademie-Verlag, Berlin, 1957.

[5] Gindikin, S. G., and F. I. Karpelevič, Plancherel measure for Riemann symmetric spaces of nonpositive curvature, *Soviet Math. Dokl.* 3(1962), 962-965.

[6] Godement, R., Sur les relations d'orthogonalité de V. Bargmann, *C. R. Acad. Sci. Paris* 225(1947), 521-523 and 657-659.

[7] Harish-Chandra, Plancherel formula for the 2×2 real unimodular group, *Proc. Nat. Acad. Sci. USA*, 38(1952), 337-342.

[8] ——————, Harmonic analysis on semisimple Lie groups, *Bull. Amer. Math. Soc.* 76(1970), 529-551.

[9] ——————, On the theory of the Eisenstein integral, *Conference on Harmonic Analysis*, Springer-Verlag Lecture Notes 266(1972), 123-149.

[10] Knapp, A. W., Commutativity of intertwining operators II, *Bull. Amer. Math. Soc.*, to appear.

[11] Knapp, A. W., and E. M. Stein, The existence of complementary series, *Problems in Analysis, a Symposium in Honor of Salomon Bochner*, R. Gunning (ed.), Princeton University Press, Princeton, N. J., 1970, 249-259.

[12] Knapp, A. W., and E. M. Stein, Intertwining operators for semisimple groups, *Ann. of Math.* (2), 93(1971), 489-578.

[13] ——————————, Singular integrals and the principal series III, *Proc. Nat. Acad. Sci. USA*, 71(1974), 4622-4624.

[14] ——————————, Singular integrals and the principal series IV, *Proc. Nat. Acad. Sci. USA*, 72(1975), 2459-2461.

[15] Kunze, R., and E. M. Stein, Uniformly bounded representations III, *Amer. J. Math.* 89(1967), 385-442.

[16] Romm, B. D., An analogue of the Plancherel formula for the 3×3 real unimodular group, *Soviet Math. Dokl.* 6(1965), 315-316.

[17] ——————, Analogue of the Plancherel formula for the real unimodular group of the nth order, *Amer. Math. Soc. Transl.* (2), 58(1966), 155-215.

[18] Schiffmann, G., Intégrales d'entrelacement et fonctions de Whittaker, *Bull. Soc. Math. France*, 99(1971), 3-72.

[19] Stein, E. M., *Singular Integrals and Differentiability Properties of Functions*, Princeton University Press, Princeton, N. J., 1970.

[20] Wallach, N., Cyclic vectors and irreducibility for principal series representations, *Trans. Amer. Math. Soc.*, 158(1971), 107-113.

[21] ——————, On Harish-Chandra's generalized C-functions, *Amer. J. Math.*, 97(1975), 386-403.

[22] Zelobenko, D. P., The analysis of irreducibility in the class of elementary representations of a complex semisimple Lie group, Izv. Akad. Nauk SSSR 32(1968), 105-128.

INEQUALITIES FOR THE MOMENTS OF THE EIGENVALUES OF THE SCHRÖDINGER HAMILTONIAN AND THEIR RELATION TO SOBOLEV INEQUALITIES

Elliott H. Lieb*
Walter E. Thirring

1. *Introduction*

Estimates for the number of bound states and their energies, $e_j \leq 0$, are of obvious importance for the investigation of quantum mechanical Hamiltonians. If the latter are of the single particle form $H = -\Delta + V(x)$ in R^n, we shall use available methods to derive the bounds

$$\sum_j |e_j|^\gamma \leq L_{\gamma,n} \int d^n x\, |V(x)|_-^{\gamma+n/2}\,, \qquad \gamma > \max(0, 1-n/2)\,. \quad (1.1)$$

Here, $|V(x)|_- = -V(x)$ if $V(x) \leq 0$ and is zero otherwise.

Of course, in many-body theory, one is more interested in Hamiltonians of the form $-\sum_i \Delta_i + \sum_{i>j} v(x_i - x_j)$. It turns out, however, that the energy bounds for the single particle Hamiltonian yield a lower bound for the kinetic energy, T, of N fermions in terms of integrals over the single particle density defined by

$$\rho(x) \equiv N \int |\psi(x, x_2, \cdots, x_N)|^2\, d^n x_2 \cdots d^n x_N\,, \qquad (1.2)$$

where ψ is an antisymmetric, normalized function of the N variables $x_i \in R^n$. Our main results, in addition to (1.1), will be of the form

*Work supported by U. S. National Science Foundation Grant MPS 71-03375-A03.

$$T \equiv \sum_{i=1}^{N} \int |\vec{\nabla}_i \psi(x_1 \cdots x_N)|^2 \, d^n x_1 \cdots d^n x_N$$
$$\geq K_{p,n} \left[\int d^n x \, \rho(x)^{p/(p-1)} \right]^{2(p-1)/n} \tag{1.3}$$

when $\max\{n/2, 1\} \leq p \leq 1 + n/2$.

For $N = 1$, $p = n/2$, (1.3) reduces to the well-known Sobolev inequalities. (1.3) is therefore a partial generalization of these inequalities, and we shall expand on this in Section 3.

Our constants $K_{p,n}$ are not always the best possible ones, but nevertheless, they may be useful for many purposes. In particular, in ref. [1], a special case of (1.3) was used to give a simple proof of the stability of matter, with a constant of the right order of magnitude. The result for q species of fermions $(2m = e = \hbar = 1)$ moving in the field of M nuclei with positive charges Z_j is

$$H \geq -1.31 \, q^{2/3} N \left[1 + \left(\sum_{j=1}^{M} Z_j^{7/3}/N \right)^{1/2} \right]^2 . \tag{1.4}$$

In particular, if $q = 2$ (spin 1/2 electrons), we have a bound $\sim N$, and if we set $q = N$, we get a bound $\sim N^{5/3}$ if no symmetry requirement is imposed on the wave function; *a fortiori* this is a bound for *bosons*. Our bound implies stability of matter in its intuitive meaning such that the volume occupied by N particles will be $\sim N$ (Bohr radius)^3. To give a formal demonstration of this fact, one might use a method which gives lower bounds for the radii of complex atoms (compare Equation (3.6, 38) of ref. [20]). As a first observation, one calculates the ground state energy of N electrons (with spin) in a harmonic potential. Filling the oscillator levels, one finds

$$\frac{1}{2} \sum_{i=1}^{N} (-\Delta_i + \omega^2 \vec{x}_i^2) \geq \omega \, N^{4/3} \, \frac{3^{4/3}}{4} \, (1 + O(N^{-1/3})) . \tag{1.5}$$

Next, take the expectation value of this operator inequality with the ground state of H, set

$$\omega = \frac{4}{3^{4/3}} \left\langle -\sum_i \Delta_i \right\rangle \frac{1}{N^{4/3}} \tag{1.6}$$

and use the virial theorem

$$\left\langle -\sum_i \Delta_i \right\rangle = -E_0 \leq 2.08\,N \left[1 + \left(\sum_{j=1}^{M} Z_j^{7/3}/N \right)^{1/2} \right]^2 . \tag{1.7}$$

Altogether we find

$$\left\langle \sum_{i=1}^{N} \vec{x}_i^{\,2} \right\rangle \geq \frac{(3N)^{8/3}}{16 \left\langle -\sum_i \Delta_i \right\rangle} \geq \frac{3^{8/3} N^{5/3}}{16 \cdot 2.08 \left[1 + \left(\sum_j Z_j^{7/3}/N \right)^{1/2} \right]^2} . \tag{1.8}$$

Thus we have proved that

$$\langle \vec{x}_i^{\,2} \rangle^{1/2} \geq c N^{1/3} , \qquad c = \frac{.75}{\left[1 + \left(\sum_j Z_j^{7/3}/N \right)^{1/2} \right]} . \tag{1.9}$$

Therefore, if the system is not compressed by other forces, so that the virial theorem is valid, it will not collapse, but will adjust its volume to a size proportional to the number of particles. Regarding the Z-dependence, we see that with $Z = Z_j = N/M$ we have (for large Z)

$$\langle \vec{x}_i^{\,2} \rangle^{1/2} \sim M^{1/3} Z^{-1/3} .$$

That is, the mean atomic radius is predicted to be $\geq Z^{-1/3}$. A better result can hardly be expected since for $M = 1$, this is the correct Z-dependence for large Z.

Although we have no results on the best possible constants, $K_{p,n}$, except in a few special cases, experience drawn from computer calculations suggests that there is a critical value $\gamma_{c.n}$ above which the classical value gives a bound:

$$\left(\sum |e_j|^\gamma\right)_{\text{classical}} = (2\pi)^{-n} \int d^np\, d^nx\, |p^2 + V(x)|_-^\gamma$$

$$\equiv L_{\gamma,n}^C \int |V(x)|_-^{\gamma+n/2}\, d^nx \ ,$$

$$\gamma \geq \gamma_{c,n} \ , \tag{1.10}$$

and where $L_{\gamma,n}^C$, given by the above integral, is

$$L_{\gamma,n}^C = 2^{-n}\pi^{-n/2}\,\Gamma(\gamma+1)/\Gamma(\gamma+1+n/2) \ . \tag{1.11}$$

We conjecture $\gamma_{c,1} = 3/2$, $\gamma_{c,3} \cong .863$ and $\gamma_{c,n} = 0$, all $n \geq 8$. If this conjecture were to be true, the constants in (1.3, 1.4) could be further improved.

In the next section we shall deduce bounds for $\sum_j |e_j|^\gamma$ and use them in Section 3 to derive (1.3). In Section 4 we shall discuss our conjectures and support them for $n = 1$ with results from the Korteweg-de Vries equation. Section 5 contains new results added in proof. In Appendix A, generously contributed by J. F. Barnes, further evidence from computer studies is presented. We are extremely grateful to Dr. Barnes for taking an interest in this problem, for without his results we would have been hesitant to put forth our conjectures.

2. *Bounds for Moments of the Eigenvalues*

In this section we shall deduce bounds of the form (1.1), and we shall compare our $L_{\gamma,n}$ with the classical values which one gets by replacing

$$\sum_j |e_j|^\gamma \qquad \text{by} \qquad (2\pi)^{-n} \int d^nx\, d^np\, |p^2 + V(x)|_-^\gamma \ .$$

For $n \sim 3$ and $\gamma \sim 1$, the latter are smaller by about an order of magnitude.

Our inequalities are based on the Birman-Schwinger [2, 3] method for estimating N_E, the number of bound states of $H = -\Delta + V(x)$ having an energy $\leq E$. Since

$$\frac{\partial}{\partial E} N_E = \sum_j \delta(E - e_j)$$

we have

$$\sum_j |e_j|^\gamma = \gamma \int_0^\infty d\alpha\, \alpha^{\gamma-1} N_{-\alpha} \,. \tag{2.1}$$

Now, according to Birman-Schwinger [2, 3], for all $\alpha \geq 0$, $m \geq 1$ and $t \in [0, 1]$,

$$N_{-\alpha} \leq \mathrm{Tr}(|V + (1-t)\alpha|_-^{1/2}(-\Delta + t\alpha)^{-1}|V + (1-t)\alpha|_-^{1/2})^m \,. \tag{2.2}$$

REMARKS ABOUT (2.2):

1. We are only interested in potentials such that $V_- \in L^{\gamma+n/2}(R^n)$ for $\gamma \geq \min(0, 1-n/2)$. For such potentials (2.2) is justified, and a complete discussion is given in Simon [4, 5]. Moreover, it is sufficient to consider $V \in C_0^\infty(R^n)$ in (2.2), and in the rest of this paper, and then to use a limiting argument. Such potentials have the advantage that they have only a finite number of bound states [5].

2. Since we are interested in maximizing $\Sigma |e_j|^\gamma / \int |V|_-^{\gamma+n/2}$, we may as well assume that $V(x) \leq 0$, i.e. $V = -|V|_-$. This follows from the max-min principle [4] which asserts that $e_j(V) \geq e_j(-|V|_-)$, all j, including multiplicity.

To evaluate the trace in (2.2), we use the inequality

$$\mathrm{Tr}(B^{1/2} A B^{1/2})^m \leq \mathrm{Tr}\, B^{m/2} A^m B^{m/2} \tag{2.3}$$

when A, B are positive operators and $m \geq 1$. When m is integral and A, B is of our special form, (2.3) is a consequence of Hölder's inequality. For completeness, we shall give a more general derivation of (2.3) in Appendix B.

To calculate

$$\mathrm{Tr}\,|V+(1-t)\alpha|_-^m(-\Delta+t\alpha)^{-m}, \tag{2.4}$$

we shall use an x-representation where $(-\Delta+t\alpha)^{-m}$ is the kernel

$$G_{t\alpha}^{(m)}(x-y) = (2\pi)^{-n}\int d^np\,(p^2+t\alpha)^{-m}\,e^{ip(x-y)} \tag{2.5}$$

if $m > n/2$. Using

$$d^np = \frac{2\pi^{n/2}}{\Gamma(n/2)}\int_0^\infty dp\;p^{n-1}, \tag{2.6}$$

we easily compute

$$G_{t\alpha}^{(m)}(0) = (2\pi)^{-n}\,\frac{2\pi^{n/2}}{\Gamma(n/2)}\,(t\alpha)^{-m+n/2}\int_0^\infty p^{n-1}(p^2+1)^{-m}\,dp$$

$$= (4\pi)^{-n/2}\,\frac{\Gamma(m-n/2)}{\Gamma(m)}\,(t\alpha)^{-m+n/2} \tag{2.7}$$

if $m > n/2$. Thus,

$$N_{-\alpha} \le (4\pi)^{-n/2}\,\frac{\Gamma(m-n/2)}{\Gamma(m)}\,(t\alpha)^{-m+n/2}\int d^nx\,|V(x)+(1-t)\alpha|_-^m. \tag{2.8}$$

Next, we substitute (2.8) into (2.1). If we impose the condition that $t < 1$, it is easy to prove that one can interchange the α and the x integration. Changing variables $\alpha \to (1-t)^{-1}\,|V(x)|_-\beta$, leads to

$$\sum |e_j|^\gamma \le \gamma(4\pi)^{-n/2}\,t^{-m+n/2}(1-t)^{m-\gamma-n/2}\,\frac{\Gamma(\gamma-m+n/2)\Gamma(m-n/2)}{\Gamma(\gamma+1+n/2)}\,m\int d^nx$$

$$\times\,|V(x)|_-^{\gamma+n/2} \tag{2.9}$$

provided $n/2 < m < n/2+\gamma$, $m \geq 1$ and $0 < t < 1$. The optimal t is $t = (m-n/2)/\gamma$.

If we put our results together, we obtain the following (see note added in proof, Section 5).

THEOREM 1. *Let* $V_- \in L^{\gamma+n/2}(R^n)$, $\gamma \geq \max(0, 1-n/2)$. *Let* $H = -\Delta + V(x)$, *and let* $e_j \leq 0$ *be the negative energy bound states of* H. *Then*

$$\sum |e_j|^\gamma \leq L_{\gamma,n} \int |V(x)|_-^{\gamma+n/2} \tag{2.10}$$

where

$$L_{\gamma,n} \leq \tilde{L}_{\gamma,n} \equiv \min_m (4\pi)^{-n/2} \gamma^{\gamma+1} \frac{m}{\Gamma\left(\gamma+\frac{n}{2}+1\right)} F\left(m-\frac{n}{2}\right) F\left(\gamma+\frac{n}{2}-m\right), \tag{2.11}$$

and where $F(x) = \Gamma(x) x^{-x}$, $\max\{1, n/2\} \leq m < n/2 + \gamma$.

REMARKS:

1. When $\gamma = 0$, $\Sigma |e_j|^0$ means the number of bound states, including zero energy states. For $n \geq 2$, our $\tilde{L}_{0,n} = \infty$. In Section 4, we shall discuss the $\gamma = 0$ case further. See also Section 5.

2. In (2.11), $\tilde{L}_{\gamma,n}$ is the bound we have obtained using the Birman-Schwinger principle. We shall henceforth reserve the symbol $L_{\gamma,n}$ for the quantity

$$L_{\gamma,n} \equiv \sup_V \sum |e_j|^\gamma \Big/ \int |V|_-^{\gamma+n/2} . \tag{2.12}$$

Optimization with respect to m in (2.11) can be done either numerically or analytically in the region where Stirling's formula

$$F(x) \sim e^{-x} \sqrt{2\pi/x} \tag{2.13}$$

can be applied. In [1], for $n = 3$, $\gamma = 1$, we used the value 2 for m. A marginal improvement can be obtained with $m = 1.9$.

If (2.13) were exact, the best m would be

$$\bar{m} = n(\gamma + n/2)/(n+\gamma) . \tag{2.14}$$

Note that as $\gamma \to \infty$, $\bar{m}$ is bounded by n. Using $\bar{m}$, together with (2.13), which is valid when $\gamma n(\gamma+n)^{-1}$ is large,

$$\tilde{L}_{\gamma,n} \sim (4\pi)^{1-n/2} \frac{\gamma^\gamma e^{-\gamma}}{\Gamma(\gamma+n/2)} \left[\frac{n/2}{\gamma+n/2}\right]^{1/2} . \qquad (2.15)$$

Finally, we want to compare our bounds with their classical values, $L^C_{\gamma,n}$. From the results of Martin [6] and Tamura [7], one has the following

THEOREM 2. *If* $V(x) \le 0$ *and* $V \epsilon C_0^\infty(R^n)$, *then*

$$\lim_{\lambda\to\infty} \sum_j |e_j(\lambda V)|^\gamma \Big/ \int |\lambda V|^{\gamma+n/2} = L^C_{\gamma,n} . \qquad (2.16)$$

COROLLARY.

$$L_{\gamma,n} \ge L^C_{\gamma,n} . \qquad (2.17)$$

Our $\tilde{L}_{\gamma,n}$ satisfies (2.17), in particular in the asymptotic region (2.15), we find

$$\tilde{L}_{\gamma,n}/L^C_{\gamma,n} \approx [4\pi n(\gamma+n/2)]^{1/2}\gamma^{-1/2} . \qquad (2.18)$$

We conjecture in Section 4 that for γ sufficiently large, the best possible $L_{\gamma,n}$ should be $L^C_{\gamma,n}$, a result which does not follow from the Birman-Schwinger method employed here. For small γ, we know that $L^C_{\gamma,n}$ is not a bound.

We conclude this section with a theorem about $L_{\gamma,n}$ which will be useful in the discussion of the one-dimensional case in Section 4.

THEOREM 3. *Let* $\gamma \ge 1 + \max(0, 1-n/2)$. *Then*

$$L_{\gamma,n} \le L_{\gamma-1,n}[\gamma/(\gamma+n/2)] . \qquad (2.19)$$

PROOF. Choose $\varepsilon > 0$. We can find a $V \epsilon C_0^\infty(R^n)$, with $V \le 0$, such that

$$L_{\gamma,n}(V) = \sum_j |e_j(V)|^\gamma \Big/ \int |V|^{\gamma+n/2} \ge L_{\gamma,n} - \varepsilon .$$

Let $g \in C_0^\infty(R^n)$ be such that $0 \le g(x) \le 1$, $\forall x$, and $V(x) \ne 0$ implies $g(x) = 1$. Let $V_\lambda(x) = V(x) - \lambda g(x)$, $\lambda \le 0$. The functions $|e_j(V_\lambda)|$ are continuous and monotone increasing in λ. Furthermore, there are a finite number of values $-\infty < \lambda_1 \le \lambda_2 \le \cdots \le \lambda_k \le 0$ with λ_j being the value of λ at which $e_j(V_\lambda)$ first appears. λ_1 is finite because V_λ is non-negative for λ sufficiently negative. $e_j(V_\lambda)$ is continuously differentiable on $\Lambda = \{\lambda | 0 \ge \lambda > \lambda_1, \lambda \ne \lambda_i,\ i = 1,\cdots,k\}$ and

$$de_j(V_\lambda)/d\lambda = -\int |\psi_j(x; V_\lambda)|^2 g(x) d^n x$$

by the Feynman-Hellman theorem. It is easy to prove that if $f, g \in L^p(R^n)$, $p > 1$, then

$$h(\lambda) \equiv \int |f(x) - \lambda g(x)|_-^p d^n x$$

is differentiable, $\forall\lambda$ and

$$dh/d\lambda|_{\lambda=0} = p \int |f(x)|_-^{p-1} g(x) d^n x .$$

Thus $L_{\gamma,n}(V_\lambda)$ is piecewise C^1 on Λ and its derivative, $\dot{L}_{\gamma,n}$, is given by

$$\dot{L}_{\gamma,n} = \left[\int |V_\lambda|_-^{\gamma+n/2}\right]^{-1} \Big\{ \gamma \sum_j |e_j(V_\lambda)|^{\gamma-1} \int g(x) \psi_j(x; V_\lambda)|^2 d^n x - (\gamma+n/2) L_{\gamma,n}(V_\lambda)$$

$$\cdot \int |V_\lambda(x)|_-^{\gamma+n/2-1} g(x) d^n x \Big\} .$$

By the stated properties of $L_{\gamma,n}$, there exists a $\lambda \in (\lambda_1, 0]$ such that

(i) $\dot{L}_{\gamma,n}(V_\lambda) \ge 0$;

(ii) $L_{\gamma,n}(V_\lambda) \ge L_{\gamma,n} - 2\varepsilon$.

Thus, using the properties of g,

$$0 \le \gamma \sum_j |e_j(V_\lambda)|^{\gamma-1} - L_{\gamma,n}(V_\lambda)(\gamma + n/2) \int |V_\lambda|_-^{\gamma+n/2-1} . \tag{2.20}$$

Since ε was arbitrary, (2.20) implies the theorem.

If we use (2.17) together with the fact that $L^C_{\gamma,n} = L^C_{\gamma-1,n}[\gamma/(\gamma+n/2)]$, we have

COROLLARY. *If for some* $\gamma \geq \max(0, 1-n/2)$, $L_{\gamma,n} = L^C_{\gamma,n}$, *then*

$$L_{\gamma+j,n} = L^C_{\gamma+j,n}, \qquad j = 0,1,2,3,\cdots .$$

REMARK. By the same proof

$$L^1_{\gamma,n} \leq L^1_{\gamma-1,n}[\gamma/(\gamma+n/2)] \tag{2.21}$$

(see (3.1) for the definition of $L^1_{\gamma,n}$).

3. *Bounds for the Kinetic Energy*

In this section, we shall use Theorem 1 to derive inequalities of the type (1.3). We recall the definition (2.14) and we further define

$$L^1_{\gamma,n} \equiv \sup_V |e_1|^\gamma \Big/ \int |V|_-^{\gamma+n/2} . \tag{3.1}$$

Clearly,

$$L^1_{\gamma,n} \leq L_{\gamma,n} . \tag{3.2}$$

If $\psi \in \mathcal{H}_{N,n,q}$ = the N-fold antisymmetric tensor product of $L^2(R^n;C^q)$, we can write ψ pointwise as $\psi(x_1,\cdots,x_N;\sigma_1,\cdots,\sigma_N)$ with $x_j \in R^n$, $\sigma_j \in \{1,2,\cdots,q\}$ and $\psi \to -\psi$ if (x_i,σ_i) is permuted with (x_j,σ_j). $q = 2$ for spin 1/2 fermions. We can extend the definition (1.2) to

$$\rho_\sigma(x) \equiv N \sum_{\sigma_2=1}^{q} \cdots \sum_{\sigma_N=1}^{q} \int |\psi(x,x_2,\cdots,x_N;\sigma,\sigma_2,\cdots,\sigma_N)|^2 d^nx_2 \cdots d^nx_N . \tag{3.3}$$

We also define

$$T_\psi \equiv \sum_{j=1}^{N} \sum_{\sigma_1=1}^{q} \cdots \sum_{\sigma_N=1}^{q} \int |\nabla_j \psi(\underline{x};\underline{\sigma})|^2 \, d^{nN}\underline{x} \,, \tag{3.4}$$

$$\|\psi\|_2^2 \equiv \sum_{\sigma_1=1}^{q} \cdots \sum_{\sigma_N=1}^{q} \int |\psi(\underline{x};\underline{\sigma})|^2 \, d^{nN}\underline{x} \,. \tag{3.5}$$

Our result is

THEOREM 4. *Let* p *satisfy* $\max\{n/2, 1\} \le p \le 1 + n/2$ *and suppose that* $L_{p-n/2,n} < \infty$. *If* $\|\psi\|_2 = 1$, *then, except for the case* $n = 2$, $p = 1$, *there exists a positive constant* $K_{p,n}$ *such that*

$$T_\psi \ge K_{p,n} \sum_{\sigma=1}^{q} \left[\int \rho_\sigma(x)^{p/(p-1)} \, d^n x\right]^{2(p-1)/n} \tag{3.6}$$

and

$$K_{p,n} \ge \frac{1}{2} n p^{-2p/n} (p-n/2)^{-1+2p/n} (L^1_{p-n/2,n}/L_{p-n/2,n})^{-1+2p/n} L^1_{p-n/2,n}{}^{-2/n}. \tag{3.7}$$

Before giving the proof of Theorem 4, we discuss its relation to the well-known Sobolev inequalities [9, 10]:

THEOREM 5 (Sobolev-Talenti-Aubin). *Let* $\nabla\psi \in L^r(R^n)$ *with* $1 < r < n$. *Let* $t = nr/(n-r)$. *Then*

$$\int |\nabla\psi|^r \ge C_{r,n} \left(\int |\psi|^t\right)^{r/t} \tag{3.8}$$

for some $C_{r,n} > 0$.

Talenti [11] and Aubin [21] have given the best possible $C_{r,n}$ (for $n = 3$, $r = 2$, $t = 6$, $C_{2,n}$ is also given in [8] and [12]):

$$C_{r,n} = n\pi^{r/2}\left(\frac{r-1}{n-r}\right)^{1-r}\left\{\frac{\Gamma(1+n-n/r)\Gamma(n/r)}{\Gamma(n)\Gamma(1+n/2)}\right\}^{r/n} . \tag{3.9}$$

Our inequality (3.6) relates only to the $r = 2$ case in (3.8), in which case $t = 2n/(n-2)$. Consider (3.8) with $r = 2$ and $\|\psi\|_2 = 1$. Using Hölder's inequality on the right side of (3.8), one gets

$$\int |\nabla\psi|^2 \geq C_{2,n}\left[\int |\psi|^{2p/(p-1)}\right]^{2(p-1)/n}\left[\int |\psi|^2\right]^{-2(p-n/2)/n} \tag{3.10}$$

whenever $n > 2$ and $p \geq n/2$. However, $C_{2,n}$ is not necessarily the best constant in (3.10) when $p \neq n/2$ ($p = n/2$ corresponds to $r = 2$ in (3.8)). Indeed, Theorem 4 says something about this question.

In the case that $N = 1$ and $q = 1$, Theorem 4 is of the same form as (3.10) (since $\rho = |\psi|^2$ and $\|\psi\|_2 = 1$). We note two things:

1. For $n > 2$ and $p = n/2$, (3.6) agrees with (3.8) except, possibly, for a different constant. We have, therefore, an alternative proof of the usual Sobolev inequality (for the $r = 2$ case). As we shall also show $K_{n/2,n} = C_{2,n}$, so we also have the best possible constant for this case.

2. If $\max\{n/2, 1\} < p \leq 1 + n/2$, Theorem 4 gives an improved version of (3.10), *even if* $n = 1$ or 2 (in which cases $C_{2,n} = 0$, but $K_{p,n} > 0$). For $p > 1 + n/2$, one can always use Hölder's inequality on the $p = 1+n/2$ result to get a nontrivial bound of the form (3.10). However, in Theorem 4, the restriction $p \leq 1 + n/2$ is really necessary. This has to do with the dependence of T_ψ on N rather than on n, as we shall explain shortly.

Next we turn to the case $N > 1$. To illustrate the nature of (3.6), we may as well suppose $q = 1$. To fix ideas, we take a special, but important form for ψ, namely

$$\psi(x_1, \cdots, x_N) = (N!)^{-1/2}\,\mathrm{Det}\,\{\phi^i(x_j)\}_{ij=1}^N \tag{3.11}$$

and where the ϕ^i are orthonormal functions in $L^2(R^n)$. Then, suppressing the subscript σ because $q = 1$,

$$\rho(x) = \sum_{i=1}^{N} \rho^i(x) \ ,$$

$$\rho^i(x) = |\phi^i(x)|^2 \ ,$$

$$T_\psi = \sum_{i=1}^{N} t^i \ ,$$

$$t^i = \int |\nabla\phi^i|^2 \ . \tag{3.12}$$

Theorem 4 says that

$$\sum_i t^i \geq K_{p,n} \left\{ \int \left[\sum_i \rho^i(x) \right]^{p/(p-1)} d^n x \right\}^{2(p-1)/n} \ . \tag{3.13}$$

If we did not use the orthogonality of the ϕ^i, all we would be able to conclude, using (3.6) with $N = 1$, N times, would be

$$\sum_i t^i \geq K_{p,n} \sum_i \left[\int \rho^i(x)^{p/(p-1)} \, d^n x \right]^{2(p-1)/n} \ . \tag{3.14}$$

If $p = n/2$, then (3.14) is better than (3.13), by convexity. In the opposite case, $p = 1 + n/2$, (3.13) is superior. For in between cases, (3.13) is decidedly better if N is large and if the ρ^i are close to each other (in the $L^{p/(p-1)}(R^n)$ sense). Suppose $\rho^i(x) = \rho(x)/N$, $i = 1, \cdots, N$. Then the right side of (3.13) is proportional to $N^{2p/n}$ while the right side of (3.14) grows only as N. This difference is caused by the orthogonality of the ϕ^i, or the Pauli principle.

In fact, the last remark shows why $p \leq 1 + n/2$ is important in Theorem 4. If $\rho^i = \rho/N$, all i, then the best bound, insofar as the N dependence is concerned, occurs when p is as large as possible. It is easy to see by example, however, that the largest growth for T_ψ due to the orthogonality condition can only be $N^{(n+2)/n}$.

PROOF OF THEOREM 4. Let $V(x) \leq 0$ be a potential in $\mathbf{R}^n$ with at least one bound state. If $e_1 = \min\{e_i\}$, then, for $\gamma \in [0,1]$,

$$\sum_j |e_j|^\gamma \geq |e_1|^{\gamma-1} \sum_j |e_j| \; .$$

Using the definition (2.14) and (3.2), we have that

$$\sum_j |e_j| \leq A_{\gamma,n} \left\{ \int |V|^{\gamma+n/2} \right\}^{1/\gamma} \tag{3.15}$$

$$A_{\gamma,n} = L_{\gamma,n}(L^1_{\gamma,n})^{-1+1/\gamma} \tag{3.16}$$

when $1 \geq \gamma \geq \max(0, 1-n/2)$. (3.15) holds even if V has no bound state. Let π_σ, $\sigma = 1,\cdots,q$, be the projection onto the state σ, i.e. for $\psi \in L^2(\mathbf{R}^n; \mathbf{C}^q)$, $(\pi_\nu\psi)(x,\sigma) = \psi(x,\nu)$ if $\sigma = \nu$ and zero otherwise. Choose $\gamma = p - n/2$. Let $\{\rho_\sigma\}_{\sigma=1}^q$ be given by (3.3) and, for $a_\sigma \geq 0$, $\sigma = 1,\cdots,q$, define

$$h = -\Delta - \sum_{\sigma=1}^{q} a_\sigma \rho_\sigma(x)^{1/(\gamma+n/2-1)} \pi_\sigma \tag{3.17}$$

to be an operator on $L^2(\mathbf{R}^n; \mathbf{C}^q)$ in the usual way. Define

$$H_N = \sum_{i=1}^{N} h_i \tag{3.18}$$

where h_i means h acting on the i-th component of $\mathcal{H}_{N,n,q}$. Finally, let $E = \inf \operatorname{spec} H_N$.

Now, by the Rayleigh-Ritz variational principle

$$E \leq (\psi, H_N \psi) = T_\psi - \sum_{\sigma=1}^{q} a_\sigma \int \rho_\sigma^{p/(p-1)} . \qquad (3.19)$$

On the other hand, $E \geq$ the sum of all the negative eigenvalues of h

$$\geq -A_{\gamma,n} \sum_{\sigma=1}^{q} a_\sigma^{p/\gamma} \left\{ \int \rho_\sigma^{p/(p-1)} \right\}^{1/\gamma} \qquad (3.20)$$

by (3.15). Combining (3.19) and (3.20) with

$$a_\sigma = \left\{ \int \rho_\sigma^{p/(p-1)} \right\}^{2(\gamma-1)/n} \left\{ \frac{\gamma}{\gamma + n/2} \frac{1}{A_{\gamma,n}} \right\}^{2\gamma/n} ,$$

the theorem is proved.

Note that when $p = 1 + n/2$ (corresponding to $\gamma = 1$ in the proof), $L_{1,n}^1$ does not appear in (3.7). In this case, the right side of (3.7) is the best possible value of $K_{1+n/2,n}$, as we now show.

Lemma 6. *From* (3.7), *define*

$$L_{1,n}^* \equiv [n/(2K_{1+n/2,n})]^{n/2} (1+n/2)^{-1-n/2} .$$

Then $L_{1,n}^* = L_{1,n}$.

Proof. By (3.7), we only have to prove that $L_{1,n}^* \geq L_{1,n}$. Let $V \leq 0$, $V \in C_0^\infty(R^n)$ and let $H = -\Delta + V$. Let $\{\phi_i, e_i\}_{i=1}^N$ be the bound state eigenfunctions and eigenvalues of H. Let ψ and ρ^i be as defined in (3.11), (3.12). Then

$$\sum_i |e_i| = -\int V\rho - T_\psi \leq \|V\|_p \|\rho\|_{p/(p-1)} - T_\psi$$

with $p = 1 + n/2$. Using Theorem 4 for T_ψ, one has that

$$\sum_i |e_i| \leq \max_{y>0} \{\|V\|_p y - K_{1+n/2,n} y^{2p/n}\} = L_{1,n}^* \|V\|_p^p \ .$$

We conclude with an evaluation of $K_{n/2,n}$ for $n > 2$ as promised. By a simple limiting argument

$$K_{n/2,n} \geq \lim_{p \downarrow n/2} \text{(right side of (3.7))} \ . \tag{3.21}$$

Our bound (2.11) on $L_{p-n/2,n}$ shows that

$$\lim_{p \downarrow n/2} (L_{p-n/2,n})^{-1+2p/n} = 1 \ . \tag{3.22}$$

Hence

$$K_{n/2,n} \geq (L_{0,n}^1)^{-2/n} \ . \tag{3.23}$$

On the other hand, by the method of Lemma 6 applied to the $N = 1$ case, $K_{n/2,n} \leq (L_{0,n}^1)^{-2/n}$. The value of $L_{0,n}^1$ is given in (4.24). To be honest, its evaluation requires the solution of the same variational problem as given in [8, 11, 12]. Substitution of (4.24) into (3.23) yields the required result

$$K_{n/2,n} = C_{2,n} = \pi n(n-2)[\Gamma(n/2)/\Gamma(n)]^{2/n} \ . \tag{3.24}$$

If we examine (3.23) when $n = 2$, one gets $K_{1,2} \geq 0$ since $L_{0,2}^1 = \infty$. This reflects the known fact [5] that an arbitrarily small $V < 0$ always has a bound state in two dimensions. This observation can be used to show that

$$K_{1,2} = 0 \ . \tag{3.25}$$

When $n = 1$, the smallest allowed p is $p = 1$. In this case, (3.6) reads

$$T_\psi \geq K_{1,1} \sum_{\sigma=1}^{q} \|\rho_\sigma\|_\infty^2 \ . \tag{3.26}$$

Using (3.7) and (4.20),

$$K_{1,1} \geq [2L_{1/2,1}]^{-1} . \tag{3.27}$$

If one accepts the conjecture of Section 4 that $L_{1/2,1} = L^1_{1/2,1} = 1/2$, then

$$K_{1,1} = 1 . \tag{3.28}$$

The reason for the equality in (3.28) is that $K_{1,1} = 1$ is well known to be the best possible constant in (3.26) when $q = 1$ and $N = 1$.

4. *Conjecture About* $L_{\gamma,n}$

We have shown that for the bound state energies $\{e_j\}$ of a potential V in n dimensions and with

$$L_{\gamma,n}(V) \equiv \sum_j |e_j|^\gamma / \int |V|^{\gamma+n/2} , \tag{4.1}$$

then

$$L_{\gamma,n} \equiv \sup_{V \in L^{\gamma+n/2}} L_{\gamma,n}(V) \tag{4.2}$$

is finite whenever $\gamma + n/2 > 1$ and $\gamma > 0$. The "boundary points" are

$$\begin{aligned} &\gamma = 1/2 && n = 1 \\ &\gamma = 0 && n \geq 2 . \end{aligned} \tag{4.3}$$

We showed that for $n = 1$, $L_{1/2,1} < \infty$. For $\gamma < 1/2$, $n = 1$, there cannot be a bound of this kind, for consider $V_L(x) \equiv -1/L$ for $|x| < L$ and zero otherwise. For $L \to 0$, this converges towards $-2\delta(x)$ and thus has a bound state of finite energy (which is -1 for $-2\delta(x)$). On the other hand,

$$\lim_{L \to 0} \int dx \, |V_L|^{1/2+\gamma} = 0 \quad \text{for} \quad \gamma < 1/2 .$$

For $n = 2$, $\gamma = 0$ is a "double boundary point" and $L_{0,2} = \infty$, i.e. there is no upper bound on the number of bound states in two dimensions. (Cf. [5].)

For $n \geq 3$, $L_{0,n}$ is *conjectured* to be finite (see note added in proof, Section 5); for $n = 3$, this is the well-known $\int |V|_-^{3/2}$ conjecture on the number, $N_0(V)$, of bound states (cf. [5]). The best that is known at present is that

$$N_0(V) \leq c\left[\int |V|_-^{3/2}\right]^{4/3} , \tag{4.4}$$

but for spherically symmetric V, a stronger result is known [8]:

$$N_0(V) \leq I\left(1 + \frac{1}{4} \ln I\right) ,$$

$$I = 4(3\pi^2 3^{1/2})^{-1} \int |V|_-^{3/2} . \tag{4.5}$$

In (1.4) and (3.1), we introduced L^C and L^1 and showed that

$$L_{\gamma,n} \geq \max(L^1_{\gamma,n}, L^C_{\gamma,n}) . \tag{4.6}$$

A parallel result is Simon's [22] for $n \geq 3$:

$$N_0(V) \leq D_{n,\varepsilon}(\|V_-\|_{\varepsilon+n/2} + \|V_-\|_{-\varepsilon+n/2})^{n/2}$$

with $D_{n,\varepsilon} \to \infty$ as $\varepsilon \to 0$.

In our previous paper [4], we conjectured that $L_{1,3} = L^C_{1,3}$, and we also pointed out that $L^1_{1,1} > L^C_{1,1}$. A remark of Peter Lax (private communication), which will be explained presently, led us to the following:

CONJECTURE. *For each* n, *there is a critical value of* $\gamma, \gamma_{c,n}$, *such that*

$$L_{\gamma,n} = L^C_{\gamma,n} \qquad \gamma \geq \gamma_c$$

$$L_{\gamma,n} = L^1_{\gamma,n} \qquad \gamma \leq \gamma_c$$

γ_c *is defined to be that* γ *for which* $L^C_{\gamma,n} = L^1_{\gamma,n}$; *the uniqueness of this* γ_c *is part of the conjecture. Furthermore,* $\gamma_{c,1} = 3/2$, $\gamma_{c,2} \sim 1.2$, $\gamma_{c,3} \sim .86$ *and the smallest* n *such that* $\gamma_{c,n} = 0$ *is* $n = 8$.

(A) Remarks on $L^1_{\gamma,n}$

We want to maximize

$$\left|\int [|\psi|^2 V + |\nabla\psi|^2] d^n x\right|^{\gamma} \Big/ \int |V|^{\gamma+n/2} \tag{4.7}$$

with respect to V, and where $\int |\psi|^2 = 1$ and $(-\Delta + V)\psi = e_1\psi$. By the variational principle, we can first maximize (4.7) with respect to V, holding ψ fixed. Hölder's inequality immediately yields

$$V(x) = -\sigma |\psi(x)|^{2/(\gamma+n/2-1)}$$

with $\sigma > 0$. The kinetic energy, $\int |\nabla\psi|^2$, is not increased if $\psi(x)$ is replaced by $|\psi(x)|$ and, by the rearrangement inequality [13], this is not increased if $|\psi|$ is replaced by its symmetric decreasing rearrangement. Thus, we may assume that $|V|$ and $|\psi|$ are spherically symmetric, non-increasing functions.

By the methods of [8] or [11], (4.7) can be shown to have a maximum when $\gamma + n/2 > 1$. The variational equation is

$$-\Delta\psi(x) - \sigma\psi(x)^{(\gamma+n/2+1)/(\gamma+n/2-1)} = e_1\psi(x) \tag{4.8}$$

with

$$\sigma = \left\{\frac{\gamma |e_1|^{\gamma-1}}{(\gamma+n/2) L^1_{\gamma,n}}\right\}^{1/(\gamma+n/2-1)} . \tag{4.9}$$

Equation (4.8) determines ψ up to a constant and up to a change of scale in x. The former can be used to make $\int \psi^2 = 1$ and the latter leaves (4.7) invariant.

Equation (4.8) can be solved analytically in two cases, to which we shall return later:

(i) $n = 1$, all $\gamma > 1/2$

(ii) $n \geq 3$, $\gamma = 0$.

(B) The One-Dimensional Case

Lax's remark was about a result of Gardner, Greene, Kruskal and Miura [14] to the effect that

$$L_{3/2,1} = L^C_{3/2,1} = 3/16 . \tag{4.10}$$

To see this, we may assume $V \in C_0^\infty(\mathbb{R})$, and use the theory of the Korteweg-de Vries (KdV) equation [14]:

$$W_t = 6WW_x - W_{xxx} . \tag{4.11}$$

There are two remarkable properties of (4.11):

(i) As W evolves in time, t, the eigenvalues of $-d^2/dx^2 + W$ remain invariant.

(ii) $\int W^2\,dx$ is constant in time.

Let $W(x,t)$ be given by (4.11) with the initial data

$$W(x,0) = V(x) .$$

Then $L_{3/2,1}(W(\cdot,t))$ is independent of t, and may therefore be evaluated by studying its behavior as $t \to \infty$.

There exist traveling wave solutions to (4.11), called solitons, of the form

$$W(x,t) = f(x-ct) .$$

Equation (4.11) becomes

$$-cf_x = -f_{xxx} + 6ff_x . \tag{4.12}$$

The solutions to (4.12) which vanish at ∞ are

$$f_a(x) = -2a^2 \cosh^{-2}(ax)$$

$$c = 4a^2 . \tag{4.13}$$

Any solution (4.13), regarded as a potential in the Schrödinger equation has, as we shall see shortly, exactly one negative energy bound state with energy and wave function

$$e = -a^2$$

$$\psi(x) = \cosh^{-1}(ax) \ . \tag{4.14}$$

Now the theory of the KdV equation says that as $t \to \infty$, W evolves into a sum of solitons (4.13) plus a part that goes to zero in $L^\infty(\mathbb{R})$ norm (but not necessarily in $L^2(\mathbb{R})$ norm). The solitons are well separated since they have different velocities. Because the number of bound states is finite, the non-soliton part of W can be ignored as $t \to \infty$. Hence, for the initial V,

$$\sum |e_j|^{3/2} = \sum_{\text{solitons}} a^3 \tag{4.15}$$

while

$$\int V(x)^2\,dx \geq \sum_{\text{solitons}} \int f_a(x)^2\,dx \ . \tag{4.16}$$

Since $4\int_{-\infty}^{\infty} \cosh^{-4}(x)\,dx = 16/3$, we conclude that

$$L_{3/2,1} = L^C_{3/2,1} = 3/16 \tag{4.17}$$

with equality if and only if $W(x,t)$ is composed purely of solitons as $t \to \infty$. For the same reason,

$$L^1_{3/2,1} = L^C_{3/2,1} \tag{4.18}$$

(cf. (4.21)).

Not only do we have an evaluation of $L_{3/2,1}$, (4.17), but we learn something more. When $\gamma = 3/2$, there is an infinite family of potentials for which $L_{3/2,1}(V) = L_{3/2,1}$, and these may have any number of bound states = number of solitons.

What we believe to be the case is that when $\gamma < 3/2$, the optimizing potential for $L_{\gamma,n}$ has only one bound state, and satisfies (4.8). When

$\gamma > 3/2$, the optimizing potential is, loosely speaking, infinitely deep and has infinitely many bound states; thus $L_{\gamma,n} = L^{C}_{\gamma,n}$.

An additional indication that the conjecture is correct is furnished by the solution to (4.8). When $\gamma = 3/2$, this agrees with (4.14). In general, one finds that, apart from scaling, the nodeless solution to (4.8) is

$$\psi_\gamma(x) = \Gamma(\gamma)^{1/2}\,\pi^{-1/4}\,\Gamma(\gamma-1/2)^{-1/2}\cosh^{-\gamma+1/2}(x)$$

$$V_\gamma(x) = -(\gamma^2 - 1/4)\cosh^{-2}(x)$$

$$e_1 = -(\gamma - 1/2)^2 \ . \tag{4.19}$$

Thus,

$$L^{1}_{\gamma,1} = \pi^{-1/2}\frac{1}{\gamma - 1/2}\frac{\Gamma(\gamma+1)}{\Gamma(\gamma+1/2)}\left(\frac{\gamma-1/2}{\gamma+1/2}\right)^{\gamma+1/2} \ . \tag{4.20}$$

When $L^{1}_{\gamma,1}$ is compared with $L^{C}_{\gamma,1}$, one finds that

$$L^{1}_{\gamma,1} \geq L^{C}_{\gamma,1} \qquad \gamma \leq 3/2$$

$$L^{1}_{\gamma,1} \leq L^{C}_{\gamma,1} \qquad \gamma \geq 3/2 \ . \tag{4.21}$$

This confirms at least part of the conjecture.

However, more is true. For $\gamma = 3/2$, V_γ has a zero energy single node bound state

$$\phi(x) = \tanh(x) \ .$$

Since V_γ is monotone in γ, it follows that V_γ has only one bound state for $\gamma < 3/2$ and at least two bound states for $\gamma > 3/2$. The (un-normalized) second bound state can be computed to be

$$\phi(x) = \sinh(x)\cosh^{-\gamma+1/2}(x)$$

$$e_2 = -(\gamma - 3/2)^2 \ . \tag{4.22}$$

In like manner, one can find more bound states as γ increases even further.

Thus we see that the potential that optimizes the ratio $|e_1|^{\gamma}/\int|V|^{\gamma+1/2}$ automatically has a second bound state when $\gamma > \gamma_c$.

Finally, we remark that Theorem 3, together with (4.10), shows that

THEOREM 7.

$$L_{\gamma,1} = L^C_{\gamma,1} \quad \textit{for} \quad \gamma = 3/2,\ 5/2,\ 7/2,\ \textit{etc.}$$

An application of Theorem 7 to scattering theory will be made in Section 4(D).

(C) Higher Dimensions

We have exhibited the solution to the variational equation (4.8) for $L^1_{\gamma,1}$. When $n > 2$ and $\gamma = 0$, we clearly want to take $e_1 = 0$ in order to maximize $L^1_{0,n}(V)$. (4.8) has the zero energy solution

$$\phi(x) = (1+|x|^2)^{1-n/2}$$

$$V(x) = \alpha\phi(x)^{2/(\gamma+n/2-1)} = n(n-2)(1+|x|^2)^{(2-n)/(n/2-1)} \qquad (4.23)$$

(note: $\phi \in L^2(R^n)$ if and only if $n > 4$, but $V \in L^{n/2}(R^n)$ always). This leads to

$$L^1_{0,n} = [\pi n(n-2)]^{-n/2}\,\Gamma(n)/\Gamma(n/2)\ . \qquad (4.24)$$

The smallest dimension for which $L^1_{0,n} \leq L^C_{0,n}$ is $n = 8$.

If we suppose that the ratio $L^1_{\gamma,n}/L^C_{\gamma,n}$ is monotone decreasing in γ (as it is when $n = 1$ and as it is when $n = 3$ on the basis of the numerical solution of (4.10) by J. F. Barnes, given in Appendix A), and if our conjecture is correct, then $L_{\gamma,n} = L^C_{\gamma,n}$ for $n \geq 8$. The value of γ_c obtained numerically is

$$\gamma_c = 1.165 \qquad n = 2$$

$$\gamma_c = .863 \qquad n = 3\ . \qquad (4.26)$$

The other bit of evidence, apart from the monotonicity of $L^1_{\gamma,n}/L^C_{\gamma,n}$, for the correctness of our conjecture is a numerical study of the energy levels of the potential

$$V_\lambda(x) = \lambda e^{-|x|}, \qquad \lambda > 0 ,$$

in three dimensions. This is given in Appendix A. The energy levels of the square well potential are given in [15, 16]. In both cases, one finds that

$$\lim_{\lambda \to \infty} L_{1,3}(V_\lambda) = L^C_{1,3}$$

and the limit is approached from below. Unfortunately, it is not true, as one might have hoped, that $L_{1,3}(V_\lambda)$ is monotone increasing in λ.

(D) Bounds on One-Dimensional Scattering Cross-Sections

In their study of the KdV equation, (4.11), Zakharov and Fadeev [17] showed how to relate the solution $W(x,t)$ to the scattering reflection coefficient $R(k)$ and the bound state eigenvalues $\{e_j\}$ of the initial potential $V(x)$. There are infinitely many invariants of (4.11) besides $\int W^2$ and these have simple expressions in terms of $R(k)$, $\{e_j\}$.

Thus, for any potential V,

$$\int V^2 = (16/3) \sum |e_j|^{3/2} - 4 \int_{-\infty}^{\infty} k^2\, T(k)\, dk \tag{4.27}$$

$$\int V^3 + \tfrac{1}{2} V_x^2 = -(32/5) \sum |e_j|^{5/2} - 8 \int_{-\infty}^{\infty} k^4\, T(k)\, dk \tag{4.28}$$

$$\int V^4 + 2VV_x^2 + \tfrac{1}{5} V_{xx}^2 = (256/35) \sum |e_j|^{7/2} - (64/5) \int_{-\infty}^{\infty} k^6\, T(k)\, dk \tag{4.29}$$

where

$$T(k) = \pi^{-1} \ln(1 - |R(k)|^2) \le 0 . \tag{4.30}$$

These are only the first three invariants; a recursion relation for the others can be found in [17].

Notice that 3/16, 5/32, 35/256 are, respectively $L^C_{3/2,1}$, $L^C_{5/2,1}$, $L^C_{7/2,1}$. Since $\int V^2 \geq \int |V_-|^2$, (4.27) establishes that $L_{3/2,1} = L^C_{3/2,1}$, as mentioned earlier. For the higher invariants, the signs in (4.28) and (4.29) are not as fortunately disposed and we cannot use these equations to prove Theorem 7. But, given that Theorem 7 has already been proved, we can conclude that

THEOREM 8. *For any nonpositive potential* V(x),

$$\int V_x^2 \geq -16 \int_{-\infty}^{\infty} k^4\, T(k)\, dk\ . \tag{4.31}$$

For any potential V(x),

$$2 \int V V_x^2 + (1/5) \int V_{xx}^2 \leq -(64/5) \int_{-\infty}^{\infty} k^6\, T(k)\, dk\ . \tag{4.32}$$

The first inequality, (4.31), is especially transparent: If V(x) is very smooth, it cannot scatter very much.

5. *Note Added in Proof*

After this paper was written, M. Cwikel and Lieb, simultaneously and by completely different methods, showed that the number of bound states, $N_0(V)$ for a potential, V, can be bounded (when $n \geq 3$) by

$$N_0(V) \leq A_n \int |V(x)|_-^{n/2}\, d^n x\ . \tag{5.1}$$

Cwikel exploits the weak trace ideal method of Simon [22]; his method is more general than Lieb's, but for the particular problem at hand, (5.1), his

A_n does not seem to be as good. Lieb's method uses Wiener integrals and the general result is the following:

$$N_{-\alpha}(V) \leq \int d^n x \int_0^\infty dt\, t^{-1} e^{-\alpha t} (4\pi t)^{-n/2} f(t|V(x)|_-) \tag{5.2}$$

for any non-negative, convex function $f:[0,\infty) \to [0,\infty)$ satisfying

$$1 = \int_0^\infty t^{-1} f(t) e^{-t} dt \ . \tag{5.3}$$

For $\alpha = 0$, one can choose $f(t) = c(t-b)$, $t \geq b$, $f(t) = 0$, $t \leq b$. This leads to (5.1), and optimizing with respect to b, one finds that

$$A_3 = 0.116 \, , \quad A_4 = 0.0191 \tag{5.4}$$

and, as $n \to \infty$,

$$A_n / L^C_{0,n} = (n\pi)^{1/2} + O(n^{-1/2}) \ . \tag{5.5}$$

Note that $A_3/L^1_{0,3} = 1.49$, i.e. A_3 exceeds $L_{0,3}$ by at most 49%.

Since $N_{-\alpha}(V) \leq N_0(-|V+\alpha|_-)$, one can use (5.1) and (2.1) to deduce that for $\gamma \geq 0$ and $n \geq 3$,

$$L_{\gamma,n} \leq L^C_{\gamma,n}(A_n/L^C_{0,n}) \ . \tag{5.6}$$

This is better than (2.11), (2.18). In particular, for $n = 3$, $\gamma = 1$, the improvement of (5.6) over (2.11) with $m = 2$ is a factor of 1.83. The factor 1.31 in Equation (1.4) can therefore be replaced by $1.31 \ (1.83)^{-2/3} = 0.87$.

APPENDIX A. NUMERICAL STUDIES

John F. Barnes
Theoretical Division
Los Alamos Scientific Laboratory
Los Alamos, New Mexico 87545

I. *Evaluation of* $L^1_{\gamma,n}$, $n = 1, 2, 3$

The figure shows the numerical evaluation of $L^1_{\gamma,n}$ as well as $L^C_{\gamma,n}$. The latter is given in (1.11)

$$L^C_{\gamma,n} = 2^{-n}\pi^{-n/2}\Gamma(\gamma+1)/\Gamma(\gamma+1+n/2) \ .$$

The former is obtained by solving the differential equation (4.8) in polar coordinates and choosing σ such that $\psi(x) \to 0$ as $|x| \to \infty$. Note that by scaling, one can take $e_1 = -1$, whence

$$(L^1_{\gamma,n})^{-1} = \sigma^{(\gamma+n/2)} \int |\psi(x)|^{(2\gamma+n)/(\gamma-1+n/2)} d^n x \ .$$

In one dimension, $L^1_{\gamma,1}$ is known analytically and is given in (4.20). Another exact result, (4.24), is

$$L^1_{0,3} = 4\pi^{-2}\, 3^{-3/2} = 0.077997 \ .$$

The critical values of γ, at which $L^1_{\gamma,n} = L^C_{\gamma,n}$ are:

$$\gamma_{c,1} = 3/2$$

$$\gamma_{c,2} = 1.165$$

$$\gamma_{c,3} = 0.8627 \ .$$

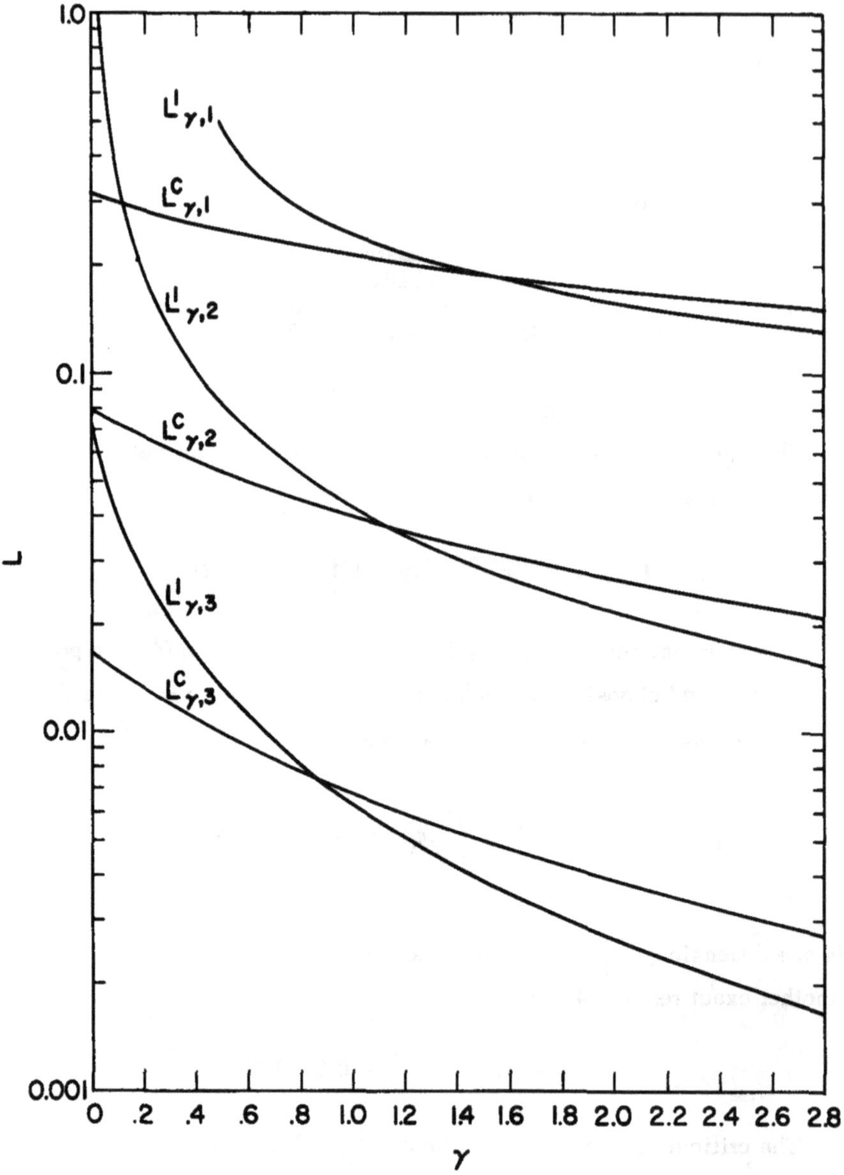
1.0
0.1
0.01
0.001
L
$L^I_{\gamma,1}$
$L^C_{\gamma,1}$
$L^I_{\gamma,2}$
$L^C_{\gamma,2}$
$L^I_{\gamma,3}$
$L^C_{\gamma,3}$
0 .2 .4 .6 .8 1.0 1.2 1.4 1.6 1.8 2.0 2.2 2.4 2.6 2.8
γ

II. *The Exponential Potential*

To test the conjecture that $L_{1,3} = L^C_{1,3}$, the eigenvalues of the potential $V_\lambda = -\lambda \exp(-|x|)$ in three dimensions were evaluated for $\lambda = 5, 10, 20, 30, 40, 50,$ and 100. These are listed in the table according to angular momentum and radial nodes. These numbers have been corroborated by H. Grosse, and they can be used to calculate $L_{\gamma,3}(V_\lambda)$ for any γ. The final column gives $L_{1,3}(V_\lambda)$, since $\int |V_\lambda|^{5/2} = \lambda^{5/2}(64\pi)/125$. It is to be noted that the classical value $L^C_{1,3} = 0.006755$, is approached from below, in agreement with the conjecture, but *not monotonically*.

$$V_\lambda = -\lambda e^{-r}$$

	ℓ	$\lvert e\rvert$	nodes	states	$\sum \lvert e\rvert$	$\dfrac{\sum \lvert e\rvert}{\lambda^{5/2}\frac{64\pi}{125}}$
$\lambda = 5$	0	0.55032	0	$\frac{1}{1}$	$\frac{0.55032}{0.55032}$	0.006120
$\lambda = 10$	0	0.06963	1			
		2.18241	0	2	2.2520	
	1	0.33405	0	$\frac{3}{5}$	$\frac{1.0022}{3.2542}$	0.006398
$\lambda = 20$	0	0.00869	2			
		1.42562	1			
		6.62410	0	3	8.0584	
	1	0.16327	1			
		2.71482	0	6	8.6342	
	2	0.43136	0	$\frac{5}{14}$	$\frac{2.1568}{18.8494}$	0.006551

$V_\lambda = -\lambda e^{-r}$ (continued)

	ℓ	$\lvert e\rvert$	nodes	states	$\sum \lvert e\rvert$	$\dfrac{\sum \lvert e\rvert}{\lambda^{5/2}\frac{64\pi}{125}}$
$\lambda = 30$	0	0.58894	2			
		3.83072	1			
		11.84999	0	3	16.270	
	1	1.39458	1			
		6.12302	0	6	22.553	
	2	0.00593	1			
		2.36912	0	10	11.875	
	3	0.07595	0	7 26	0.532 51.230	0.006461
$\lambda = 40$	0	0.07676	3			
		1.86961	2			
		6.88198	1			
		17.53345	0	4	26.362	
	1	0.41991	2			
		3.35027	1			
		10.13596	0	9	41.718	
	2	0.93459	1			
		5.03378	0	10	29.842	
	3	1.54738	0	7 30	10.832 108.754	0.006682

$V_\lambda = -\lambda e^{-r}$ (continued)

	ℓ	$\|e\|$	nodes	states	$\sum\|e\|$	$\dfrac{\sum\|e\|}{\lambda^{5/2}\frac{64\pi}{125}}$
$\lambda = 50$	0	0.60190	3			
		3.66447	2			
		10.39110	1			
		23.53215	0	4	38.190	
	1	1.43321	2			
		5.81695	1			
		14.56904	0	9	65.458	
	2	0.07675	2			
		2.45887	1			
		8.19840	0	15	53.670	
	3	0.26483	1			
		3.61626	0	14	27.168	
	4	0.49009	0	9	4.411	
				51	188.897	0.006643
$\lambda = 100$	0	0.39275	5			
		2.91408	4			
		8.29231	3			
		17.44909	2			
		32.07168	1			

$V_\lambda = -\lambda e^{-r}$ (continued)

ℓ	$\|e\|$	nodes	states	$\sum \|e\|$	$\dfrac{\sum \|e\|}{\lambda^{5/2}\,\frac{64\pi}{125}}$
	56.28824	0	6	117.41	
1	1.10170	4			
	4.76748	3			
	11.62740	2			
	22.79910	1			
	40.45495	0	15	242.25	
2	0.02748	4			
	2.04022	3			
	6.85633	2			
	15.22147	1			
	28.46495	0	25	263.05	
3	0.22692	3			
	3.14743	2			
	9.13429	1			
	19.04073	0	28	220.85	
4	0.52962	2			
	4.37856	1			
	11.56470	0	27	148.26	

$V_\lambda = -\lambda e^{-r}$ (continued)

ℓ	$\lvert e\rvert$	nodes	states	$\sum \lvert e\rvert$	$\dfrac{\sum \lvert e\rvert}{\lambda^{5/2}\,\frac{64\pi}{125}}$
5	0.88997	1			
	5.69707	0	22	72.46	
6	1.26789	0	13 136	16.48 1080.76	0.006719

APPENDIX B: PROOF OF (2.3)

THEOREM 9. *Let* $\mathcal{H}$ *be a separable Hilbert space and let* A, B *be positive operators on* $\mathcal{H}$. *Then, for* $m \geq 1$,

$$\operatorname{Tr}(B^{1/2} A B^{1/2})^m \leq \operatorname{Tr} B^{m/2} A^m B^{m/2} . \tag{B.1}$$

REMARK. When $\mathcal{H} = L^2(R^n)$ and A is a kernel $a(x-y)$ and B is a multiplication operator $b(x)$ (as in our usage (2.2)), Seiler and Simon [19] have given a proof of (B.1) using interpolation techniques. Simon (private communication) has extended this method to the general case. Our proof is different and shows a little more than just (B.1).

PROOF. For simplicity, we shall only give the proof when A and B are matrices; for the general case, one can appeal to a limiting argument. For $m = 1$, the theorem is trivial, so assume $m > 1$. Let $C = A^m$ and $f(C) \equiv g(C) - h(C)$, where $g(C) \equiv \operatorname{Tr}(B^{1/2} C^{1/m} B^{1/2})^m$ and $h(C) = \operatorname{Tr} B^{m/2} C B^{m/2}$. Let M^+ be the positive matrices. Clearly $M^+ \ni C \to h(C)$ is linear. Epstein [18] has shown that $M^+ \ni C \to g(C)$ is concave (actually, he showed this for m integral, but his proof is valid generally for $m \geq 1$). Write $C = C^D + C^0$ where C^D is the diagonal part of C in a basis in which B is diagonal. $C_\lambda \equiv C^D + \lambda C^0 = \lambda C + (1-\lambda) C^D$ is in M^+ for $\lambda \in [0, 1]$, because $C^D \in M^+$. Then $\lambda \to f(C_\lambda) \equiv R(\lambda)$ is concave on $[0,1]$. Our goal is to show that $R(1) \leq 0$. Since $[C^D, B] = 0$, $R(0) = 0$ and, by concavity, it is sufficient to show that $R(\lambda) \leq 0$ for $\lambda > 0$ and λ small. $h(C_\lambda) = h(C^D)$ for $\lambda \in [0,1]$. Since $f(C)$ is continuous in C, we can assume that C^D is nondegenerate and strictly positive, and that C_λ is positive when $\lambda \geq -\varepsilon$ for some $\varepsilon > 0$. Then $R(\lambda)$ is defined and concave on $[-\varepsilon, 1]$. $\lambda \to C_\lambda^{1/m}$ is differentiable at $\lambda = 0$ and its derivative at $\lambda = 0$ has zero diagonal elements. (To see this, use the representation $C^{1/m} = K\int_0^\infty dx\, x^{-1+1/m} \cdot C(C+xI)^{-1}$.) Likewise, the derivative of $(B^{1/2}(D + \lambda O)B^{1/2})^m$ at $\lambda = 0$ has zero diagonal elements when O has and when D is diagonal. Thus

$$dR(\lambda)/d\lambda|_{\lambda=0} = 0 .$$

Acknowledgment

One of the authors (Walter Thirring) would like to thank the Department of Physics of the University of Princeton for its hospitality.

ELLIOTT H. LIEB
DEPARTMENTS OF MATHEMATICS AND PHYSICS
PRINCETON UNIVERSITY
PRINCETON, NEW JERSEY

WALTER E. THIRRING
INSTITUT FÜR THEORETISCHE PHYSIK
DER UNIVERSITÄT WIEN, AUSTRIA

REFERENCES

[1] E. H. Lieb and W. E. Thirring, Phys. Rev. Lett. *35*, 687 (1975). See Phys. Rev. Lett. *35*, 1116 (1975) for errata.

[2] M. S. Birman, Mat. Sb. *55 (97)*, 125 (1961); Amer. Math. Soc. Translations Ser. 2, *53*, 23 (1966).

[3] J. Schwinger, Proc. Nat. Acad. Sci. *47*, 122 (1961).

[4] B. Simon, "Quantum Mechanics for Hamiltonians Defined as Quadratic Forms," Princeton University Press. 1971.

[5] B. Simon, "On the Number of Bound States of the Two Body Schrödinger Equation – A Review," in this volume.

[6] A. Martin, Helv. Phys. Acta *45*, 140 (1972).

[7] H. Tamura, Proc. Japan Acad. *50*, 19 (1974).

[8] V. Glaser, A. Martin, H. Grosse and W. Thirring, "A Family of Optimal Conditions for the Absence of Bound States in a Potential," in this volume.

[9] S. L. Sobolev, Mat. Sb. *46*, 471 (1938), in Russian.

[10] ―――――, Applications of Functional Analysis in Mathematical Physics, Leningrad (1950), Amer. Math. Soc. Transl. of Monographs, 7 (1963).

[11] G. Talenti, Best Constant in Sobolev's Inequality, Istituto Matematico, Universitá Degli Studi Di Firenze, preprint (1975).

[12] G. Rosen, SIAM Jour. Appl. Math. *21*, 30 (1971).

[13] H. J. Brascamp, E. H. Lieb and J. M. Luttinger, Jour. Funct. Anal. *17*, 227 (1974).

[14] C. S. Gardner, J. M. Greene, M. D. Kruskal and R. M. Miura, Commun. Pure and Appl. Math. *27*, 97 (1974).

[15] S. A. Moszkowski, Phys. Rev. *89*, 474 (1953).

[16] A. E. Green and K. Lee, Phys. Rev. *99*, 772 (1955).

[17] V. E. Zakharov and L. D. Fadeev, Funkts. Anal. i Ego Pril. *5*, 18 (1971). English translation: Funct. Anal. and its Appl. *5*, 280 (1971).

[18] H. Epstein, Commun. Math. Phys. *31*, 317 (1973).

[19] E. Seiler and B. Simon, "Bounds in the Yukawa Quantum Field Theory,' Princeton preprint (1975).

[20] W. Thirring, T7 Quantenmechanik, Lecture Notes, Institut für Theoretische Physik, University of Vienna.

[21] T. Aubin, C. R. Acad. Sc. Paris *280*, 279 (1975). The results are stated here without proof; there appears to be a misprint in the expression for $C_{r,n}$.

[22] B. Simon, "Weak Trace Ideals and the Bound States of Schrödinger Operators," Princeton preprint (1975).

ON THE NUMBER OF BOUND STATES OF TWO BODY SCHRÖDINGER OPERATORS – A REVIEW

Barry Simon*

Given a measurable function V on R^n, consider the operator $-\Delta + V$ on $L^2(R^n)$. Under wide circumstances, this operator is known to be essentially self-adjoint on $C_0^\infty(R^n)$ (see [1] for a review) and under more general circumstances, it can be defined as a sum of quadratic forms [2, 3, 4]. Physically, it represents the Hamiltonian (energy) operator of the particles in nonrelativistic quantum mechanics after the center of mass motion has been removed. For this reason, $-\Delta + V$ is called a two-body Schrödinger operator. We will denote by $N(V)$ the dimension of the spectral projection for $-\Delta + V$ associated with $(-\infty, 0)$; physically the number of bound states. If V is spherically symmetric, we abuse notation and use V also as the symbol for the obvious function on $[0, \infty)$, i.e. the one with $V(x) = V(|x|)$. $n_\ell(V)$ for $\ell \geq 0$ will denote the number of bound states of the operator $-d^2/dx^2 + \ell(\ell+1)r^{-2} + V(r)$ on $L^2(0, \infty)$ (with the boundary condition $u(0) = 0$ if $\ell = 0$). Of course, for $n = 3$, one has the well-known partial wave expansion which yields

$$N(V) = \sum_{\ell=0}^{\infty} (2\ell+1)\, n_\ell(V) \; .$$

For $n > 3$, similar expansions exist but are associated with some non-negative nonintegral ℓ. (For $n = 2$, $\ell = -1/2$ enters.) It is an interesting

---

*A. Sloan Fellow; research partially supported by USNSF under Grant GP-39048.

question to relate qualitative properties of V to N(V) and $n_\ell(V)$. Results of this kind go back to Jost-Pais [5] who proved that $N(V) = 0$ if $\int_0^\infty r|V(r)|\,dr < \infty$ and Bargmann [6] who proved the celebrated bound:

$$n_\ell(V) \le (2\ell+1)^{-1} \int_0^\infty r|V(r)|\,dr \ .$$

Stimulated by Bargmann's paper, something of an industry has developed and we will review some of the results and methods that have emerged. Throughout we will be cavalier about self-adjointness questions, but we emphasize that these kind of details can easily be filled in by following e.g. [7].

§1. *The Methods and Bounds of Bargmann and Calegero*

As a common thread running through all work on the properties of N(V) is the min-max principle of Weyl, Fisher and Courant which takes the following general form:

THEOREM 1.1. *Let* A *be self-adjoint operator which is bounded below, and let* Q(A) *be its quadratic form domain. Let*

$$\mu_n(A) \equiv \max_{\psi_1,\dots,\psi_{n-1}} \left[\min_{\substack{\phi\in[\psi_1,\dots,\psi_{n-1}]^\perp;\ \|\phi\|=1 \\ \phi\in Q(A)}} (\phi, A\phi) \right] .$$

Then either:

(a) $\mu_n(A)$ *is the* nth *eigenvalue from bottom of the spectrum of* A *counting multiplicity and* A *has purely discrete spectrum in* $(-\infty, \mu_n(A))$ *or*

(b) μ_n *is the bottom of the essential spectrum of* A. *If* (b) *holds, then* A *has at most* n−1 *eigenvalue in* $(-\infty, \mu_n)$ *and* $\mu_n(A) = \mu_{n+1}(A) = -$

For a proof and further discussion, see [7]. A major corollary of the min-max principle is the following:

COROLLARY 1.2 (Comparison Theorem). *Let* A *and* B *be self-adjoint operators with* $A \leq B$ *in the sense that* $Q(A) \supset Q(B)$ *and* $(\psi, A\psi) \leq (\psi, B\psi)$ *for all* $\psi \in Q(A)$. *Then* $\mu_n(A) \leq \mu_n(B)$ *for all* n *and, in particular,*

$$\dim P_{(-\infty,a)}(A) \leq \dim P_{(-\infty,a)}(B)$$

for all a.

The proof is immediate. Since one has the following (see e.g. [7]):

THEOREM 1.3. *Let* $V \in L^{n/2}(R^n) + L^\infty_\varepsilon(R^n)$ [*for any* ε, $V = V_1^{(\varepsilon)} + V_2^{(\varepsilon)}$ *with* $V_1^{(\varepsilon)} \in L^{n/2}$ *and* $\|V_2^{(\varepsilon)}\|_\infty < \varepsilon$]. *Then* $\sigma_{ess}(-\Delta + V) = [0, \infty)$ $(n \geq 3)$.

One can apply Corollary 1.2 to Schrödinger operators:

THEOREM 1.4.

(a) *Let* $V \in L^{n/2} + L^\infty_\varepsilon$ *and let* V_- *be its negative part, i.e.* $V_- = \max(-V, 0)$. *Then*

$$N(V) \leq N(-V_-)$$

$$n_\ell(V) \leq n_\ell(-V_-) \text{ if } V \text{ is central .}$$

(b) *Let* $V, W \in L^{n/2} + L^\infty_\varepsilon$ *with* $V \leq W$ *pointwise. Then*

$$N(W) \leq N(V)$$

$$n_\ell(W) \leq n_\ell(V) \text{ if } V \text{ and } W \text{ are central.}$$

In addition to this result, the main input used by Bargmann is the following:

THEOREM 1.5. *Let* $V \in C_0^\infty(R^n)$ *be centrally symmetric. Fix* $\ell \geq 0$. *Let* u *be a solution of*

$$-u'' + \ell(\ell+1)r^{-2}u + Vu = 0; \quad u(0) = 0 .$$

Then $n_\ell(V)$ *is the number of zeroes of* u *on* $(0, \infty)$.

REMARKS.

1. This result is true under much greater generality than $V \in C_0^\infty$. However, for our purposes, this is enough. For a bound of the form $n_\ell(V) \leq (2\ell+1)^{-1} \int_0^\infty r|V(r)|\,dr$, once proven for $V \in C_0^\infty$ extends to all V by a simple limiting argument.

2. Theorem 1.5 follows by a simple min-max principle which exploits the Sturm comparison theorem; see [7]. Alternatively Theorem 1.5 can be proven by combining Levinson's theorem [8] with the method of variable phases [9]; see [7, 9].

Martin [10] has remarked on an interesting "local" comparison theorem:

THEOREM 1.6 (Martin's local comparison theorem). *Let* u *be any solution of* $-u'' + \ell(\ell+1)r^{-2}u + Vu = 0$. *Suppose that* $V \geq W$ *on* (a,b) *and that* u *has* n *zeroes in* (a,b). *Then* $n_\ell(W) \geq n-1$.

REMARKS.

1. The proof is simple. By a Sturm comparison theorem, any other solution of $-u'' + \ell(\ell+1)r^{-2}u + Vu = 0$ has at least $n-1$ zeroes in (a,b) and therefore by another Sturm argument, any solution of $-u'' + \ell(\ell+1)r^{-2} + Wu = 0$ has at least $n-1$ zeroes in (a,b).

2. As a typical application of this result, we note that so long as W is strictly negative in some interval (a,b), $\varliminf_{\lambda\to\infty} \lambda^{-1/2} n_\ell(\lambda W) > 0$, for compare with a square wall.

3. One can use Martin's principle to prove [52]: If $V(r)$ is a continuous function on $(0,\infty)$ and $\ell_{max}(\lambda)$ is the largest angular momentum for which $-\Delta+\lambda V$ has bound states on $\mathbf{R}^3$, then

$$\lim_{\lambda\to\infty} \ell_{max}(\lambda V)/\lambda^{1/2} = [-\min(r^2V)]^{1/2} .$$

Calegero [11] invented a very elegant method for exploiting Theorem 1.5. In case $\ell = 0$, it goes as follows: Let u solve $-u'' + Vu = 0$. Define $a(r)$ by

$$(a(r)+r)u'(r) = u(r) . \qquad (1)$$

Then, $a(r)$ obeys the Riccati equation

$$a'(r) = -V(r)[r+a(r)]^2 . \qquad (2)$$

Now, by (1), $a(r) \to 0$ as $r \to 0$, in fact, $a(r) = o(r)$. Moreover, if $V \leq 0$, (2) says that a is monotone increasing. A simple geometric argument (Figure 1) shows that the number of zeroes of u is identical to the number of "poles" of a. The idea is to introduce an auxiliary function which is a function of $a(r)$, use (2) to get a differential inequality which can be integrated.

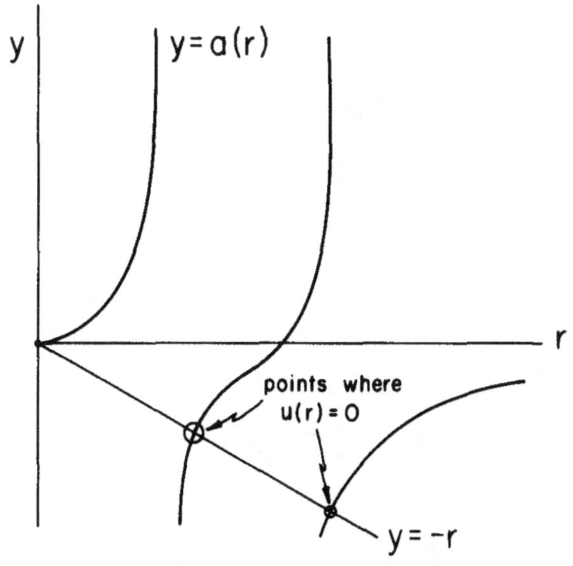

Fig 1

EXAMPLE 1. Let $b(r) = r^{-1}a(r)$. Then

$$b'(r) = -rV(r)[1+b(r)]^2 - r^{-1}b(r) . \qquad (3)$$

If $n_0(V) = n$, then $b(r)$ has poles at $p_1, \cdots, p_n$ and zeroes at $z_1 = 0$, $z_2, \cdots, z_n$ (and perhaps at a z_{n+1}) with $z_1 < p_1 < z_2 < p_2 < \cdots < p_n$. On

(z_i, p_i), b is positive, so by (3)

$$b'(r) \le r|V(r)|\,(b(r)+1)^2$$

so that

$$1 = \int_{z_i}^{p_i} -\frac{d}{dr}\left(\frac{1}{b+1}\right) dr \le \int_{z_i}^{p_i} r|V(r)|\,dr \ .$$

Summing over i, we get the $\ell = 0$ Bargmann bound [6]:

$$n_0(V) \le \int_0^\infty r|V(r)|\,dr \ .$$

EXAMPLE 2. Suppose that $V \le 0$ is smooth and $V'(r) \ge 0$. Define $\nu(r)$ by

$$\tan \nu(r) = (-V(r))^{1/2}\,(a(r)+r) \ .$$

Then ν obeys:

$$\nu'(r) = |V(r)|^{1/2} - \frac{1}{2}\,(V'(r)/|V(r)|)\,(\cos^2 \nu(r) \tan \nu(r)) \ .$$

Now, if $n_0(V) = n$, $(a(r)+r)$ has zeroes $z_1 = 0$, $z_2, \cdots, z_n$ and poles, $p_1, \cdots, p_n$ with $z_1 < p_1 < z_2 < \cdots < p_n$. In (z_i, p_i), $\tan \nu > 0$, so $\nu'(r) \le |V(r)|^{1/2}$. Thus

$$\frac{\pi}{2} = \int_{z_i}^{p_i} \nu'(r)\,dr \le \int_{z_i}^{p_i} |V(r)|^{1/2}\,dr \ .$$

Summing over i, we get Calegero's bound [11]:

$$n_0(V) \le \frac{2}{\pi} \int_0^\infty |V(r)|^{1/2} \ .$$

REMARKS.

1. For $\ell \neq 0$, one defines a_ℓ by:

$$u'(r)[r^{\ell+1} + a_\ell(r)\, r^{-\ell}] = u(r)[(\ell+1)r^{\ell} - \ell a_\ell(r)\, r^{-\ell-1}] .$$

Then a_ℓ obeys the Riccati equation:

$$a_\ell' = -(2\ell+1)^{-1} V(r)[r^{2\ell+1} + a_\ell(r)]^2 .$$

Bargmann's bound is proven by using $b_\ell = r^{-2\ell-1} a_\ell$; see [9], pp. 182-184.

2. There is a connection between $a(r)$ and the scattering length; in particular, $\lim_{r\to\infty} a(r)$ is the scattering length [11, 12].

We close this section by stating formally some of the bounds on $n_\ell(V)$:

THEOREM 1.7 (Bargmann [6])

$$(2\ell+1)\, n_\ell(V) \leq \int_0^\infty r|V_{-}(r)|\, dr .$$

Calegero [9, 11, 13] has proven a variety of bounds on $n_\ell(V)$ among which we mention:

THEOREM 1.8 (Calegero [11], Cohn [14]). *Suppose that* V *is negative and monotone increasing. Then:*

$$n_\ell(V) \leq \frac{2}{\pi} \int_0^\infty |V(r)|^{1/2}\, dr .$$

THEOREM 1.9 (Calegero [11]). *Let* $I_\rho = \int_0^\infty dr\, r^\rho |V(r)|$. *Then:*

$$n_\ell(V) \leq \frac{1}{2} + \frac{2}{\pi}(I_0 I_2)^{1/2}$$

$$n_\ell(V) \leq 1 + \frac{2}{\pi}(I_0 I_2 - I_1^2)^{1/2} .$$

Glaser et al., have proven:

THEOREM 1.10 ([15]). *For* $1 < p < \infty$:

$$(2\ell+1)^{2p-1} n_\ell(V) \le \frac{(p-1)^{p-1}}{p^p} \frac{\Gamma(2p)}{[\Gamma(p)]^2} \int_0^\infty r^{2p-1} |V(r)|^p \, dr \, .$$

Notice that as $p \downarrow 1$, this bound goes over to Bargmann's bound.

§2. *The Method of Birman and Schwinger*

In 1961, the Russian mathematician, M. Birman, and the American physicist, J. Schwinger, independently published almost identical proofs of the following theorem:

THEOREM 2.1 (Birman [16] - Schwinger [17]). *On* $\mathbf{R}^3$:

$$N(V) \le \frac{1}{(4\pi)^2} \int dx \, dy \, |x-y|^{-2} |V(x)| \, |V(y)| \, . \tag{4}$$

The first step in the proof is to note that:

LEMMA 2.2. E *is an eigenvalue of* $-\Delta + \lambda V$ *with* $V \le 0$, $\lambda > 0$, $E < 0$, *if and only if* λ^{-1} *is an eigenvalue of* $K_E = |V|^{1/2}(-\Delta - E)^{-1}|V|^{1/2}$ *and the multiplicities are equal.*

REMARK. *Formally* $(-\Delta + \lambda V)\psi = E\psi$ if and only if $K_E\phi = \lambda^{-1}\phi$ where $\phi = |V|^{-1/2} V\psi$. For a careful proof, see [3].

The second step is the simple but elegant:

LEMMA 2.3. *The number of eigenvalues of* $-\Delta + V$ *less than* $E < 0$, *is the number of* $\lambda \in (0, 1]$ *for which* E *is an eigenvalue of* $-\Delta + \lambda V$.

PROOF. Let $\mu_n(\lambda)$ be given by the min-max principle for $-\Delta + \lambda V$. Then $\mu_n(\lambda) \le 0$ for all λ and decreases as λ increases. Moreover, $\mu_n(\lambda) \uparrow 0$

as $\lambda \downarrow 0$. Thus the number of $\mu_n(1)$ less than E is identical to the number of μ_n for which $\mu_n(\lambda) = E$ for some $\lambda \in (0,1]$.

The two lemmas immediately imply the following basic "Birman-Schwinger" principle.

THEOREM 2.4. *Let* $V \leq 0$, $E < 0$. *The number of eigenvalues of* $-\Delta+V$ *in* $(-\infty, E]$ *is the same as the number of eigenvalues of* $K_E = |V|^{1/2}(-\Delta-E)^{-1}|V|^{1/2}$ *in* $[1,\infty)$ *(counting multiplicity).*

PROOF OF THEOREM 2.1. The number of eigenvalues of K_E larger than 1 is clearly dominated by the sum of the squares of the eigenvalues which equals $\mathrm{Tr}(K_E^2)$. Since $(-\Delta-E)^{-1}$ has an integral Kernel $(4\pi)^{-1}|x-y|^{-1}\exp(-\sqrt{-E}|x-y|)$, we see that:

$$N_E(V) \leq \frac{1}{(4\pi)^2}\int dx\, dy\, |x-y|^{-2}|V(x)|\,|V(y)|e^{-2k|x-y|} \tag{5}$$

where $E = -k^2$ and $N_E(V)$ is the number of eigenvalues in $(-\infty, E]$; (5) is also a bound of Birman-Schwinger. Taking $E \uparrow 0$, (4) results.

REMARKS.

1. By a classical inequality of Sobolev (see e.g. [1]), $\int dx\, dy\, |x-y|^{-2}|V(x)|\,|V(y)| \leq C\|V\|_{3/2}^2$ where $\|V\|_p = (\int |V|^p dx)^{1/2}$. Thus

$$N(V) \leq C\|V\|_{3/2}^2 \tag{6}$$

We return to this in Section 3C below.

2. See Ghirardi-Rimini [18], Fonda-Ghirardi [19] and Konno-Kuroda [20] for modification of the Birman-Schwinger bound.

§3. *Further Applications of the Birman-Schwinger Principle*

A. *Recovery of Bargmann's Bound* [16, 17]. Let h_0 be the operator $-\frac{d^2}{dr^2} + r^{-2}\ell(\ell+1)$ on $L^2(0,\infty)$ (with boundary condition $u(0) = 0$ if $\ell = 0$).

Then, as in the proof of Theorem 2.1,

$$n_\ell(V) \le \lim_{E\uparrow 0} \mathrm{Tr}(|V_-|^{1/2}(h_0-E)^{-1}|V_-|^{1/2})$$
$$= \mathrm{Tr}(|V_-|^{1/2}h_0^{-1}|V_-|^{1/2}) \ . \tag{7}$$

Now h_0^{-1} is an (unbounded) integral operator with kernel $(2\ell+1)^{-1}[\min(x,y)]^{\ell+1}[\max(x,y)]^{-\ell}$, so the trace in (7) is $\int_0^\infty (2\ell+1)^{-1}x|V_-(x)|dx$ which gives Bargmann's bound.

B. *Low-Dimensions.* Students in a first quantum mechanics course, learn that if V is a negative spherical square wall in R^n, then $-\Delta+\lambda V$ has bound states for all positive λ if $n = 1$ and has no bound states for small λ if $n = 3$. What about $n = 2$? There is some confusion about this question in the published and preprint literature – we first learned the correct answer from M. Kac. The Birman-Schwinger principle is ideal for studying this question:

THEOREM 3.1. *Consider* $-\Delta+\lambda V$ *on* $L^2(R^n)$ *for* $n = 1$ *or* 2. *Suppose* $V \le 0$ *and* $V \in L^p + L^q$ *with* $1 < q < \infty$ *and* $p > 1$ *if* $n = 2$, $p = 1$ *if* $n = 1$ [*in this case* $-\Delta+\lambda V$ *can be defined as a sum of forms,* K_E *is a bounded, compact operator and* $\sigma_{ess}(-\Delta+\lambda V) = [0,\infty)$]. *Then* $N(\lambda V) > 0$ *for all* $\lambda > 0$.

PROOF. By the Birman-Schwinger principle, we must show that for any $\lambda > 0$, there is $E < 0$, so that K_E has an eigenvalue larger than λ^{-1}. Since K_E is positive and compact, it clearly suffices to prove that $\lim_{E\uparrow 0} \|K_E\| = \infty$. This follows if we prove that $\lim_{E\uparrow 0} (\eta, K_E\eta) = \infty$ for some fixed $\eta \in L^2$. Let η be the characteristic function of some bounded set on which V obeys $a < V(x) < a^{-1}$ for some $a > 0$. Let $f = |V|^{1/2}\eta \in L^1 \cap L^2$, $f \ge 0$ so $\hat{f}$ is nonvanishing near 0. Thus

$$\lim_{E\uparrow 0} (\eta, K_E \eta) = \lim_{E\uparrow 0} \int |\hat{f}(p)|^2 (p^2 - E)^{-1} d^n p$$

$$= \int |\hat{f}(p)|^2 p^{-2} d^n p$$

diverges if $n = 1$ or 2.

REMARKS.

1. As we shall see in Section 5, if $V \in L^{n/2}(R^n)$ and $n \geq 3$, then $N(\lambda V) = 0$ if λ is small.

2. If $V \in C_0^\infty(R^n)$, $n = 1, 2$ with $V \leq 0$, then $N(\lambda V) = 1$ for λ small by the following argument: Place Neumann boundary conditions on a sphere, S, containing suppV. This can only increase $\dim P_{(-\infty,0)}$ (see, e.g. [7]). But S breaks R^n into a ball B and an exterior E. $-\Delta_N$ is positive on $L^2(E)$ and since $-\Delta_N$ on $L^2(B)$ has an isolated simple eigenvalue at 0, $-\Delta_N + \lambda V$ can have at most one negative eigenvalue for λ small.

3. Let $\|-\|$ be a translation invariant norm on $C_0^\infty(R^n)$ $(n = 1, 2)$. Then, given any m, ε, we can find $V \in C_0^\infty$ with $\|V\| < \varepsilon$ and $N(V) \geq m$. For pick any $f \in C_0^\infty$ with $f \leq 0$ and $\|f\| \leq \varepsilon/m$. Since $-\Delta + f$ has at least one eigenvalue, $-\Delta + V$ will have at least m if V is the sum of m translates of f all sufficiently far from one another. Thus, there is no bound if $n = 1, 2$ of the form $N(V) \leq$ function of a translation invariant norm. The situation is very different if $n \geq 3$ (see Section 3C).

4. If $n = 1$, we have the bound

$$N(V) \leq 2 + \int_{-\infty}^{\infty} |x|\, |V_-(x)|\, dx \tag{8}$$

for, let $-\Delta_D$ be the operator with Dirichlet boundary conditions at $x = 0$. Then, by Bargmann's bounds in each half-space, $-\Delta_D + V$ has at most $\int_{-\infty}^{\infty} |x|\, |V_-(x)|\, dx$ eigenvalues in $(-\infty, 0)$. Thus, since $-\Delta_D + V$ and $-\Delta + V$ are self-adjoint extensions of a common operator with deficiency indices $(2, 2)$, (8) follows.

5. Theorem 3.1 illustrates that Calegero's bound, Theorem 1.8 does not hold for all V.

6. It is false that if V is negative somewhere on $\mathbb{R}$, then $-d^2/dx^2+V$ has negative bound states. For example, if $V(x) = -1$ for $|x| < \pi/4$, $V(x) = 4$ for $\pi/4 \leq |x| \leq \pi/4+2$; then $-d^2/dx^2+V$ has no bound states. One can show [21] if $n = 1,2$ and $\int |x|^2 |V(x)| dx < \infty$ if $n = 1$ ($\int (1 + |x|^2)^\delta |V(x)| d^2x < \infty$ and $\int |V(x)|^{1+\delta} dx < \infty$ if $n = 2$), then $-\Delta + \lambda V$ has a bound state for all small λ if and only if $\int V(x) d^n x \leq 0$.

C. *Quasi-Classical and Almost Quasi-Classical Bounds.* The basic principle of the quasi-classical limit to quantum mechanics is that each bound state requires a volume h^3 in phase space. Thus, in units with $\hbar = 1 = 2m$, on $\mathbb{R}^n$

$$N_{c\ell}(V) = (2\pi)^{-n} \tau_n \int (V_-(x))^{n/2} dx \tag{9}$$

where τ_n is the volume of the unit ball in $\mathbb{R}^n$. $\tau_n \int (V_-(x))^{n/2} dx$ is the volume of phase space $\langle p, x \rangle$ where $p^2 + V(x) < 0$. As we discuss in Section 4, there is a sense in which $N(V)$ and $N_{c\ell}(V)$ become equal when V is large. There is a general conjecture which has been made by Glaser et al., [15], Simon [22] and E. Lieb [23]:

CONJECTURE. *Let* $n \geq 3$. *There is a constant* C_n *so that*

$$N(V) \leq C_n N_{c\ell}(V) \tag{10}$$

for all $V \in L^{n/2}$.

REMARKS.

1. In fact, Glaser et al., [15] suggest that $C_3 = 8/\sqrt{3}$ and prove (10) for $n = 3$ whenever $N(V) \leq 2$.

2. For $n = 1, 2$, (10) fails by our remarks in 3B.

3. As we shall see, (10) holds as $N(V) \to \infty$, in the sense that $\lim_{\lambda \to \infty} N(\lambda V)/N_{c\ell}(\lambda V) = 1$.

4. In general, for suitable C_n, (10) holds with $N(V) = 1$ (see Section 5) and then, by an argument of Glaser et al., [15], for $N(V) \leq 2$.

5. Simon [22] has proven (10) is equivalent to a natural conjecture involving "weak trace ideals."

6. By a limiting argument, (10) need only be proven for $V \in C_0^\infty$.

(10) says $N(V) \leq C_n' \|V_-\|_{n/2}^{n/2}$ where $\|V_-\|_p^p = \int |V_-|^p dx$. Using the Birman-Schwinger principle and interpolation theory for weak trace ideals (developed in [22]), one can prove:

THEOREM 3.2 ([22]). *Let* $n \geq 3$, $\varepsilon > 0$. *Then, there exists a constant* D_n, *so that*

$$N(V) \leq D_{n,\varepsilon}(\|V\|_{\frac{1}{2}n+\varepsilon} + \|V\|_{\frac{1}{2}n-\varepsilon})^{n/2} . \tag{11}$$

REMARK. As we discuss in Section 4, this estimate, unlike those of Bargmann and Birman-Schwinger, has the proper large coupling constant behavior.

D. *The Lieb-Thirring Bound.* In their beautiful paper on the stability of matter, Lieb and Thirring use the Birman-Schwinger bound (5), to prove:

THEOREM 3.3 ([24]). *Let* $n = 3$. *Let* $V \in L^{5/2}(R^3)$. *Let* $e_1(V) \leq e_2(V) \leq \cdots$ *be the negative eigenvalues of* $-\Delta + V$. *Then*

$$\sum_i |e_i(V)| \leq \frac{4}{15\pi} \int V^{5/2}(x) d^3x . \tag{12}$$

SKETCH OF PROOF. By (5) and the comparison theorem:

$$N_E(V) \leq N_{E/2}((V-E/2)_-)$$

$$\leq (4\pi)^{-2} \int (V-E/2)_-(x)(V-E/2)_-(y)|x-y|^{-2} e^{-\sqrt{2E}|x-y|}$$

$$\leq (4\pi\sqrt{2E})^{-1} \int |(V-\tfrac{1}{2}E)_-(x)|^2 dx$$

by Young's inequality. Now:

$$\sum_i |e_i(V)| = \int_{-\infty}^{0} |E|\, dN_E$$

$$= \int_{-\infty}^{0} N_E\, dE$$

$$\leq (4\pi\sqrt{2})^{-1} \int dx \int_0^{\infty} a^{-1/2}\, |(V+1/2a)\_(x)|^2\, da$$

which yields (12).

REMARKS.

1. Similar results hold for sums $\sum |e_i(V)|^{\nu}$ for other ν and for n different from 3, see [25].

2. This theorem is especially interesting since the quasi-classical value for $\sum |e_i(V)|$ is $(15\pi^2)^{-1} \int |V\_(x)|^{5/2}\, d^3x$.

§4. *Large Coupling Constant: The Quasi-Classical Limit*

The number of bound states of $-\Delta + \lambda V \cdot$ is the same as that for $-\lambda^{-1}\Delta + V$, so that large λ is the same as small $\hbar$. Thus one expects the quasi-classical approximation to be good. Martin [26] has proven:

THEOREM 4.1 ([26]). *If* V *is a Hölder continuous function of compact support, then,*

$$\lim_{\lambda \to \infty} N(\lambda V)/N_{c\ell}(\lambda V) = 1\ . \tag{13}$$

Martin uses the method of Dirichlet-Neumann bracketing [27, 7]. Independently of Martin, Tamura [28] proved (13) for a wider class of V.

Since $N_{c\ell}(\lambda V) = \lambda^{n/2} N_{c\ell}(V)$, (13) gives the large λ behavior of $N(\lambda V)$. It shows that for λ large, the Birman-Schwinger bound which $\sim c\lambda^2$ is not good. The advantage of the bound (11) is that it gives the proper large λ behavior. Using (11), one can prove:

THEOREM 4.2 ([22]). *Let* $n \geq 3$. *Let* $V \in L^{n+\varepsilon} \cap L^{n-\varepsilon}$. *Then* (13) *holds.*

§5. *Small Coupling Constant; When is* $N = 0$?

The question of when $N = 0$ was first asked by Jost and Pais [5]; more recently, it has been deeply studied by Glaser, Martin, Grösse and Thirring [15]. In the one-dimensional case, special interest is connected with this problem because of the following remark of Glaser et al., [15]:

THEOREM 5.1. *Let* $a(V) = \int_0^\infty f(x)|V(x)|^p$ *for* $f \geq 0$. *If* $a(V) < 1$ *implies that* $0 = N(V)$, *then, for any* V,

$$N(V) \leq a(V) .$$

PROOF. Let $n = N(V)$. Let u be the zero energy solution of the Schrodinger equation. Let $x_0 = 0,\ x_1, \cdots, x_n$ be its zeroes. Let $V_i = V\chi_i$ where χ_i is the characteristic function of (x_{i-1}, x_i). Then $u_i \equiv u\chi_i \in Q(-d^2/dx^2 + V_i)$, $(u_i, (-d^2/dx^2 + V_i)u_i) = 0$, so $N(V_i) \geq 1$. Thus $a(V_i) \geq 1$, so $a(V) \geq \sum_i a(V_i) \geq n$.

Thus the Jost-Pais result [5] implies Bargmann's bound [6]. The bounds of Glaser et al., [15] quoted as Theorem 1.10 are proven by using Theorem 5.1 and the results below.

THEOREM 5.2. *If* $\| |V|^{1/2}(-\Delta)^{-1}|V|^{1/2}\| < 1$, *then* $N(V) = 0$.

PROOF. This follows from the Birman-Schwinger principle, but also by an alternate proof [15]: If $\| |V|^{1/2}(-\Delta)^{-1}|V|^{1/2}\| < 1$, then $|V| \leq -\Delta$ so $-\Delta + V \geq 0$ which implies $N(V) = 0$.

THEOREM 5.3. *Let* $n \geq 3$ *and* $p \geq n/2$. *Then, there is a constant* $C_{n,p}$ *so that if*

$$C_{n,p} \int |x|^{2p-n} |V_-(x)|^p \, d^n x < 1 , \tag{14}$$

then $N(V) = 0$.

PROOF. It is well known that $r^{-2} \leq C_n p^2$ if $n \geq 3$. Thus $r^{-1}p^{-2}r^{-1}$ is bounded from L^2 to L^2. By Sobolev's inequality p^{-2} is bounded from L^{q_0} to $L^{q_0'}$ where $q_0^{-1} = 1/2 + 1/n$ and $p' = (1-p^{-1})^{-1}$. Thus, by the Stein interpolation theorem, $r^{-\alpha}p^{-2}r^{-\alpha}$ is bounded from L^{q_α} to $L^{q_\alpha'}$ where $q_\alpha^{-1} = 1/2 + n^{-1}(1-\alpha)$. As a result, if $r^{-\alpha}V^{1/2} \in L^{n/(1-\alpha)}$, $V^{1/2}(-\Delta)^{-1}V^{1/2}$ is bounded and if $\|r^{-\alpha}V^{1/2}\|_{n/1-\alpha} \leq D_{n,\alpha}$, then $N(V) = 0$. This is just (14).

REMARKS.

1. For $n = 3$, this result is independently due to Glaser et al., [15].

2. That $r^{-\alpha}p^{-2}r^{-\alpha}$ is bounded from L^{q_α} to $L^{q_\alpha'}$ is a result of Strichartz [29].

3. One can use the same argument if $n = 1$, $p \geq 1$ if we deal with the operator $-d^2/dx^2$ on $L^2(0,\infty)$ with boundary condition $u(0) = 0$: one interpolates between $r^{-1}p^{-2}r^{-1}$ bounded from L^2 to L^2 and that $r^{-1/2}p^{-2}r^{-1/2}$ is bounded from L^1 to L^∞: The later follows from the explicit kernel $x^{-1/2}\min(x,y)y^{-1/2}$ of the integral operator $r^{-1/2}p^{-2}r^{-1/2}$.

The above theorem leaves open the question of the best value of the constant $C_{n,p}$. Glaser et al., [15] proves:

THEOREM 5.4. $C_{3,p} = (p-1)^{p-1}\Gamma(2p)/4\pi p^p[\Gamma(p)]^2$. *For* $n = 1$ *with Dirichlet boundary conditions*:

$$\tilde{C}_{1,p} = (p-1)^{p-1}\Gamma(2p)/p^p[\Gamma(p)]^2 \ .$$

In the above, we have considered the question of when $N(V)$ is zero. If $N(V)$ is zero, there is some possibility that $-\Delta$ and $-\Delta+V$ are unitarily equivalent. This problem is discussed by Kato [30] (small coupling) and Lavine [31] (repulsive interactions).

§6. *Other Results*

We want to briefly describe some further results about $N(V)$ giving references to additional literature. The really interesting open questions involve n-body problems. There are no known general bounds on the number of eigenvalues below the continuum limit: As we shall see (C and D below), there are various pathologies which make obtaining such bounds difficult.

A. *How does* $N_E(V)$ *approach infinity if* $N(V) = \infty$? We have already seen that if $V \in L^{n/2}$ $(n \geq 3)$, then $N(V) < \infty$. As a complimentary result, one has the following:

THEOREM 6.1. *If* $V(r) \leq -C r^{-2+\varepsilon}$ *for* $|r| > R_0$ *for some* $C, \varepsilon > 0$, *then* $\mu_n(-\Delta+V) < 0$ *for all* n. *In particular,* $N(V) = \infty$.

REMARKS.

1. Of course, if $V \geq -C r^{-2-\varepsilon}$, then $N(V) < \infty$ by the $L^{n/2}$-bound.
2. One can actually show $V \leq -(c_n+\varepsilon) r^{-2}$ implies $N(V) = \infty$ and $V \geq -(c_n-\varepsilon) r^{-2}$ implies $N(V) < \infty$ for suitable c_n. $c_n = 1/4$ if $n = 1,3$.
3. For more details of the proof, see [27, 7].

SKETCH. Let ψ be a fixed function with support in $\{x \mid 1 \leq |x| \leq 2\}$. Let $\psi_n(x) = \psi(2^{-n}x)$. Then, for n large, $(\psi_n, (-\Delta+V)\psi_n) \leq c_1 n^{-2} - c_2 n^{-2+\varepsilon} < 0$ so one can find an infinite orthonormal set with $(\phi_n, H\phi_m) = 0$ if $n \neq m$ and $(\phi_n, H\phi_n) < 0$. It follows that $\mu_n(H) < 0$ for all n.

Suppose now that $\sigma_{ess}(-\Delta+V)$ is $[0, \infty)$. Then $N_E(V) \uparrow \infty$ as $E \uparrow 0$. One can ask how. As one might expect, this limit is one where quasi-classical (i.e. phase space) consideration predominate; see Brownell-Clark [32], McLeod [33], Tamura [34] or Reed-Simon [7].

B. N-*Body; Small Coupling*. For small coupling N-Body systems, one can show that N(V) is zero by the general principle:

THEOREM 6.2. *Let* V_{ij} $(1 \leq i < j \leq n)$ *be potentials on* R^3 *with* $N(\tfrac{1}{2}(n-1) V_{ij}) = 0$ *for all* i, j. *Then*

$$H = -\sum_{i=1}^{n} \Delta_i + \sum_{i<j} V_{ij}(r_i - r_j)$$

has spectrum contained in $[0, \infty)$.

PROOF. Since $-\Delta_i - \Delta_j = -1/2\Delta_{[ij]} - 2\Delta_{(ij)}$ where $[ij]$ is associated with $1/2(r_i + r_j)$ and (ij) with $r_j - r_i$. Thus

$$\sum -\Delta_i + \sum V_{ij} \geq \sum_{i<j} \frac{2}{n-1}(-\Delta_{(ij)} + V_{ij})$$

$$= \sum_{i<j} \frac{2}{n-1}\left(-\Delta_{(ij)} + \frac{n-1}{2} V_{ij}\right) \geq 0$$

by hypothesis.

For results on when $-\sum_i \Delta_i$ and $-\sum \Delta_i + \sum V_{ij}$ are unitarily equivalent, see Orio-O'Carroll [35] (small coupling) and Lavine [31] (repulsive potentials).

C. *The Effimov Effect.* Effimov [36] has suggested the following: Let V be a fixed short range spherically symmetric potential. Let

$$\tilde{H}(\lambda) = -\Delta_1 - \Delta_2 - \Delta_3 + \lambda[V(r_{21}) + V(r_{31}) + V(r_{23})]$$

and let $H(\lambda)$ be $\tilde{H}(\lambda)$ with the center of mass motion removed. Fix λ to be the coupling constant at which $-2\Delta + \lambda V$ has its first s-wave resonance. At this value of λ, it is claimed that $H(\lambda)$ has infinitely many bound states.

The point of this prediction is that the occurrence of an s-wave resonance sets up an effective long range force. The occurrence of such an effect shows the difficulty of establishing bounds on the number of bound states in the N-body case.

For further discussion of this effect, see Amado-Noble [37] and Yafeev [38].

D. *Two-Body Continuum Limit.* It is a basic theorem of the spectral theory of multiparticle Schrödinger operators that for a large class of potentials, the continuous spectrum of the Schrödinger operator is $[\Sigma, \infty)$ where Σ is determined as follows: let $C_1, \cdots, C_k$ be a breakup of the n-particles into clusters. Let $E_1, \cdots, E_k$ be eigenvalues of the Hamiltonian associated to clusters $C_1, \cdots, C_k$ with their center of mass motion removed (if $\#(C_i) = 1$, E_i must be 0). Then $\Sigma = \min(E_1 + \cdots + E_k)$ where the minimum is over all breakups into clusters and all choices of eigenvalues. This is a result of Hunziker [39], Van Winter [40] and Zhislin [41]; see also Jörgons-Weidmann [42] or Reed-Simon [7].

There is considerable information available about N(V) in case Σ is determined by a breakup into two clusters. An example of this is the case of atoms where one has the result of Zhislin [41].

THEOREM 6.3. *Atoms have an infinite number of bound states.*

REMARK. For Helium with an infinitely heavy nucleus, this is a result of Kato [43].

In case Σ is determined by a two cluster breakup, one has the following *intuition* [44]: If the sum of the potentials between the cluster is a long range two-body potential (i.e. $r^{-2+\varepsilon}$ at infinity), then the number of bound states is infinite. If this sum is a short range (i.e. $r^{-2-\varepsilon}$ at infinity), then there are only finitely many eigenvalues in $(-\infty, \Sigma)$. Under suitable technical hypothesis, the former was proven by Simon [44] and the latter by Combes [45]. One result of this is the following [44]:

THEOREM 6.4. *There exist three two body potentials* V_1, V_2, V_{12} *so that*

$$H(\lambda) = -\Delta_1 - \Delta_2 + \lambda V_1(r_1) + \lambda V_2(r_2) + \lambda V_{12}(r_1 - r_2)$$

has the following property. There are $0 < \lambda_1 < \lambda_2 < \cdots$ *so that if* $\lambda \in [0, \lambda_1) \cup (\lambda_2, \lambda_3) \cup \cdots \cup (\lambda_{2n}, \lambda_{2n+1}) \cdots$ $H(\lambda)$ *has finitely many eigen-*

values in $(-\infty, \Sigma(\lambda))$ *and if* $\lambda \in [\lambda_1, \lambda_2] \cup \cdots \cup [\lambda_{2n-1}, \lambda_{2n}] \cup \cdots$ *then* $H(\lambda)$ *has infinitely many eigenvalues in* $(-\infty, \Sigma(\lambda))$.

For additional information on bound states of N-body systems, see [46-51].

ADDED NOTE.

During the production of this book, three interesting new papers on N(V) appeared. A. Martin (CERN preprint) has proven that

$$N(V) \leq (2\pi)^{-1} (\|V\|_1^{1/2} \|V\|_2)$$

in three dimensions. E. Lieb (Princeton Preprint) and M. Cwikel (IAS Preprint) have proven the general conjecture discussed in the text, that for $n \geq 3$

$$N(V) \leq c_n \|V\|_{n/2}^{n/2} .$$

Both bounds have the correct large coupling constant behavior.

DEPARTMENTS OF MATHEMATICS AND PHYSICS
PRINCETON UNIVERSITY

REFERENCES

[1] M. Reed and B. Simon, *Methods of Modern Mathematical Physics, II, Fourier Analysis, Self-Adjointness*, Academic Press, 1975.

[2] W. Faris, Pac. J. Math. *22* (1967), 47-70.

[3] B. Simon, *Quantum Mechanics for Hamiltonians Defined as Quadratic Forms*, Princeton University Press, 1971.

[4] W. Faris, *Self-Adjoint Operators*, Springer Lecture No. 433, 1975.

[5] R. Jost and A. Pais, Phys. Rev. *82* (1951), 840.

[6] V. Bargmann, Proc. Nat. Acad. Sci. *38* (1952), 961-966.

[7] M. Reed and B. Simon, *Methods of Modern Mathematical Physics, III, Analysis of Operators*, Academic Press, in preparation.

[8] N. Levinson, Kgl. Dansk. Vid. Selsk. Mat. Fys. Medd. (1949) *25* No. 9.

[9] F. Calegero, *The Variable Phase Approach to Potential Scattering*, Academic Press, 1967, New York.

[10] A. Martin, Unpublished.

[11] F. Calegero, Comm. Math. Phys. *1* (1965), 80.

[12] R. Dashen, J. Math. Phys. *4* (1963), 388.

[13] F. Calegero, Nuovo Cimento *36* (1965), 199.

[14] P. Cohn, J. London Math. Soc. *41* (1966), 474.

[15] V. Glaser, A. Martin, H. Grösse and W. Thirring, this volume.

[16] M. S. Birman, Mat. Sb. *55* (1961), 125-174; Eng. trans. AMS Trans. *53* (1966), 23-80.

[17] J. Schwinger, Proc. Nat. Acad. Sci. *47* (1961), 122-129.

[18] G. C. Ghirardi and A. Rimini, J. Math. Phys. *6* (1965), 6.

[19] L. Fonda and G. C. Ghirardi, Nuovo Cimento *46* (1966), 47-58.

[20] R. Konno and S. T. Kuroda, J. Fac. Sci., Univ. Tokyo *13* (1966), 55-63.

[21] B. Simon, "The Bound State of Weakly Coupled Schrödinger Operators in One and Two Dimensions," Ann. Phys., to appear.

[22] B. Simon, "Weak Trace Ideals and the Bound States of Schrödinger Operators," Trans. AMS, to appear.

[23] E. Lieb, private communication.

[24] E. Lieb and W. Thirring, Phys. Rev. Lett. *35* (1975), 687.

[25] E. Lieb and W. Thirring, this volume.

[26] A. Martin, Helv. Phys. Acta. *45* (1972), 140-148.

[27] R. Courant and D. Hilbert, *Methods of Mathematical Physics*, Vol. I, Interscience, 1953.

[28] H. Tamura, Proc. Japan Acad. *50* (1974), 19-22.

[29] R. Strichartz, J. Math. Mech. *16* (1967), 1031-1060.

[30] T. Kato, Math. Ann. *162* (1966), 258-279.

[31] R. Lavine, Comm. Math. Phys. *20* (1971), 301-323.

[32] F. Brownell-C. Clark, J. Math. Mech. *10* (1961), 31-70.

[33] J. McLeod, Proc. London Math. Soc. *11* (1961), 139-158.

[34] H. Tamura, Proc. Japan Acad. *50* (1974), 185-187.

[35] R. J. Iorio and M. O'Carroll, Comm. Math. Phys. *27* (1972), 137.

[36] V. Effimov, Phys. Lett. *33B* (1970), 563; Sov. J. Phys. *12* (1971), 589.

[37] R. Amado and J. Noble, Phys. Lett. *35B* (1971), 25, Phys. Rev. *D5* (1972), 1992-2002.

[38] P. Yafeev, Math. Sb. *94* (1974), 567-593.

[39] W. Hunziker, Helv. Phys. Acta *39*, (1966), 451.

[40] C. Van Winter, Kgl. Donsk. Vid. Selsk, *2* No. 8, 1964.

[41] G. M. Zhislin, Trudy Mosk. Math. Obsc. *9* (1960), 82.

[42] K. Jörgens and J. Weidmann, *Spectral Properties of Hamiltonian Operators*, Springer Lecture No. 313, 1973.

[43] T. Kato, Trans. Am. Math. Soc. *70* (1951), 212.

[44] B. Simon, Helv. Phys. Acta, *43* (1970), 607-630.

[45] J. M. Combes, in preparation.

[46] G. M. Zhislin and A. G. Sigalov, Izv. Akad. Nauk SSR *29* (1965), 834. (AMS. Trans. *91*, 297).

[47] E. Balslev, Ann. of Phys. *73* (1972), 49-107.

[48] I. Uchiyama, Publ. Res. Inst. Math. Sci. Kyoto *5* (1969), 61-62.

[49] G. M. Zhislin, AMS Trans. -Math USSR Izv. *3* (1969), 559 (Russian original: Izv. Nauk. SSSR *33* (1969), 590).

[50] G. M. Zhislin, J. Theo. Math. Phys. (Russian) *7* (1971).

[51] A. G. Sigalov, Russian Math Surveys *22* (1967), #2, 1-18.

[52] B. Simon, unpublished.

QUANTUM DYNAMICS:
FROM AUTOMORPHISM TO HAMILTONIAN

Barry Simon*

We describe the mathematical arguments involved in passing from a one-parameter measurable group of automorphisms of the basic quantum structures to the Schrödinger equation.

§1. *Introduction*

Every student of quantum mechanics learns in a first course that quantum dynamics is governed by the Schrödinger equation $i\hbar\dot{\psi} = H\psi$. However, even professional quantum mechanics who have delved into the axiomatic foundations of quantum theory are sometimes unaware of the full chain of argument leading from the primitive version of dynamics as a one-parameter continuous (or measurable) group of automorphisms of the axiomatic structure to the Schrödinger equation. Our goal in this note is to put down in one place this full chain of argument. We expect the experts will find nothing new here and we do not intend a review of the literature. This note will have served its purpose if the student of the foundations of quantum theory is able to find here, in one place, things which would have formerly taken him into five or six research articles. The traditional route from continuous automorphisms takes the following steps:

(1) *Wigner's Theorem.* Every automorphism is induced by a unitary or antiunitary, uniquely determined up to a phase.

(2) *Bargmann-Wigner Theorem.* Given a one-parameter continuous group of automorphisms, the phases of step (1) can be chosen so that the corresponding family of unitaries depends continuously on the parameter.

*A. Sloan Fellow; research supported in part by USNSF under Grant GP 39048.

(3) *Multiplier Problem for* R. Steps (1) and (2) lead to a family of unitaries U with $U(a)U(b) = \omega(a,b)U(a+b)$. There is a function $\lambda(a)$ with $\omega(a,b) = \lambda(a+b)\lambda(a)^{-1}\lambda(b)^{-1}$ so that $\tilde{U}(a) = U(a)\lambda(a)$ obeys $\tilde{U}(a)\tilde{U}(b) = \tilde{U}(a+b)$.

(4) *Stone's Theorem.* Every one-parameter strongly continuous group of unitaries is of the form $U(a) = \exp(-iaA)$ for some self-adjoint operator A.

We intend to follow a slightly longer route to prove a result slightly stronger in two ways. First we wish to consider several different meanings of automorphism which *a priori* might be very different. Step (1) is then several theorems including results of Wigner [1] and Kadison [2]. Secondly, we wish to assume *a priori* only that the automorphisms are (Borel) measurable so that steps (2) and (3) take on a slightly different content. Before step (4) we must insert

(3½) *Von Neumann's Theorem.* Every weakly measurable unitary representation of R is strongly continuous.

§2. *What is a Quantum Automorphism?*

Let $\mathcal{H}$ be a *separable* Hilbert space. Depending on which axiom scheme one adopts, one is led to various *a priori* notions of automorphism:

(1) *Wigner automorphism*: Let $P(\mathcal{H})$ be the complex projective space for $\mathcal{H}$, i.e. identify $\psi, \eta \in \mathcal{H}$ with $\|\psi\| = \|\eta\| = 1$ if $\psi = a\eta$ for some $a = e^{i\theta} \in \mathbb{C}$. $P(\mathcal{H})$ is the family of equivalence classes under this relation. By definition, $<[\psi],[\eta]> = |<\psi,\eta>|$. A *Wigner automorphism* is a bijection $a : P(\mathcal{H}) \to P(\mathcal{H})$ such that $<a[\psi],a[\eta]> = <[\psi],[\eta]>$ · $t \to a_t$ is called *measurable* if $t \to <a_t[\psi],[\eta]>$ is measurable for all $[\psi],[\eta] \in P(\mathcal{H})$. An alternative way of describing $P(\mathcal{H})$ is as the space of all projections P_ψ of rank one. Then $<[\psi],[\eta]> = \mathrm{Tr}(P_\psi P_\eta)^{1/2}$.

(2) *Kadison automorphism.* Let $\mathcal{S}(\mathcal{H})$ denote the set of density matrices on $\mathcal{H}$, i.e. trace class operators ρ with $\mathrm{Tr}(\rho) = 1$, $\rho \geq 0$. $\mathcal{S}(\mathcal{H})$ is a convex set and a *Kadison automorphism* is a map $\beta : \mathcal{S}(\mathcal{H}) \to \mathcal{S}(\mathcal{H})$ which is a bijection and which is affine, i.e.

$$\beta(t\rho_1 + (1-t)\rho_2) = t\beta(\rho_1) + (1-t)\beta(\rho_2)$$

all $\rho_1, \rho_2 \in \mathcal{S}(\mathcal{H})$, $0 \le t \le 1$. A family $t \to \beta_t$ is called measurable if and only if $t \to \mathrm{Tr}(\beta_t(\rho)A)$ is measurable for all $\rho \in \mathcal{S}$, $A \in \mathcal{B}(\mathcal{H})$, the bounded operators on $\mathcal{H}$.

(3) *Segal automorphism.* Let $\mathcal{B}_r(\mathcal{H})$ denote the bounded self-adjoint operators on $\mathcal{H}$ endowed with the natural linear structure and the Jordan product:

$$A \circ B = 1/2\,(AB + BA) \ .$$

A Segal automorphism is a bijection $\gamma : \mathcal{B}_r(\mathcal{H}) \to \mathcal{B}_r(\mathcal{H})$ so that γ is linear and

$$\gamma(A \circ B) = \gamma(A) \circ \gamma(B) \ .$$

If this condition is only assumed for commuting A and B, we call γ a weak Segal automorphism. $t \to \gamma_t$ is called measurable, if and only if $t \to (\psi, \gamma_t(A)\psi)$ is measurable for all $A \in \mathcal{B}_r(\mathcal{H})$ and $\psi \in \mathcal{H}$.

Wigner automorphisms arise in a framework of analyzing quantum mechanics in terms generalized Stern-Gerlach experiments and overlap probabilities (see Ax [3] for a recent treatment). Kadison automorphisms arise in a framework describing general states and observables (see von Neumann [4] or Mackey [5]). Segal automorphisms arise in a framework that emphasizes the structure of observables (see e.g. Segal [6]).

A priori the four types of automorphisms appear very different although each does seem to capture an important property of a symmetry. Wigner automorphisms preserve the basic objects of the theory as viewed from an overlap probability point of view. Kadison automorphisms are based on the interpretation of convex combinations of states as statistical mixtures. Weak Segal automorphisms are based on the notion of expectations of commuting observables. I see no simple physical reason why these families of automorphisms are essentially the same; in fact, if "automorphism" is replaced by "endomorphism" (bijection no longer required), then the families are no longer the same! Nevertheless, one has:

THEOREM 2.1 (Wigner's Theorem). *Every Wigner automorphism is of the form*

$$\alpha[\psi] = [U\psi]$$

where U *is either unitary or antiunitary and is uniquely determined up to one overall phase, i.e. if* $\alpha[\psi] = [U'\psi]$, then $U' = aU$ with $a = e^{i\theta}$.

THEOREM 2.2 (Kadison's Theorem). *Every Kadison automorphism is of the form:*

$$\beta(\rho) = U\rho U^*$$

for a unitary or antiunitary map U *uniquely determined up to a phase.*

THEOREM 2.3. *Every weak Segal automorphism is a strong Segal automorphism and is of the form*

$$\gamma(A) = U^*AU$$

for a unitary or antiunitary map U *uniquely determined up to a phase.*

We remark that Wigner's Theorem looks more like the others if we write it in terms of projections:

$$P_{\alpha(\psi)} = UP_\psi U^* .$$

We write Theorems 2.2 and 2.3 as $U\rho U^*$ and U^*AU so that $\mathrm{Tr}(\beta(\rho)A) = \mathrm{Tr}(\rho\gamma(A))$, i.e. the β and γ associated to a fixed U are related by the distinction between the Schrödinger and Heisenberg pictures.

These theorems have been compared (e.g. [5]) to the fundamental theorem of projective geometry [7]; this theorem says that given any map $\alpha : P(V) \to P(V)$, the projective space of a finite dimensional vector space V over a field F, which takes lines in P(V) into lines we can find $U : V \to V$ so that $\alpha([\psi]) = [U\psi]$. U will be additive and obey $U(a\psi) = m(a)(U\psi)$ for some automorphism m of F. Galois theory yields many automorphisms of C other than the identity and complex conjugation, so the theorems, while related, are certainly not equivalent; we discuss this further in Section 4.

The main "super theorem" of this note is the following:

THEOREM 2.4. *Let* $t \to \alpha_t$ (resp. β_t, γ_t) *be a map from* R *to the Wigner (resp. Kadison, Segal, weak Segal) automorphisms which is measurable and obeys* $\alpha_t \alpha_s = \alpha_{t+s}$ *(resp. for* β *or* γ*). Then, there exists a self-adjoint operator* H *(not necessarily bounded) unique up to an overall additive constant so that*

$$\alpha_t[\psi] = [e^{-iHt}\psi]$$

$$(\text{resp. } \beta_t(\rho) = e^{-itH}\rho\, e^{+itH}\,;\ \gamma_t(A) = e^{itH} A\, e^{-itH})\ .$$

§3. *The Two-Dimensional Case*

The two-dimensional case of Theorems 2.1-3 is not only a preliminary step in the general case, but provides a simple insight into the theorems: the unitary vs. antiunitary choice is related to the fact that isometries of Euclidean three space are either orientation preserving ("pure rotations") or orientation reversing ("reflections"). This remark and much of our discussion in this section follows Hunziker [8].

All three types of automorphisms involve self-adjoint operators, so we will exploit the fact that the matrices

$$1 = \begin{pmatrix} 1 & 0 \\ 0 & 1 \end{pmatrix}; \qquad \sigma_3 = \begin{pmatrix} 1 & 0 \\ 0 & -1 \end{pmatrix}$$

$$\sigma_1 = \begin{pmatrix} 0 & 1 \\ 1 & 0 \end{pmatrix}; \qquad \sigma_2 = \begin{pmatrix} 0 & i \\ -i & 0 \end{pmatrix}$$

are a basis for $\mathcal{B}_r(C^2)$, as a *real* vector space. If

$$A = a + \vec{a}\cdot\vec{\sigma}$$

then

$$UAU^* = a + \overrightarrow{R(U)a}\cdot\vec{\sigma}$$

where R(U) is an orientation preserving (resp. reversing) isometry if U is unitary (resp. antiunitary). The map R is onto all isometries and $R(U) = R(U')$ if and only if $U = e^{i\theta}U'$. We will not prove all these facts from scratch, but assume the reader is familiar with the two-to-one map of SU(2) onto SO(3). Given this, the general unitary case follows from the fact that any unitary U is of the form $e^{i\theta}\tilde{U}$ with $\tilde{U} \in SU(2)$ and $e^{i\theta}$ determined up to ± 1. The antiunitary case follows by noting that if U is the natural complex conjugation on $\mathbf{C}^2$, then R(U) is just reflection in the plane, orthogonal to the 2 axis and that any antiunitary is the product of a unitary and the complex conjugation.

(1) *Wigner Automorphisms*. The rank one projections are precisely of the form

$$P(\vec{a}) = 1/2(1+\vec{a}\cdot\vec{\sigma}); \quad |\vec{a}| = 1$$

so any Wigner automorphism is associated with a bijection of the unit sphere in $\mathbf{R}^3$. Since

$$2\mathrm{Tr}(P(\vec{a})P(\vec{b})) = 1 + \vec{a}\cdot\vec{b},$$

this map is an isometry of the sphere and hence a "pure rotation" or "relection."

(2) *Kadison Automorphisms*. The $\mathcal{S}(\mathbf{C}^2)$ operators are precisely of the form

$$\frac{1}{2}(1+\vec{a}\cdot\vec{\sigma}); \quad |\vec{a}| \leq 1$$

so any Kadison automorphism is associated with an affine map of the unit sphere *onto* itself. Such a map is isometry of $\mathbf{R}^3$.

(3) *Segal Automorphisms*. Let γ be a weak-Segal automorphism. Then γ clearly takes the orthogonal projections onto themselves since it preserves $P^2 = P$. Since 1 is the unique projection of the form $P_1 + P_2$ for non-zero projections P_1, P_2; $\gamma(1) = 1$. Thus γ also induce in a map of the rank one projections to themselves and so of the unit sphere in $\mathbf{R}^3$ to itself, i.e.,

$$\gamma(a+\vec{a}\cdot\vec{\sigma}) = a + \overrightarrow{Ma}\cdot\vec{\sigma}$$

where M is a linear map taking the unit sphere to itself. M is thus an isometry of R^3. Since every weak-Segal automorphism is of the requisite form, a fortiori every Segal automorphism is of the requisite form.

§4. *Wigner's Theorem*

In this section we prove Theorem 2.1. We use the letters $p, q, \cdots$ to denote points in $P(\mathcal{H})$, i.e. "rays" in $\mathcal{H}$. The linear structure of $\mathcal{H}$ induces a geometric structure on $P(\mathcal{H})$, as is well-known [7]; if $M \subset \mathcal{H}$ is a subspace of $\mathcal{H}$, we denote by $P(M)$, those $p \in P(\mathcal{H})$ which arise from rays in M.

LEMMA 4.1. *If a is a Wigner automorphism and $M \subset \mathcal{H}$ is a k-dimensional subspace, then there is an M' (denoted by $a(M)$) also of dimension k so that $a(p) \in M'$ if and only if $p \in M$.*

PROOF. Let $\phi_1, \cdots, \phi_k$ be an orthonormal basis for M (we write $M = [\phi_1, \cdots, \phi_k]$). Pick representatives $\psi_1, \cdots, \psi_k$ for $a([\phi_1]), \cdots, a([\phi_k])$ and let M' be their span. Since the ψ_i are orthonormal, M' has dimension k. Now $p \in M$ if and only if $\sum_{i=1}^{k} <p, [\phi_i]>^2 = 1$ if and only if $\sum_{i=1}^{k} <a(p), a[\phi_i]>^2 = 1$ if and only if $a(p) \in M'$. ∎

REMARK. In the language of projective geometry [7], we have just proven that a defines a collinealition. Below we reduce Wigner's Theorem to the three dimensional case. We could prove this three dimensional case by appealing to the fundamental theorem of projective geometry. This would leave open an arbitrary automorphism of C which we could prove was either the identity or the conjugation by appealing to the two-dimensional case (Section 3).

LEMMA 4.2. *For any two dimensional $M \subset \mathcal{H}$ and any Wigner automorphism, a, there exists a unitary or antiunitary map $U : M \to a(M)$, unique up to phase, so that*

$$a[\phi] = [U\phi]$$

for all $\phi \in M$.

PROOF. Let ϕ_1, ϕ_2 be an orthonormal basis for M and choose ψ_1, ψ_2 so that $a[\phi_i] = [\psi_i]$. Define a unitary $V: a(M) \to M$ by $V\psi_i = \phi_i$. Then $\beta: P(M) \to P(M)$, given by $\beta = a_V \circ a$ with $a_V[\psi] = [V\psi]$ is a Wigner automorphism. The lemma follows by appealing to the two dimensional case (Section 3). ∎

LEMMA 4.3. *If the map of Lemma 4.2 is unitary (resp. antiunitary) for one* $M \subset \mathcal{H}$, *it is unitary (resp. antiunitary) for all two dimensional* $M' \subset \mathcal{H}$.

PROOF. There is a direct algebraic proof (see page 89 of [7]), but we prefer to develop a suggestion of Bargmann. Let us topologize $P(\mathcal{H})$ by putting the metric

$$\rho(p,q) = 1 - \langle p,q \rangle$$

on it (to check it is a metric, we note for any $\psi \in p$, there is an $\eta \in q$ so that $\frac{1}{2}\|\psi - \eta\| = \rho(p,q)$ and use the triangle inequality on $\mathcal{H}$). Clearly, any Wigner automorphism is an isometry on $P(\mathcal{H})$ and so continuous.

Given $p, q, r \in P(\mathcal{H})$, following Bargmann [9], we define a number $\chi(p,q,r)$ by choosing $\psi \in p$, $\eta \in q$, $\gamma \in r$ and letting

$$\chi(p,q,r) = \langle \psi, \eta \rangle \langle \eta, \gamma \rangle \langle \gamma, \psi \rangle$$

and noting that χ is independent of choice. χ is jointly continuous, since given $p_n \to p$, we can find $\psi_n \in p_n$, $\psi \in p$ so that $\|\psi_n - \psi\| \to 0$.

Let ψ and ϕ be orthonormal vectors and let p, q, r be given by

$$p = [\psi]; \quad q = [\eta], \quad \eta = \frac{1}{\sqrt{2}}(\psi - \phi); \quad r = [\gamma], \quad \gamma = \frac{1}{\sqrt{2}}(\psi - i\phi) .$$

Then $\chi(p,q,r) = \frac{1}{4}(1+i)$. Let M' be the two dimensional space spanned by ψ, ϕ. Then the map induced by a via Lemma 4.2 is unitary (resp. antiunitary) if and only if $\chi(a(p), a(q), a(r)) = \frac{1}{4}(1+i)$ (resp. $\frac{1}{4}(1-i)$). Given another M'' we can continuously vary ψ and ϕ to ψ', ϕ' generating M'' and so p, q, r to p', q', r' so by continuity $\chi(a(p), a(q), a(r))$ will be $\frac{1}{4}(1+i)$ if and only if $\chi(a(p'), a(q'), a(r')) = \frac{1}{4}(1+i)$. ∎

LEMMA 4.4. *Let $\mathcal{H}$ have dimension 3. Let α be a Wigner automorphism and ϕ_1, ϕ_2, ϕ_3 a basis for $\mathcal{H}$. If α leaves the subsets $P([\phi_1, \phi_2])$ and $P([\phi_1, \phi_3])$ pointwise fixed, then α is the identity.*

PROOF. Let $\eta = a\phi_1 + b\phi_2 + c\phi_3$ with $a \neq 0$ and real. We will first prove that $\alpha([\eta]) = [\eta]$. Pick $\eta' \in \alpha[\eta]$ with $(\phi_1, \eta') = a$ which fixes η'. Set $b' = (\phi_2, \eta')$. Now since $([\eta], [\psi]) = ([\eta'], [\psi'])$ for $\psi = \phi_2$, $\frac{1}{\sqrt{2}}(\phi_1 + \phi_2)$ and $\frac{1}{\sqrt{2}}(\phi_1 + i\phi_2)$, we have

$$|b| = |b'|\,; \quad |a+b| = |a+b'|\,; \quad |a-ib| = |a-ib'|\ .$$

For fixed b, the last three equations on b' require b' to lie on three circles which intersect only in one point. Thus $b' = b$. Similarly $c' = c$ so $\eta' = \eta$. The $a = 0$ case follows by continuity. ∎

LEMMA 4.5. *Wigner's theorem holds in case* $\dim \mathcal{H} = 3$.

PROOF. We only prove existence; uniqueness follows as in the general proof below. Given U unitary or antiunitary, let α_U be the induced Wigner automorphism. Given α, a Wigner automorphism, it clearly suffices to find $U_1, \cdots, U_k$ so that $\alpha_{U_k} \circ \cdots \circ \alpha_{U_1} \circ \alpha = \mathrm{id}$ since then $\alpha = \alpha_U$ with $U = U_1^{-1} \cdots U_k^{-1}$. Pick a basis ϕ_1, ϕ_2, ϕ_3 for $\mathcal{H}$ and ψ_1, ψ_2, ψ_3 so that $\alpha([\phi_i]) = [\psi_i]$. Define U_1 to be the unique unitary (resp. antiunitary) with $U_1\psi_i = \phi_i$ if α is unitary (resp. antiunitary) on two dimensional subspaces. Let $\alpha_1 = \alpha_{U_1} \circ \alpha$. Then $\alpha_1([\phi_i]) = [\phi_i]$ and is unitary on two dimensional subspaces. Now, by the two dimensional theorem, we can find a unitary $V_1 : [\phi_1, \phi_2] \to [\phi_1, \phi_2]$ so that $\alpha_1 \restriction [\phi_1, \phi_2] = \alpha_{V_1}$. Let $U_2 = V_1^{-1}$ on $[\phi_1, \phi_2]$ and the identity on $[\phi_3]$. Then $\alpha_2 = \alpha_{U_2} \circ \alpha_1$ is the identity on $[\phi_1, \phi_2]$ and on $[\phi_3]$. Applying the two dimensional case to $[\phi_1, \phi_3]$, we can find $V_2 : [\phi_1, \phi_3] \to [\phi_1, \phi_3]$ so that $\alpha_2 \restriction [\phi_1, \phi_3] = \alpha_{V_2}$ and we can fix the phase so that $V_2\phi_1 = \phi_1$. Let $U_3 = V_2^{-1}$ on $[\phi_1, \phi_3]$ and the identity on $[\phi_2]$ (and so on all of $[\phi_1, \phi_2]$). Then $\alpha_{U_3} \circ \alpha_2$ is the identity on $[\phi_1, \phi_2]$, and $[\phi_1, \phi_3]$ and so the identity by Lemma 4.4. ∎

LEMMA 4.6. *For any three dimensional* $M \subset \mathcal{H}$ *and Wigner automorphism* α, *there is a unitary or antiunitary* $U : M \to \alpha(M)$ *so that* $\alpha[\psi] = [U\psi]$ *for all* $\psi \in M$.

PROOF. As in Lemma 4.2.

PROOF OF THEOREM 2.1. Without loss, suppose α is unitary on two-dimensional spaces. Fix $\phi \in \mathcal{H}$ and $\psi \in \alpha[\phi]$. Given any $\eta \in \mathcal{H}$, let M be the span of ϕ and η. By Lemma 4.2, find $V_M : M \to \alpha(M)$ inducing α on M with its phase determined by $V_M\phi = \psi$. Define $U\eta$ to be $V_M\eta$. We must prove U is linear, so given $\eta_1, \eta_2 \in \mathcal{H}$, let N be the span of ϕ, η_1, η_2. By Lemma 4.6 we can find a unitary $W : N \to \alpha(N)$ (it can't be antiunitary since its restrictions to two dimensional spaces must be unitary!) inducing α on N. By change of phase we can suppose $W\phi = \psi$. Since the restriction of W to any two dimensional $M \subset N$ is unitary, $W \restriction M = V_M$ and so $U \restriction N = W$. Since $W(a\eta_1 + b\eta_2) = aW(\eta_1) + bW(\eta_2)$, U is linear. By construction U is norm preserving. In the above construction only the phase of ψ is arbitrary. ∎

§5. *Kadison's Theorem*

Following Roberts-Roepstorff [10], we prove Theorem 2.2 by reducing it to Theorem 2.1. This reduction is essentially part of an argument of Hunziker [8]. For $M \subset \mathcal{H}$, a subspace, $\mathcal{S}(M)$ is the subset of $\mathcal{S}(\mathcal{H})$ of those ρ with $\operatorname{Ran} \rho \subset M$.

LEMMA 5.1. *Let* β *be a Kadison automorphism. Then for any two dimensional subspace* $M \subset \mathcal{H}$, *there is a two-dimensional subspace* $\beta(M) \subset \mathcal{H}$ *so that* $\beta[\mathcal{S}(M)] = \mathcal{S}(\beta(M))$.

PROOF. $\mathcal{S}(M)$ is a face of the convex set $\mathcal{S}(\mathcal{H})$ with the property: there exist two extreme points $u, v \in \mathcal{S}(\mathcal{H})$ so that $\mathcal{S}(M)$ is the smallest face containing u and v. Moreover any such face with more than one point is $\mathcal{S}(M')$ for some M' (for if $u = P_\phi$, $v = P_\psi$, take $M' = [\phi, \psi]$). Since β

is a convex automorphism, it preserves the structure of faces and in particular $\beta[\mathcal{S}(M)]$ is $\mathcal{S}(M')$ for some M'. ∎

REMARKS.

1. This proof extends to all finite-dimensional subspaces.
2. For an alternate proof, see Hunziker [8].

LEMMA 5.2. *Given any Kadison automorphism* β, *there is a Wigner automorphism* α *with* $\beta(P_\psi) = P_{\alpha(\psi)}$ *for every one dimensional projection* $P_\psi \in \mathcal{S}(\mathcal{H})$.

PROOF. Given β, we note that since the P_ψ are the extreme points of $\mathcal{S}(\mathcal{H})$, $\beta(P_\psi) = P_{\alpha(\psi)}$ for some map α on the rays. We must prove that α preserves the inner product on $P(\mathcal{H})$. Given $\phi, \psi \in \mathcal{H}$, let M be span of ϕ and ψ. By composing β with the β induced by a unitary U which maps $\beta(M)$ into M, we obtain $\tilde{\beta} = \beta_u \circ \beta$ leaving $\mathcal{S}(M)$ invariant. By the two dimensional case, $\tilde{\beta}$ is induced by a unitary, so there is a unitary or antiunitary $V: M \to \beta(M)$ so that $\beta(\rho) = V\rho V^*$ for $\rho \in \mathcal{S}(M)$ and thus β preserves $\mathrm{Tr}(P_\psi P_\phi)$, i.e. $(\alpha[\psi], \alpha[\phi]) = ([\psi], [\phi])$. ∎

LEMMA 5.3. *If* β *is a Kadison automorphism which leaves each extreme point fixed, then* β *is the identity.*

PROOF. We need only prove that β is continuous in trace-norm topology since any $\rho \in \mathcal{S}(\mathcal{H})$ has an expansion $\sum_{i=1}^{\infty} t_i P_{\psi_i}$ converging in trace norm and so is a limit of finite convex combinations of the P_{ψ_i}. Now β extends to the positive trace class operators by defining $\beta_{ext}(A) = \mathrm{Tr}(A)\beta(A/\mathrm{Tr}(A))$. β_{ext} obeys $\beta_{ext}(A+B) = \beta_{ext}(A) + \beta_{ext}(B)$ (since β is affine), $\beta_{ext}(\lambda A) = \lambda\beta_{ext}(A)$ for $\lambda \geq 0$ and

$$\mathrm{Tr}(\beta_{ext}(A)) = \mathrm{Tr}(A) .$$

We define β on all self-adjoint trace class operators by $\tilde{\beta}(A) = \beta_{ext}(A_+) - \beta_{ext}(A_-)$ where A_+ and A_- are the positive and negative parts of A.

Thus letting $\|A\|_1 = \mathrm{Tr}(|A|)$:

$$\|\tilde{\beta}(A)\|_1 \leq \|\beta_{ext}(A_+)\|_1 + \|\beta_{ext}(A_-)\|_1$$
$$= \mathrm{Tr}(A_+) + \mathrm{Tr}(A_-) = \|A\|_1 .$$

Since $\tilde{\beta}$ is linear, β is continuous in $\|-\|_1$ and so β is continuous on $\mathcal{S}(\mathcal{H})$. ∎

PROOF OF THEOREM 2.2. By Wigner's theorem and Lemma 5.2, we can find U, unitary or antiunitary so that $\beta_U \circ \beta$ is the identity on all P_ψ. Thus by Lemma 5.3, $\beta_U \circ \beta = \mathrm{id}$, i.e. $\beta = \beta_{U^{-1}}$. ∎

§6. *The Structure of Segal Automorphisms*

We will prove Theorem 2.3 by reducing it to Wigner's theorem. We first note:

LEMMA 6.1. *Let γ be a weak Segal automorphism. Then γ is order preserving (i.e. $A \geq B$ implies $\gamma(A) \geq \gamma(B)$), γ takes projections into projections, $\gamma(1) = 1$ and $\|\gamma(A)\| = \|A\|$.*

PROOF. Since γ is linear, we need only prove $C \geq 0$ implies $\gamma(C) \geq 0$ to conclude that γ is order preserving. But $\gamma(C) = \gamma(C^{1/2} \circ C^{1/2}) = \gamma(C^{1/2}) \circ \gamma(C^{1/2}) \geq 0$. γ clearly takes projections into themselves since $P^2 = P$ implies $\gamma(P) = \gamma(P \circ P) = \gamma(P)^2$. Since 1 is the unique maximal projection, $\gamma(1) = 1$. Finally $1\|A\| \pm A \geq 0$ implies $\pm\gamma(A) + \|A\|1 \geq 0$ implies $\|\gamma(A)\| \leq \|A\|$. Since γ is invertible and γ^{-1} is a Segal automorphism $\|A\| = \|\gamma^{-1}(\gamma(A))\| \leq \|\gamma(A)\|$. ∎

LEMMA 6.2. *Any weak Segal automorphism γ takes one dimensional projections onto one dimensional projections. γ thus induces a map $a : P(\mathcal{H}) \to P(\mathcal{H})$ so that $\gamma(P_{[\psi]}) = P_{a[\psi]}$. a is a Wigner automorphism.*

PROOF. One dimensional projections are minimal (non-zero) projections, so, by Lemma 6.1, γ must take one into another. By a similar argument, γ must take two dimensional projections into themselves, so, as in Sections 4, 5, the two dimensional analysis of weak Segal automorphisms implies for any two dimensional M, there is a unitary or antiunitary map $U : M \to \mathcal{H}$ so that $\gamma(P_{[\psi]}) = P_{[U\psi]}$ so that α will preserve inner products. ∎

REMARK. Given this theorem, one might expect that $|<\psi,\phi>|$ should be expressible in terms of P_ψ, P_ϕ and the Jordan product. In fact:

$$P_\psi \circ (P_\psi \circ P_\phi) - \frac{1}{2}(P_\psi \circ P_\phi) = \frac{1}{2}|(\psi,\phi)|^2 P_\psi .$$

LEMMA 6.3. *If* γ *is a weak Segal automorphism and* $\gamma(P_\psi) = P_\psi$ *for all one dimensional projections, then* γ = *identity.*

PROOF. Let P be any projection. Since $\psi \in \operatorname{Ran} P$ if and only if $P_\psi \leq P$ and γ is order preserving, $\operatorname{Ran} \gamma(P) = \operatorname{Ran} P$, i.e. γ leaves all projections invariant. Since γ is continuous in norm and any $A \in \mathcal{B}(\mathcal{H})$ is a norm limit of finite linear combinations of projections (by the Spectral Theorem), γ is the identity. ∎

Given these lemmas, Theorem 2.3 follows from Wigner's theorem in just the way that Theorem 2.2 followed. ∎

§7. *Lifting Measurability*

DEFINITION. A map $t \to U(t)$ from the reals to the unitaries is called weakly measurable if $t \to (\phi, U(t)\psi)$ is measurable for all $\phi,\psi \in \mathcal{H}$. A map $t \to \phi(t)$ from the reals to $\mathcal{H}$ is called weakly measurable if $(\psi,\phi(t))$ is measurable for each $\psi \in \mathcal{H}$.

In this section we prove:

THEOREM 7.1. *If* $t \to \alpha_t$ *is a measurable family of Wigner automorphisms obeying* $\alpha_{t+s} = \alpha_t \alpha_s$, *then there is a family of unitaries* $U(t)$ *so that* $\alpha_t[\psi] = [U(t)\psi]$ *and so that* $U(t)$ *is weakly measurable.*

REMARKS.

1. We emphasize that we are dealing with everywhere defined functions, not merely almost everywhere defined functions.

2. $\alpha_{t+s} = \alpha_t \alpha_s$ plays no critical role in the proof. We include it for convenience.

3. That continuity lifts (i.e. the analog of Theorem 7.1 with continuity replacing measurability) is a result of Wigner [11], generalized (with local continuity only) by Bargmann [12] to more general groups. Given Section 8, Section 9, we actually prove their result.

4. Since $\mathcal{H}$ is separable, weak measurability is equivalent to apparently stronger notions, see [13].

5. A similar result holds for β_t and γ_t given their relation to α's.

LEMMA 7.2. *If* $U_1(t)$ *and* $U_2(t)$ *are weakly measurable, so is* $U_1(t)U_2(t)$. *If* $\psi(t)$ *is a weakly measurable vector valued function, so is* $U_1(t)\psi(t)$. *If* $\psi(t)$ *and* $\eta(t)$ *are measurable, so is* $(\psi(t), \eta(t))$.

PROOF. Let $\{\phi_n\}_{n=1}^{\infty}$ be an orthonormal basis. Since

$$(\psi, U_1(t)U_2(t)\psi) = \sum_{n=1}^{\infty} (\psi, U_1(t)\phi_n)(\phi_n, U_2(t)\psi) ,$$

it is measurable as the limit of measurable functions. The second and third statements follow similarly. ∎

LEMMA 7.3. *There exists a unitary operator valued function* $U(\psi, \eta)$ *defined on pairs of unit vectors so that:*

(1) *If* ϕ *is orthogonal to* ψ *and* η, *then* $U(\psi, \eta)\phi = \phi$

(2) $U(\psi,\eta)\eta = \psi$

(3) *If* $\psi(t)$ *and* $\eta(t)$ *are weakly measurable, then* $U(\psi(t),\eta(t))$ *is weakly measurable.*

PROOF. If $\psi = a\eta$ for some complex a, then define $U(\psi,\eta) = 1+(a-1)P_\eta$. If ψ is orthogonal to η, define $U(\psi,\eta)$ by

$$U(\psi,\eta)\phi = \phi + (\psi,\phi)(\eta-\psi) + (\eta,\phi)(\psi-\eta) .$$

If $(\psi,\eta) \neq 0$, write $(\psi,\eta) = |(\psi,\eta)|e^{i\theta}$, $0 \leq \theta < 2\pi$ and $U(\psi,\eta)$ so that (1) and (2) holds and $U(\psi,\eta)\psi = e^{-2i\theta}\eta$. Measurability (property (3)) is easy to check. ∎

LEMMA 7.4. *If* $t \to \alpha_t$ *is a measurable family of Wigner automorphisms obeying* $\alpha_{t+s} = \alpha_t\alpha_s$, *then each* α_t *is induced by a unitary. Moreover, for any* ϕ, *we can choose* $\eta(t)$ *weakly measurable so that* $\alpha_t([\phi]) = [\eta(t)]$.

PROOF. Since $\alpha_t = (\alpha_{t/2})^2$, α_t is induced by a unitary. Let $\{\psi_m\}_{m=1}^\infty$ be an orthonormal basis. Let $X_k = \{t \mid <[\psi_i],\alpha_t[\phi]> = 0,\ i=1,\cdots,k-1;\ <[\psi_k],\alpha_t[\phi]> \neq 0\}$. Each X_k is measurable, so we need only choose $\eta(t)$ measurable on each X_k. Choose $\eta(t)$ on X_k so that $<\psi_k,\eta(t)> > 0$ and $\alpha_t([\phi]) = [\eta(t)]$. Let $f_j(t) = <\psi_j,\eta(t)>$. We must show that each $f_j(t)$ is measurable. As in Lemma 4.4, f_j is determined by $|<\psi_j,\eta(t)>|$, $|<\psi_k+\psi_j,\eta(t)>|/\sqrt{2}$ and $|<\psi_k+i\psi_j,\eta(t)>|/\sqrt{2}$, so f_j is measurable. ∎

LEMMA 7.5. *Let* $\dim \mathcal{H} = 2$ *and* $\phi \in \mathcal{H}$ *fixed. Let* $t \to \alpha_t$ *be measurable and induced by a unitary with* $\alpha_t([\phi]) = [\phi]$ *for all* t. *Choose* $U(t)$ *inducing* α_t *so that* $U(t)\phi = \phi$. *Then* $U(t)$ *is measurable in* t.

PROOF. Choose an isomorphism of $\mathcal{H}$ and C^2 so that ϕ corresponds to $\begin{pmatrix}1\\0\end{pmatrix}$. In terms of Section 3, α_t corresponds to a rotation by angle $\theta(t)$ in the 1-2 plane. Thus

$$U(t) = \begin{pmatrix} 1 & 0 \\ 0 & \exp(i\theta(t)) \end{pmatrix}$$

is measurable. ∎

PROOF OF THEOREM 7.1. Choose $\phi \in \mathcal{H}$. By Lemma 7.4, $\eta(t)$ measurable can be found so that $[\eta(t)] = a_t([\phi])$. Let $\tilde{a}_t = a_{U(\phi,\eta(t))} a_t$. Then $\tilde{a}_t$ is measurable and $\tilde{a}_t[\phi] = [\phi]$. Choose $\tilde{U}(t)$ inducing $\tilde{a}_t$ so that $\tilde{U}(t)\phi = \phi$. Since $U(\phi,\eta(t))^{-1}\tilde{U}(t)$ induces a_t, we need only prove $\tilde{U}(t)$ measurable. It suffices to show $\tilde{U}(t)\psi$ is measurable for each ψ orthogonal to ϕ. Choose $\kappa(t)$ measurable so that $\tilde{a}_t([\psi]) = [\kappa(t)]$ and let $\beta_t = a_{U(\psi,\kappa(t))}\tilde{a}_t$. Then β_t is a measurable family leaving $[\phi,\psi]$ invariant. Thus there is measurable $V(t)$ on $[\phi,\psi]$ so that $V(t)\phi = \phi$. Then, since $U(\psi,\kappa(t))\tilde{U}(t)\phi = \phi$, we conclude that $\tilde{U}(t)\psi = U(\psi,\kappa(t))^{-1}V(t)\psi$, so $\tilde{U}(t)\psi$ is measurable. ∎

The measurable choice $t \to U(t)$ must obey $U(t)U(s) = \omega(t,s)U(t+s)$ for some $\omega(t,s) \in \mathbf{C}$ with modulus 1 on account of uniqueness up to phase and $a_{t+s} = a_t a_s$.

§8. *Multipliers for* R

At this stage, we have a map $a \mapsto U(a)$ from $\mathbf{R}$ to unitary operators which is weakly measurable and obeys

$$U(a)U(b) = \omega(a,b)U(a+b)$$

where $\omega(a,b)$ is a measurable function from $\mathbf{R}\times\mathbf{R}$ to $\{\alpha \in \mathbf{C} \mid |\alpha| = 1\}$. The associative law easily implies that ω is a multiplier, where

DEFINITION. A (Borel) multiplier on $\mathbf{R}$ is a measurable map $\omega : \mathbf{R}\times\mathbf{R} \to \{\alpha \mid |\alpha| = 1\}$ so that for all $a, b, c \in \mathbf{R}$

$$\omega(a,b)\omega(a+b,c) = \omega(a,b+c)\omega(b,c) .$$

DEFINITION. Given a measurable function $\lambda : \mathrm{R} \to \{a \mid |a| = 1\}$, we define $\partial\lambda$ by

$$\partial\lambda(a, b) = \lambda(a+b)\lambda(a)^{-1}\lambda(b)^{-1} .$$

Each $\partial\lambda$ is easily seen to be a multiplier. The next step in the proof of Theorem 2.4 is:

THEOREM 8.1. *Every Borel multiplier on* R *is of the form* $\omega = \partial\lambda$ *for some* λ.

The point of Theorem 8.1 is that $\tilde{U}(a) = \lambda(a)U(a)$ is a weakly measurable map from R to the unitarities obeying $\tilde{U}(a+b) = \tilde{U}(a)\tilde{U}(b)$. The proof we give will basically follow Wigner [11] although we have been influenced by lectures of Mackey [14]. Before giving the proof, we note:

REMARKS.

1. The symbol $\partial\lambda$ is motivated by the connection with cohomology of groups. Theorem 8.1 is essentially a statement that the two-dimensional (Borel) group cohomology of R with coefficients in the circle group is trivial.

2. Theorem 8.1 represents a subtle interplay between the group and measure structure of R. For R^2 is isomorphic to R as a group (both are vector spaces over the rationals of the same dimension) and as a measure space (the Borel structures of any two complete separable metric spaces are isomorphic) but not as a group with a Borel structure. While R has no nontrivial multiplier, R^2 does have nontrivial multipliers, e.g.

$$\omega(\langle a, b\rangle, \langle c, d\rangle) = \exp\left[\frac{1}{2}(ad - bc)\right]$$

the multiplier associated to the Heisenberg group $\{\exp[i(aP - aQ)]\}$.

3. For the structure of multipliers of more general groups, see Bargmann [12] or Mackey [15].

LEMMA 8.2. *Without loss of generality, we may suppose that* $\omega(a, 0) = 1 = \omega(0, a)$.

Proof. Taking $b = c = 0$ in the definition of a multiplier, we see that $\omega(a,0) = \omega(0,0)$ for all a so $\omega(a,0)$ is a constant d. Similarly $\omega(0,a) = \omega(0,0)$. Let $\tilde{\omega} = (\omega)\partial\tilde{\lambda}$ where $\tilde{\lambda}(a) = d$. Then $\tilde{\omega}$ is a multiplier with $\tilde{\omega}(a,0) = 1 = \tilde{\omega}(0,a)$. If $\tilde{\omega} = \partial\lambda'$, then $\omega = \partial[\lambda'\tilde{\lambda}^{-1}]$. ∎

Henceforth we suppose $\omega(a,0) = \omega(0,a) = 1$.

Lemma 8.3. *Without loss of generality, we may suppose that* $\omega(a,-a) = 1$ *for all* a.

Proof. First note that taking $b = -a$, $c = a$ in the definition of multiplier

$$\omega(a,-a)\omega(0,a) = \omega(a,0)\omega(-a,a)$$

so that $\omega(a,-a) = \omega(-a,a)$. Define $\lambda(a) = [\omega(a,-a)]^{1/2}$ where we take the square root with argument in $[0,\pi)$. Then

$$\partial\lambda(a,-a) = \lambda(0)\lambda(a)^{-1}\lambda(-a)^{-1} = \omega(a,-a)^{-1}$$

so that $(\omega)(\partial\lambda) = \tilde{\omega}$ obeys $\tilde{\omega}(a,-a) = 1$. ∎

Henceforth we suppose that $\omega(a,-a) = 1$.

Lemma 8.4. *For any multiplier, there is a map* $a \to U(a)$ *to the unitaries on some Hilbert space so that*

$$U(a)U(b) = \omega(a,b)U(a+b) .$$

Proof. Let $\mathcal{H} = L^2(\mathbf{R}, d\mu)$ where μ is Lebesgue measure. Define $U(a)$ by

$$(U(a)f)(b) = \omega(b,a)f(a+b) .$$

An immediate computation shows that

$$U(a)U(b) = \omega(a,b)U(a+b) . \;\blacksquare$$

Lemma 8.5. *If* ω *is a multiplier, then* $\omega(a,b) = \omega(b,a)$ *for all* $a, b \in \mathbf{R}$.

PROOF. Let $q(a,b) = \omega(a,b)/\omega(b,a)$ and let U be some ω-representation. Then

$$U(a)\,U(b)\,U(a)^{-1} = q(a,b)\,U(b) .$$

Moreover, $\omega(a,b)\omega(-b,-a) = 1$ since by $\omega(a,-a) = 1$, $U(a)^{-1} = U(-a)$ whence $(U(a)\,U(b))^{-1} = U(-b)\,U(-a) = \omega(-b,-a)\,U(-a-b)$ on the one hand and $= (\omega(a,b)\,U(a+b))^{-1} = \omega(a,b)^{-1}U(-a-b)$ on the other hand. From the last two formulae, we conclude that $q(a+b,c) = q(a,c)\,q(b,c)$ so that the measurability of q implies that $q(a,b) = \exp(2\pi i\ a\ f(b))$ [to see this, just follow the arguments of Sections 9, 10 with $\mathcal{H} = \mathbf{C}$!]. Since $q(a,a) = 1$, $f(a) = a^{-1}n(a)$ where $n(a)$ is an integer. Clearly $q(a,b)q(b,a) = 1$ so $ab^{-1}n(b) + ba^{-1}n(a) = n(a,b)$ for integers $n(a)$, $n(b)$, $n(a,b)$. Let $b = a\sqrt[3]{2}$. Since $\sqrt[3]{2}$, $(\sqrt[3]{2})^{-1}$ and 1 are independent over Z, we conclude $n(a) = 0$ for all a, i.e. $q(a,b) = 1$ for all a, b. ∎

LEMMA 8.6. *For any multiplier, there is an irreducible family* $\{U(a)\}$ *of unitaries on some Hilbert space so that* $U(a)\,U(b) = \omega(a,b)\,U(a+b)$.

PROOF. If ω were continuous, we could form a locally compact group so that representations of the group with an additional property were in one-one correspondence to ω-representations and then appeal to the theory of such representations. Since this is not available to us, we borrow the relevant argument!

Call a function ϕ on R, ω-positive definite if and only if $\phi \in L^\infty(\mathbf{R})$ and

$$\int \overline{f(a)}\,f(b)\phi(b-a)\omega(-a,b)\,da\,db \geq 0$$

for all $f \in L^1(\mathbf{R})$. Given any ω-rep, $\phi(a) = (\psi, U(a)\psi)$ is ω-positive definite, so by appealing to Lemma 8.4, there do exist ω-positive definite functions.

Given ϕ ω-positive definite, we put an inner product

$$(f, g) = \int \overline{f(a)}\, g(b) \phi(b-a) \omega(-a, b)\, da\, db$$

on L^1 and form a Hilbert space in the obvious way. The maps $(U(a)f)(b) = \omega(a, b-a) f(b-a)$ are easily seen to obey $(U(a)U(b)f) = \omega(a,b)(U(a+b)f)$, $U(0) = 1$ $(f, U(a)g) = (U(-a)f, g)$ so that U is an ω-representation.

The set of ω-positive definite functions is a compact convex subset of L^∞ in the weak $*$ (L^1) topology, so, by the Kerin Millman theorem, there exist extreme points. Such extreme points are seen to lead to irreducible ω-representations. ∎

PROOF OF THEOREM 8.1. By Lemma 8.5, $\{U(a)\}$ is a commuting family, so the irreducible representation of Lemma 8.6 is one dimensional by Schur's lemma. Thus, if this representation is multiplication by λ:

$$\lambda(a)\lambda(b) = \omega(a, b)\lambda(a+b)$$

i.e. $\omega = \partial\lambda$. ∎

§9. *Von Neumann's Theorem*

In this section and the next, we complete the proof of Theorem 2.4 by proving that any weakly measurable family of unitary operators with $U(a)U(b) = U(a+b)$ is of the form $U(a) = e^{-iaH}$ for some self-adjoint H. Our two step proof consists in demonstrating two classical theorems. Our proofs follow those of Reed-Simon [13] to which the reader is referred for more information concerning the spectral theorem, self-adjointness, etc. Here we prove:

THEOREM 9.1 (von Neumann's Theorem [16]). *Let* $t \to U(t)$ *be a weakly measurable map from* $\mathbb{R}$ *to the unitary operators on a separable Hilbert space,* $\mathcal{H}$. *Suppose* $U(t+s) = U(t)U(s)$. *Then* $U(t)$ *is strongly continuous.*

PROOF. Since the $U(t)$'s are uniformly bounded, it suffices to find a total set (subset whose finite linear combinations are dense) of ψ's so that $t \to U(t)\psi$ is continuous. Pick an orthonormal subset $\{\phi_n\}_{n=1}^{\infty}$ of $\mathcal{H}$ and for $a > 0$, define $\phi_n(a)$ as follows:

$$\eta \to \int_0^a (\eta, U(t)\phi_n)\,dt \equiv \ell_{n,a}(\eta)$$

defines a conjugate linear function with norm $\leq a$, so there is a vector $\phi_n(a)$ with $(\eta, \phi_n(a)) = \ell_{n,a}(\eta)$. For obvious reasons we denote $\phi_n(a)$ as

$$\int_0^a U(t)\phi_n\,dt\ .$$

A simple argument proves that

$$U(s)\int_0^a U(t)\phi_n\,dt = \int_s^{a+s} U(t)\phi_n\,dt$$

and thus that

$$\|(U(s)-U(s'))\phi_n(a)\| \leq \|\int_s^{s'} U(t)\phi_n\,dt\| + \|\int_{a+s}^{a+s'} U(t)\phi_n\,dt\| \leq 2|s-s'|\,.$$

It follows that $t \to U(t)\phi_n(a)$ is continuous for all n and a, so we need only prove the $\{\phi_n(a)\}$ total by the remarks above. Suppose ψ is orthogonal to all the $\{\phi_n(a)\}$. Then, for each n, $(\psi, U(t)\phi_n) = 0$ a.e. in t, so there must be a t_0 with $(\psi, U(t_0)\phi_n) = 0$ for all n. It follows that $U(t_0)^{-1}\psi = 0$, so $\psi = 0$. Thus the $\{\phi_n(a)\}$ are total. ∎

§10. *Stone's Theorem*

We complete the proof of Theorem 2.4 with:

THEOREM 10.1 (Stone's Theorem [17]). *Let* $t \to U(t)$ *be a strongly continuous map from* $\mathbb{R}$ *to the unitaries operators so* $U(t+s) = U(t)U(s)$. *Then* $U(t) = e^{-iHt}$ *for a unique self-adjoint operator* H.

REMARKS.

1. Since e^{-iHt} (which can be defined by the spectral theorem) is a strongly continuous unitary group, this sets up a one-one correspondence between such groups and *unbounded* self-adjoint operators.

2. If both $U(t)$ and $U'(t)$ generate the automorphisms $\alpha(t)$, then $U(t)U'(t)^{-1}$ is a numerical representation of $\mathbb{R}$ and so $U(t) = U'(t)e^{i\alpha t}$ for some real α, giving the uniqueness aspect of Theorem 2.4.

PROOF. Define an operator H with domain:

$$D(H) = \{\phi \mid U(t)\phi \text{ is differentiable at } t=0\}$$

and

$$H\phi = i\frac{d}{dt}[U(t)\phi]|_{t=0} .$$

A simple argument shows that H is symmetric. Moreover, if $U(t) = e^{-iAt}$ for some self-adjoint A, then H is closed and $H = A$ (this yields uniqueness).

For any $f \in C_0^\infty$, the C^∞ functions of compact support and $\phi \in \mathcal{H}$, let $\phi_f = \int f(t)U(t)\phi dt$ and let G, the Garding domain, be the finite span of the ϕ_f. By an elementary computation $\phi_f \in D(H)$ and $H\phi_f = -i\phi_{f'}$ so $G \subset D(H)$. Moreover, $U(s)\phi_f = \phi_{f^{(s)}}$ where $f^{(s)}(t) = f(t-s)$, so G is left invariant by the $U(t)$'s. In addition, G is dense, since $\phi_f \to \phi$ as f approaches a δ-function in a suitable way (this uses the strong continuity!).

Suppose ψ is orthogonal to $(H+i)[G]$. Then, for any $\eta \in G$, $(\psi, U(t)\eta) = f(t)$ obeys:

$$\begin{aligned} \frac{d}{dt}f(t) &= (\psi, (-iH)U(t)\eta) \qquad \text{(G is invariant!)} \\ &= (i(H)^*\psi, U(t)\eta) \\ &= -f(t) . \end{aligned}$$

Thus $f(t) = f(0)e^{-t}$, since $|f(t)| \leq \|\psi\|\,\|\eta\|$; taking $t \to -\infty$, we see that $f(0) = 0$, i.e. ψ is orthogonal to G and so zero. Similarly, $(H-i)[G]$ is dense, so H is essentially self-adjoint on G.

Let $A = \overline{H}$. To complete the proof we need only show that $U(t) = e^{-iAt}$. Let $\psi, \phi \in G$. Then since $\psi \in D(A)$ which is left invariant by e^{-iAt} and $U(t)\phi \in G$: $\frac{d}{dt}(e^{-iAt}\psi, U(t)\phi) = 0$, so $U(t) = e^{-iAt}$. ∎

DEPARTMENTS OF MATHEMATICS AND PHYSICS
PRINCETON UNIVERSITY

REFERENCES

[1] E. Wigner, *Group Theory and its Applications to the Quantum Theory of Atomic Spectra*, Academic Press, 1959; German original: 1931.

[2] R. Kadison, Topology *3*, Suppl. *2* (1965), 177-198.

[3] J. Ax, Stony Brook preprint, 1975.

[4] J. von Neumann, *Mathematische Grundlagen der Quantenmechanik*, Springer, 1932.

[5] G. Mackey, *The Mathematical Foundations of Quantum Mechanics*, Benjamin, 1963.

[6] I. Segal, Ann. Math. *48* (1947) 930-940.

[7] E. Artin, *Geometric Algebra*, Wiley, 1957.

[8] W. Hunziker, Helv. Phys. Acta. *45* (1972), 233-236.

[9] V. Bargmann, J. Math. Phys. *5* (1964), 862-868.

[10] J. Roverts and G. Roepstorff, Comm. Math. Phys. *11* (1969), 321-338.

[11] E. Wigner, Ann. Math. *40* (1939), 149-204.

[12] V. Bargmann, Ann. Math. *59* (1954), 1-46.

[13] M. Reed and B. Simon, *Methods of Modern Mathematical Physics, Vol. I Functional Analysis*, Academic Press, 1972.

[14] G. Mackey, Unpublished lectures at Harvard University, 1965; see also *Induced Representations of Groups and Quantum Mechanics*, Benjamin, 1968 and Oxford University Lectures.

[15] G. Mackey, Acta Math. *99* (1958), 265-311.

[16] J. von Neumann, Ann. Math. *33* (1932), 567-573.

[17] M. Stone, Proc. Nat. Acad. Sci. *15* (1929), 198-200.

SEMICLASSICAL ANALYSIS ILLUMINATES THE CONNECTION BETWEEN POTENTIAL AND BOUND STATES AND SCATTERING*

John Archibald Wheeler

The Marchenko integral equation for finding the interaction potential V(r) from information about scattering and bound states of a single specified angular momentum $\ell\hbar$ is compared and contrasted with workaday semiclassical or JWKB methods easily adapted for use or already in use for dealing with the "inverse scattering problem," particularly in the context of atom-atom collisions. Topics considered include: One ℓ-value or many; one E-value or many; and why. The Marchenko equation and the physics that is telescoped into it. Preview. Principal features of the JWKB semiclassical approximation recapitulated. From ladder of levels to potential: semi-integration gives the "inclusion" and then a differentiation gives the "excursion." The "inverse scattering problem" via JWKB: "no slideability above the line." The difficulty of resolving slideability below the line. The Rydberg-Klein-Rees method. Semiclassical analysis of scattering in a broader context. Nucleus-nucleus interactions and the semiclassical description of scattering. The Madelung rule for the filling of atomic levels, the necklace orbit, and the Demkov-Ostrovski atomic potential. The superdirectional antenna. "Electron leakage" out of an atom. Cases where the interaction depends on velocity: their prevalence and catastrophic consequences for "inverse scattering theory." The black bird and the crimson bird: instability. The double-minimum potential and the unravelling of overlapping spectra. Barrier penetrability revealed by energy level perturbations. Inverting the barrier penetration integral to find the distance through the barrier. Refinements required for dealing with the regime of level "crossover" and the regime of "spectrum squeeze." Above the barrier. The physics hidden in the Marchenko equation, waiting to be revealed.

The Crimson Bird and the Black Bird

Not only on the Galapagos Islands has a single bird species separated into now distinct species that no longer interbreed. A similar speciation is close to completion in scattering theory. In that territory, what are the two birds that once intermingled but now so often turn the other way when they see each other? The crimson bird with the bright plumage of

---

*Preparation for publication assisted in part by NSF Grant GP30799X to Princeton University.

mathematical exactitude occupies one ecological niche; the black bird of semiclassical approximation methods, another. They get their nourishment in the same domain of nonrelativistic scattering physics. There two interacting centers, of reduced mass $\mu = M_1 M_2/(M_1 + M_2)$ and center-of-mass energy E, couple by central forces. In the Schroedinger wave equation* for the probability amplitude, $\psi(r, \theta, \phi)$, as a function of the relative coordinates, r, θ, ϕ,

$$\nabla^2\psi + (2\mu/\hbar^2)[E - V(r)]\psi = 0 , \tag{1}$$

both birds focus attention initially on a solution of angular momentum $\ell\hbar$, regular at the origin,

$$\psi = (1/r)\, f_\ell(r, E)\, Y_\ell^{(m)}(\theta, \phi) , \tag{2}$$

and on the radial equation satisfied by the radial factor,

$$(d^2 f/dr^2) + (2\mu/\hbar^2)[E - V_\ell(r)]\, f = 0 ; \tag{3}$$

or, more briefly, in a region of positive radial kinetic energy (oscillating f(r))

$$(d^2 f/dr^2) + k^2(r) f = 0 ; \tag{4}$$

and in a region of negative radial kinetic energy (barrier penetration)

$$(d^2 f/dr^2) - \kappa^2(r) f = 0 . \tag{5}$$

Both species make their living by producing knowledge about the interaction, V(r), out of information about the bound states,

$$E_{\ell 0}, E_{\ell 1}, E_{\ell 2}, \cdots, E_{\ell n}, \cdots ; \tag{6}$$

---

*The factor $(2\mu/\hbar^2)$ or $(\hbar^2/2\mu)$ is left in this and subsequent equations, in order to keep completely obvious how in applications to use the c.g.s. units of everyday physics. Anyone not so immediately concerned with applications can adopt without difficulty the frequent practice to "set $\hbar^2/2\mu$ equal to 1."

or from information for states of positive energy,

$$E = \hbar^2 k^2(\infty)/2\mu \equiv \hbar^2 k^2/2\mu \ , \tag{7}$$

about the phase shift, $\eta_\ell(E)$, as defined by the asymptotic form of the radial factor at large r values,

$$f_\ell(r, E) \sim \sin(kr - \ell\pi/2 + \eta_\ell) \tag{8}$$

[$\eta_\ell(E)$ being determined by the interaction potential $V(r)$, and being equal to zero for the case of zero interaction]; or from both kinds of information together.

One ℓ-Value or Many; One E-Value or Many; and Why

Why limit attention to data for one ℓ-value? Why not capitalize on phase shifts and bound states for all relevant ℓ-values and energies? After all, experiments hardly ever measure, and hardly ever can measure, the scattering amplitude itself,

$$A(E,\theta)(\mathrm{cm}) = \sum_{\ell=0}^{\infty} [(2\ell+1)/2ik](e^{2i\eta_\ell} - 1)P_\ell(\cos\theta) \ , \tag{9}$$

for scattering at the angle θ, easy though it would then be by spherical-harmonic analysis to focus on the one desired ℓ-value, the one desired partial wave, to the exclusion of all the others. Instead, one measures the differential cross section for scattering into each unit solid angle,

$$(d\sigma/d\Omega)(\mathrm{cm}^2/\mathrm{steradian}) = |A(E,\theta)|^2 \ , \tag{10}$$

a quantity in which the contributions of all the various ℓ-values are mixed up together. It is hardly possible to sort out the phase shift for one partial wave without at the same time getting similar information for all the other relevant partial waves [see for example Sabatier (1974 b)]. Having these η_ℓ, why not use them? And similarly the data on the bound states?

The answer is simple. In those practical connections where the semiclassical approximation is applicable, the black bird does go to work and does use every available reliable phase shift and every available reliable energy value. It feeds on experiment. It shows nowhere to better advantage than in molecular physics. Out of measurements on differential scattering of atom by atom as a function of angle and energy for energies of a fraction of an electron volt or a few electron volts, it produces an interatomic potential energy curve. It makes no call on any information about any bound state. This circumstance does not lead to indeterminateness in the potential energy curve. On the contrary, the potential energy is way overdetermined. It is a single function, $V(r)$, of a single parameter, r. Yet it has to reproduce a single function, $\eta_\ell(E)$, of two parameters, the angular momentum and the energy. It is a testimonial to the very validity of the concept of potential energy that a single "infinitude" of numbers can reproduce a double "infinitude" of numbers.

The overdeterminateness of the many-ℓ, many-E problem explains why there are at least three different ways to choose the data one will use, and corresponding methods to go from there unambiguously to the potential: the RKR method [Rydberg (1931, 1933), Klein (1932), and Rees (1947)] used by very considerable numbers of workers to evaluate an even larger number of atom-atom interaction potentials, impressively compiled in Bernstein and Muckerman (1967); the additional developments and methods contributed by Miller (1969, 1971); and the approach of Flügge and Vollmer (1971), related to, but alternative to the RKR method. However, to many a mathematical physicist of this century, an overdetermined scattering problem is as repulsive as an overdetermined bridge structure was to the engineer of the last century. The crimson bird wants to trim the problem down to a package of data precisely tailored to guarantee existence and uniqueness of a solution, free of all issues of "incompatible data."

This bright bird would be happiest if it could make do with data on phase shift at one energy for all ℓ-values. But that won't work. The black bird of semiclassical theory sees why it won't. The particle counted on to probe the potential won't reach in to a small enough r value. No matter with what angular momentum or impact parameter it enters, it has only the once-and-for-all fixed energy value, and with that energy it ordinarily won't be able to probe any reasonable repulsive potential all the way in to $r = 0$. It is not enough for the black bird to point out to the crimson bird that the region from that critical r-value on out is often the only region in which the physicist takes much interest. It is not enough to show that the semiclassical method directly yields the potential in that region from the phase shifts [Sabatier (1966, 1972, 1974 a, 1974 b, 1974 c); Wheeler (1955 b); and see in particular the reviews by Pauly and Toennies (1965), Bernstein (1966), Mason and Monchick (1967), Bernstein and Muckerman (1967), Berry and Mount (1972), Child (1974) and Buck (1974)], as well as telling how to get these phase shifts from the scattering [Sabatier (1974 b); the just cited reviews; and Ford and Wheeler (1959)] and even how to go directly from the scattering at a given energy to the potential without the intermediary of any phase shifts [Hoyt (1939); Firsov (1953); Sabatier (1974 a); and the cited reviews of Bernstein (1966) and Buck (1974)]. Nor is it enough to show [Sabatier (1972)] that exact theory "construct[s] all the potentials ... which fit this [given] set of phase shifts [and that] they depend on an arbitrary function [of one variable]." [See also Jean and Sabatier (1973).] None of this is compatible with the crimson bird's ideals of existence, uniqueness and exactness. Therefore it throws away all the data except those belonging to one ℓ-value. On the fulness of these data, however, it is insistent: everything or nothing! The phase shift must be supplied for all energies up to $E = \infty$; and for every bound state must be given a number that tells something about the effective outward reach of the potential at that energy; not the easiest data in the world for the observer to acquire!

The Marchenko Equation and the Physics that is Telescoped Into It

This information granted, the clever crimson bird performs a beautiful feat. It produces the potential energy of interaction, $V(r)$, between the two centers (on the assumption that there *is* such a potential). The prescription for this potential – provided by Marchenko (1955) – is summarized nowhere more compactly than in these words from the standard text on quantum mechanics of Landau and Lifshitz (1958 b): "We form the function

$$F(r) = \sum_n a_n^2 e^{-\kappa_n r} - (1/2\pi) \int_{-\infty}^{+\infty} \left[e^{2i\eta_0(k)} - 1\right] e^{ikr} dk \quad (105.14)$$

[here the quantity κ_n is connected with the energy of the nth bound state by the formula

$$E_n = -(\hbar^2/2\mu)\kappa_n^2 \,, \quad (11)$$

and the bound state wave function behaves asymptotically for large r as

$$(a_n/r) e^{-\kappa_n r}] \quad (12)$$

and from it the linear integral equation

$$F(x+y) + A(x,y) + \int_x^{\infty} A(x,t) F(y+t) dt = 0 \quad (105.15)$$

in which the unknown function is A(x, y), containing two variables. If this equation is solved, the required function V(r) can be found from the formula

$$V(r) = -(h^2/\mu)(d/dr) A(r,r) \text{."} \quad (105.16)$$

It is impressive that such a difficult problem in mathematical physics should have a so neatly defined exact solution. Dyson (1976) gives an attractive summary of this problem and its history [see also the book of Agranovich and Marchenko (1963) and the more recent review of Newton (1972)]. In that history, Bargmann's (1949 a, 1949 b) discovery of "phase-equivalent potentials" – distinct potentials giving the same functional dependence of the scattering phase shift on energy, $\eta = \eta(E)$, – played no small part; Gel'fand and Levitan (1951) and many others paved the way for Marchenko (1955); and Faddeev (1962) brought the whole subject into a good order.

Nothing about Marchenko's prescription stands out with more force than this, that it towers above all circumstance. It does not make any reference to big terms or little terms. It treats with equal composure the case where the wave length is small in comparison with the size of the zone of interaction and the case where the wave length is large. Not one mention does it make of such phenomena as resonance, barrier penetration, "level crossing" and "orbiting." Every one of these effects nevertheless shows up in the same context at some point or in some problem when the semiclassical analysis is employed. Would it be fair then to say that the black bird looks at the physics and the crimson bird doesn't? Not at all. To turn to Marchenko's equation is no more overlooking the physics than to turn to Schrödinger's. The physics is telescoped into the Schrödinger equation. The *same* physics is telescoped into the Marchenko equation. One knows how to "pull out the telescope full length" in the case of the Schrödinger equation. To bring out that physics is the natural daily work of the black bird. One does not yet know how to "pull out the telescope" that is compressed into the Marchenko package. But the same physics is compacted into the one package as into the other. That it can be got out and some day will be got out is the thesis of this article. To testify for this thesis, nothing better can be done here than to display some of the cited effects as they show up in the problem of "fixed ℓ-value and all E-values." The black bird will do almost all the speaking and will be the "we" of what follows.

Preview

We shall first get the semiclassical method "on the road" by looking at a small but simple problem: what information can be deduced about the potential in the case of a single smooth potential well from a knowledge of bound state energies alone. It was already clear long ago [Rydberg

(1931, 1933); Klein (1932)] from the semiclassical analysis that the given data enable one to figure out the difference, $r_2(V) - r_1(V)$, between the two turning points of a particle with energy $E = V$; that is to say, the "excursion," $X(V)$; but nothing about either turning point individually. Pöschl and Teller (1933) did more. They constructed a continuous one-parameter family of potentials, solved the Schrödinger equation in this potential for the eigenvalues, and showed that the eigenvalues remained unaffected by the change in the shape of the potential (exact quantum mechanical result on "energy-level equivalent potentials"). They also calculated for each potential the "excursion function" $X(V)$ (width as a function of height above the minimum) and showed that these functions, while not exactly identical, were very nearly so, thus harmonizing with the exact equality demanded by the semiclassical analysis. In contrast, in a square well potential, with a secondary well of half its width sitting at its bottom, a significant difference was found in calculated energy level spectra, according as the subsidiary well sat in the middle of the "floor" or next to one wall. This result is warning enough that the semiclassical analysis can fail for a potential with sharp corners or sudden changes.

After going through the semiclassical analysis of spectrum-equivalent potentials, we take a joy-ride through some of the applications of the semiclassical method to problems outside the strict "ℓ = constant" context. The purpose is to see a little of the interesting physics that can take place and that can be interpreted, if one looks out for it, with the help of the semiclassical approximation. This flavor of physics once being established, we come back at the end to look at the ℓ = constant problem to see some of the many effects that can occur there and that one will surely some day learn how to extract out of the Marchenko equation.

It is not a part of our plan to prove anything. Instead, we ask how one tells in advance, via JWKB analysis, what to expect to be provable: provable because it is true in the semiclassical approximation. If in this enterprise we turn out merely to be revealing standard trade secrets of colleagues far more knowledgeable than the author, we can offer two defenses. First, not one expert would stoop to offer as "insights" such well known conclusions as will make up this account. Second, there is not one expert whose trade secrets about scattering will be anywhere near exhausted by the present miserly give-away.

Principal Features of the JWKB Semiclassical Approximation Recapitulated

A few equations and a little terminology are unavoidable. It is typical of the semiclassical analysis to use very different mathematical expressions for the radial factor on the positive and negative kinetic energy sides of a classical turning point, a distinction accentuated by the circumstance that both expressions, one oscillatory, one exponential, diverge at the turning point itself. One knows, not least through the work of Langer (1932, 1934, 1937), and the Uppsala group of Fröman and Fröman and associates, and others, how to exploit the Airy function to make a smooth transition between the oscillatory and exponential domains. [For general treatments, in addition to the good few-page older treatments of Pauli (1933) and Kemble (1937), see the books of Heading (1962) and Fröman and Fröman (1965) and the recent review of Berry and Mount (1972). Systematic justification of the JWKB approximations, level by level, is a feature of the Uppsala investigations, among which are these: N. Fröman (1966a, 1966b, 1970, 1974); P. O. Fröman (1974); N. Fröman and P. O. Fröman (1970, 1974a, 1974b); Fröman, Fröman, Myhrman and Paulson (1972); and Karlsson (1975). Numerous special points are investigated by Yost, Wheeler and Breit (1936), Miller and Good (1953), Marchi and Mueller (1963), Miller and Marchi (1963), Smith (1964), Krainov (1971), Richardson (1971), Lu and Wald (1972), Newman and Thorson (1972), Wald and Lu (1974), and Porter, Raff, and Miller (1975). See Smith (1964) for the Gauss-Mehler technique as applied to integrating JWKB formulas near a turning point. A comprehensive review on solitons by Scott, Chu and McLaughlin (1975) discusses their relation to the JWKB approximation. For the relation to "complex trajectories,'' see Knoll and Schaeffer (1975); and for progress towards a more-than-one dimensional JWKB analysis, see, for example, Van Horn and Salpeter (1967), Doll and Miller (1972) Pechukas (1972) and Miller (1975).] However, the focus of attention here is not the wave function itself, but the energies of bound states and the phase shift associated with states in the continuum. For neither the E_n nor the $\eta(E)$ does it matter that the JWKB wave function diverges at a turning point. Moreover, the Langer continuation procedure applied, as it normally is, in lowest order gives the standard JWKB values for these observables. Therefore, it is reasonable to take the JWKB wave function as convenient shorthand – and nothing but shorthand – for the more nearly accurate but also more complicated Langer wave function that gives the same energies and phase shifts.

At small distances "centrifugal forces" dominate and the radial kinetic energy increases with increasing r. To the right of the turning point, the solution is oscillatory and the shorthand of the JWKB approximation gives for the radial factor (up to a multiplicative normalization constant)

$$f(r) = k^{-\frac{1}{2}}(r) \sin\left(\pi/4 + \int_{TP}^{r} k(r)\, dr\right) . \tag{13}$$

To the left of the turning point, with the same normalization, the symbolic expression for the radial part of the wave function is

$$f(r) = \frac{1}{2}\, \kappa^{-\frac{1}{2}}(r) \exp\left(-\int_{r}^{TP} \kappa(r)\, dr\right) . \tag{14}$$

This expression rises with increasing r, but it does not rise at the right rate (as $r^{\ell+1}$) for small r unless and until the effective potential is modified from

$$V(r) + (\hbar^2/2\mu)\ell(\ell+1)/r^2$$

to the value used now and hereafter,

$$V_\ell(r) = V(r) + (\hbar^2/2\mu)\left(\ell + \frac{1}{2}\right)^2/r^2 \ . \tag{15}$$

[This value makes the exponential damping factor $\kappa(r)$ behave for small r as $\kappa \simeq \left(\ell + \frac{1}{2}\right)/r$.] This is the reason that led Kramers (1926) to make the nowadays standard "$\left(\ell + \frac{1}{2}\right)$" modification in the potential, a modification subsequently supplied with a deeper mathematical justification by Langer (1932, 1934, and especially 1937). Physical applications supply two more reasons to make the Kramers-Langer modification: This readjustment forces the phase of the wave function (13) for the case of zero force [$V(r) = 0$] to take on the correct value, $k(\infty)r - \ell\pi/2$, at large r [zero phase shift, η_ℓ, in (8)]. It also gives improved values for the wave function itself [see Yost, Wheeler, and Breit (1936) and the cited papers of the Uppsala group].

For a state bound in a potential minimum, there are two turning points, $r = TP_1$ and $r = TP_2$. The radial wave factor for the nth quantum state of oscillation ($n = 0, 1, 2, \cdots$) has n nodes in addition to the node at the origin. It is "deformation-equivalent" to a wave function in a "square-well potential" over the interval $(0, L)$, the wave function $\sin[(n+1)\pi x/L]$: equivalent in the sense that the function $f(r)$ undergoes a total effective phase change of $(n+1)\pi$ between $r = 0$ and $r = \infty$, as follows,

$\pi/4$, *equivalent* phase change through the zone of exponential rise from $r = 0$ to TP_1, as evidenced in the phase, $\pi/4$, with which the wave function (8) starts off in the zone of oscillation;

$\left(n + \frac{1}{2}\right)\pi$, phase change in the zone of oscillation itself, as expressed in the standard *quantum condition*

$$\int_{TP_1}^{TP_2} k(r, E_n)\,dr = \left(n + \frac{1}{2}\right)\pi \ , \tag{16}$$

a condition for fixing the eigenvalue, E_n, equivalent to the Bohr-Sommerfeld condition

$$\oint \begin{pmatrix} \text{radial component} \\ \text{of momentum} \end{pmatrix} dr = \left(n + \frac{1}{2}\right) h \tag{17}$$

("phase diagram encompassing a half-integral number of quanta of action"); and

$\pi/4$, *equivalent* phase change through the zone of exponential fall from $\gamma = TP_2$ to $r = \infty$, as evidenced in the phase with which the JWKB wave function,

$$f = (-1)^n k(r)^{-\frac{1}{2}} \sin\left[(\pi/4) + \int_r^{TP_2} k(r)\,dr\right], \tag{18}$$

ends up at the right hand limit of the zone of oscillation.

Almost everything we seek to do, and everything we can do, with the aid of the semiclassical approximation is founded on Equation (16) for the phase *difference*, $\left[n(E) + \frac{1}{2}\right]\pi$, between turning point and turning point for bound states, and on the corresponding formula

$$\eta_\ell(E) = \lim_{R \to \infty}\left[\int_{TP}^{R} k(r)\,dr - \int_{TP_{free}}^{R} k_{free}(r)\,dr\right] \tag{19}$$

for the phase *shift* for states in the continuum. Here k_{free} refers to the comparison problem where the interaction potential $V(r)$ is wiped out:

$$(\hbar^2/2\mu)\,k_{free}^2 = \begin{pmatrix} \text{radial kinetic energy} \\ \text{in absence of interaction} \end{pmatrix} = E - (\hbar^2/2\mu)\left(\ell + \frac{1}{2}\right)^2/r^2 \,. \tag{20}$$

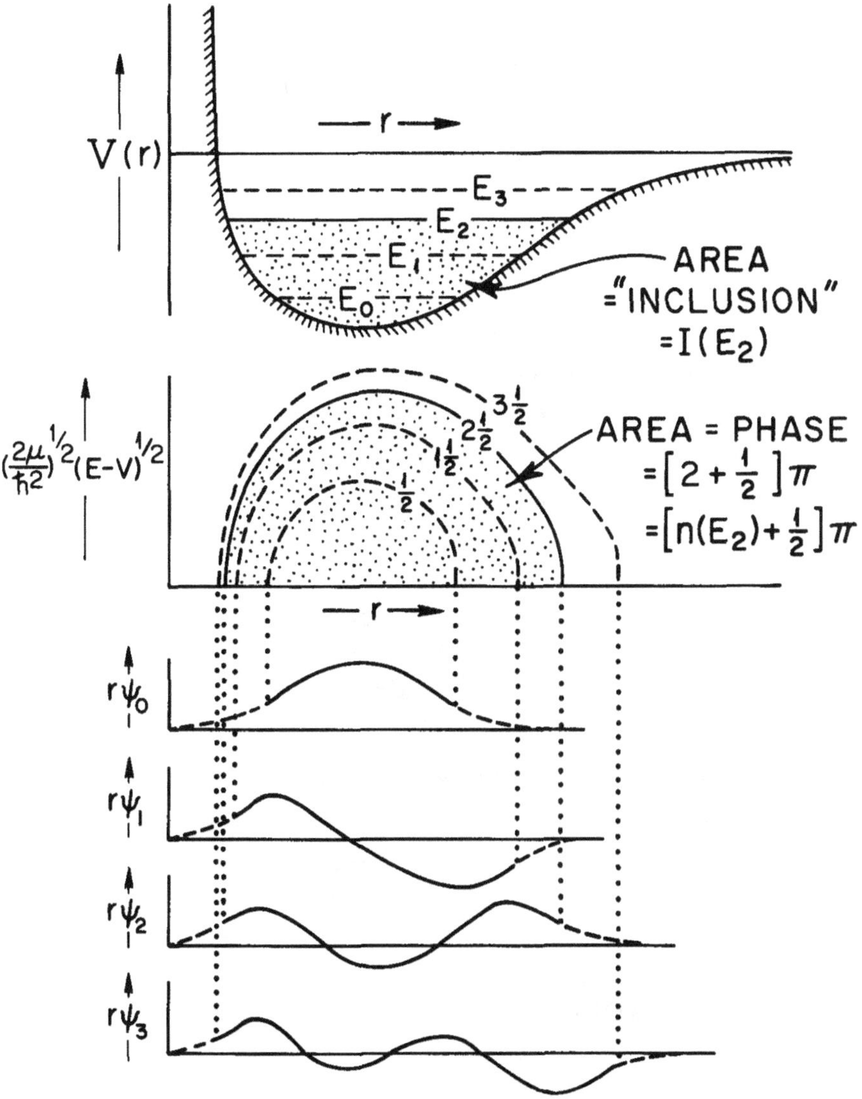

Figure 1. Four important features of the interaction potential V(r) show up at the semiclassical level of analysis. Each is a function of energy. Three of the four are illustrated here for the quantum state with radial quantum number $n_r = 2$. First is the inclusion, $I(E_2)$, shown as the dotted area in the upper diagram. Second is the phase change in the wave function (or, better, in $r\psi$) in the lower diagram, between the inner and outer turning points – a phase change which is given by the dotted area in the middle "phase diagram." Third is the actual separation between these turning points, denoted as the "excursion," X(E). Fourth (not shown) is the period, T(E), for one complete vibration with energy E.

From Ladder of Levels to Potential: A Semi-integration Gives the "Inclusion" and then a Differentiation Gives the "Excursion"

The four horseman of the semiclassical analysis are (Figure 1): the "*inclusion,*" I(E), defined as the area

$$I(E) = \int_{TP_1}^{TP_2} [E - V(r)]\, dr \tag{21}$$

included between the potential energy curve and the available energy, E, and terminated by the two turning points, $r = TP_1$ and $r = TP_2$, where the difference, $E - V(r)$, vanishes (we defer the case where the potential has more than one minimum);

the *phase change* of the wave function between the one turning point and the other,

$$\left[n(E) + \frac{1}{2}\right]\pi = (2\mu/\hbar^2)^{\frac{1}{2}} \int_{TP_1}^{TP_2} [E - V(r)]^{\frac{1}{2}}\, dr \; ; \tag{22}$$

the "*excursion*" of the r-coordinate in periodic motion at energy E,

$$X(E) = \int_{TP_1}^{TP_2} dr = r_2 - r_1 \; ; \text{ and} \tag{23}$$

the *period* of this motion,

$$T(E) = (2\mu)^{\frac{1}{2}} \int_{TP_1}^{TP_2} [E - V(r)]^{-\frac{1}{2}}\, dr \; . \tag{24}$$

Given the ladder of energy levels, $E_0, E_1, \cdots$, we know E as a function of n. We therefore also know the inverse quantity: n itself – or, better, the phase change $\left[n(E) + \frac{1}{2}\right]\pi$ – as a function of E. From this phase change we go to the inclusion, I(E), by semi-integration; and from the

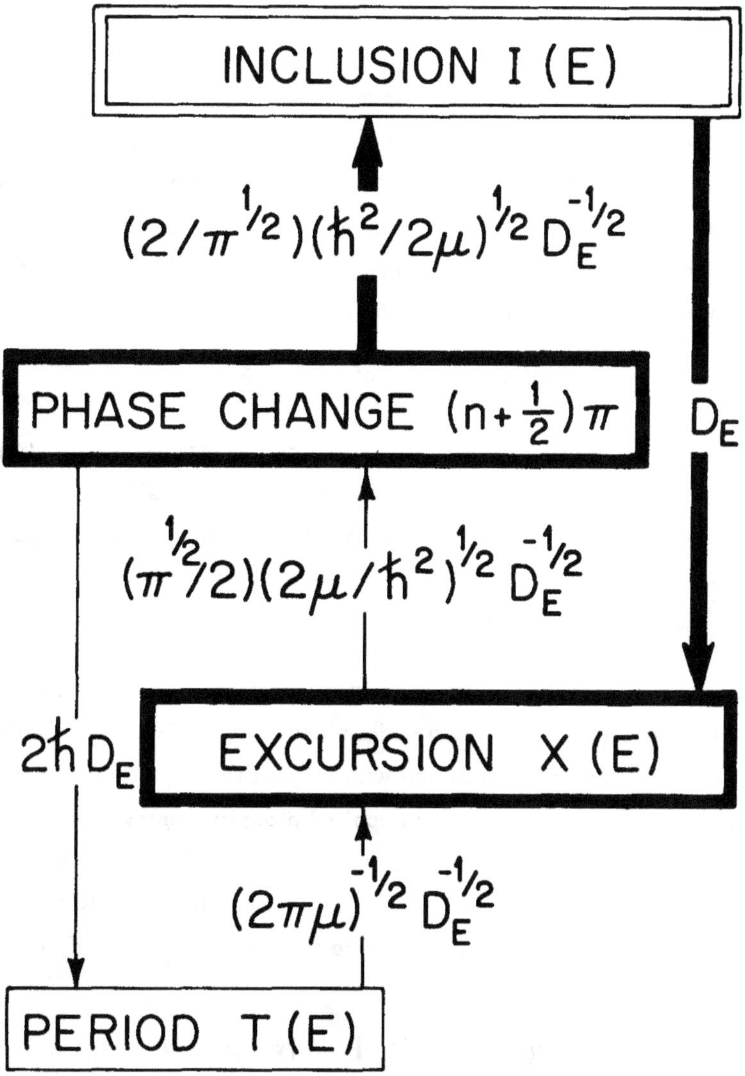

Figure 2. The relation between the four horsemen of the semiclassical analysis as expressed in terms of semi-integration, integration, and differentiation. No use is made of the operation of semidifferentiation, $D^{\frac{1}{2}}$, because the kernel associated with this operator, $\sim (E-E')^{-\frac{3}{2}}$, is too strongly singular to be acceptable in most applications. Let it be known how many levels, n, there are up to each energy, E. Let it be asked how much excursion the particle separation makes in the potential at each energy E, X(E) = r (outer turning point) − r(inner turning point). The answer is obtained (heavy arrows) by one semi-integration (see text for details) followed by one differentiation.

inclusion to the excursion, $X(E)$, by differentiation, as symbolized in the flow diagram of Figure 2. This is the result in brief. Now for commentary.

First notation. We use here the notation of "semi-integration" instead of the more familiar "Abelian integral" in order to make the systematics of Figure 2 stand out as clearly as possible. As for ordinary integration, so for semi-integration (and for fractional-order integration and differentiation in general) definiteness demands the statement of a lower limit. Here that lower limit is the minimum of the potential energy curve:

$$\text{minimum value of } V(r) \equiv E_{min} \ . \tag{25}$$

It is determined as that value of E at which the phase change between turning point and turning point (determined semiclassically and defined observationally in the first instance only for $n = 0, 1, 2, \cdots$) extrapolates to zero:

$$\left[n(E_{min}) + \frac{1}{2}\right]\pi = 0 \ . \tag{26}$$

Conforming to the notation proposed in a recent book on the fractional calculus [Ross (1975); see also Oldham and Spanier (1974)], we preface the label D_E for "differentiation with respect to E" by the subscript prefix E_{min} except when D_E is applied a positive integral number of times ("first derivative," "second derivative," etc.) or when the prefix must be omitted to save space (as in Figure 2 and from time to time in the text below). Thus semi-integration here means

$$_{E_{min}}D_E^{-\frac{1}{2}} f(E) = (1/\pi^{\frac{1}{2}}) \int_{E_{min}}^{E} (E - E')^{-\frac{1}{2}} f(E')\, dE' \ . \tag{27}$$

More generally [Ross (1975)],

$$_{E_{min}}D_E^{s} f(E) \equiv [\Gamma(-s)]^{-1} \int_{E_{min}}^{E} (E - E')^{-s-1} f(E')\, dE' \ . \tag{28}$$

Semidifferentiation we shall not use because in it the kernel $(E-E')^{-3/2}$ is so strongly singular. That is why we do not express X(E) directly in terms of $D_E^{\frac{1}{2}}\left[n(E)+\frac{1}{2}\right]$. That is why instead we apply $D_E^{-\frac{1}{2}}$ to $\left[n(E)+\frac{1}{2}\right]$ to get the intermediate quantity I(E) ("the inclusion") and then differentiate, via D_E, to get X(E).

As an illustration of how semi-integration "works," apply the operator $D_E^{-\frac{1}{2}}$ of (9) to both sides of (4); thus,

$$
\begin{aligned}
{}_{E_{min}}D_E^{-\frac{1}{2}}\left[n(E)+\frac{1}{2}\right]\pi &= {}_{E_{min}}D_E^{-\frac{1}{2}}(2\mu/\hbar^2)^{\frac{1}{2}}\int_{TP_1}^{TP_2}[E-V(r)]^{\frac{1}{2}}\,dr \\
&= (1/\pi^{\frac{1}{2}})(2\mu/\hbar^2)^{\frac{1}{2}}\int_{E_{min}}^{E}(E-E')^{-\frac{1}{2}}\left\{\int_{TP_1(E')}^{TP_2(E')}[E'-V(r)]^{\frac{1}{2}}\,dr\right\}dE' \,.
\end{aligned}
\tag{29}
$$

Rather than formalistically invert the order of integration on the right hand side of (29), ask if there is any contribution to the r-integration from a specified infinitesimal range of integration, r to r + dr. The answer is "no" if r lies outside the limits of excursion, $TP_1(E')$ to $TP_2(E')$. That means, "No, if V(r) exceeds E'." The consequence for the integration over E' is simple. There is no contribution to the integral when E' is less than V(r). Still keeping r fixed, we therefore find for the integration over E' the result

$$
\int_{E_{min}}^{V(r)} 0\,dE' + \int_{V(r)}^{E}\frac{(E'-V)^{\frac{1}{2}}}{(E-E')^{\frac{1}{2}}}\,dE' = (\pi/2)(E-V) \,. \tag{30}
$$

Finally we do the integration over r. Taking all numerical factors to the left hand side of the equation, we get

$$(2/\pi^{\frac{1}{2}})(\hbar^2/2\mu)^{\frac{1}{2}}\,{}_{E_{min}}D_E^{-\frac{1}{2}}\left[n(E)+\frac{1}{2}\right]\pi = \int_{TP_1(E)}^{TP_2(E)} [E-V(r)]\,dr$$

$$\equiv \text{“the inclusion,”} \qquad (31)$$

in agreement with the "flow diagram" of Figure 2. Other results in that diagram are verified in a similar way.

A second commentary has to do with history. Klein (1932) already had the key result when he showed how to find out something about the potential energy curve for a diatomic molecule from its spectrum of vibrational levels by differentiating an Abelian integral (the operation $D_E D_E^{-\frac{1}{2}}$ indicated by heavy arrows in Figure 2). It is also evident from the literature how to go from spectrum to excursion by the reverse order of operations, $D_E^{-\frac{1}{2}} D_E\, n(E)$ (indicated by lighter arrows in Figure 2). Thus the correspondence principle [Bohr (1913)] tells one that the classical frequency of radial oscillation is to be identified in the limit of large quantum numbers with the quantum value for the frequency, $(E_{n+1} - E_n)/h$; or in the continuum treatment adopted in the semiclassical analysis, $T(E) = 2\hbar\pi\, dn(E)/dE$. Moreover, it is well known in classical mechanics how to go from a knowledge of the period as a function of excitation to a knowledge of the excursion as a function of excitation by an Abelian integral [see for example Landau and Lifshitz (1958a)]. In the present notation of the fractional calculus, that integral is the $D_E^{-\frac{1}{2}}$ in the operator product $D_E^{-\frac{1}{2}} D_E$.

Third, and implicit in this earlier work, and explicit in the paper of Pöschl and Teller (1933), there exists an infinite family of spectrum-equivalent potentials. This result foreshadows at the semiclassical level of analysis Bargmann's (1949a, 1949b) famous discovery of a many-parameter family of potentials that (1) give identical spectra of bound states, and (2) for states in the continuum give identical expressions for phase shift as a function of energy. Thus here, where there is no continuum, only bound states, two quite different potentials can give the

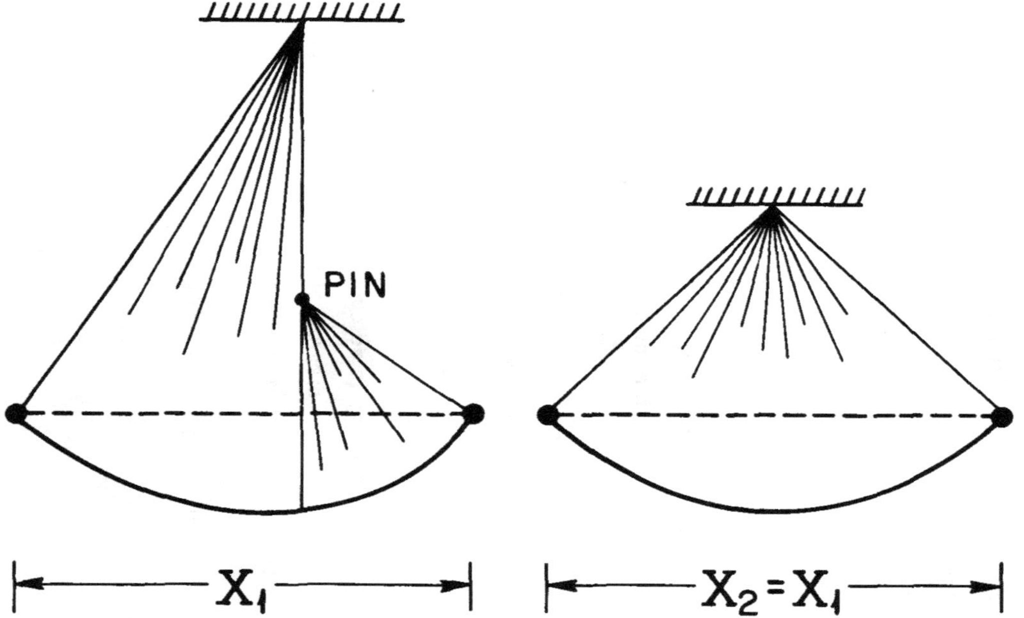

Figure 3. Elementary example of spectrum-equivalent potentials. The pendulum bob-plus-string at the left, intercepted by the pin, has the same spectrum of equally spaced levels (quantum description) or the same period of vibration (classical description) as the equally massive but unintercepted bob-plus-string at the right. That this is so follows from the circumstance that the masses have equal excursions (Figure 1) at equal excitations. Each pendulum here is to be regarded as the symbol for a two-body system of reduced mass $\mu = m_1 m_2/(m_1 + m_2)$ undergoing oscillations in the *linear* separation coordinate r. All corrections for the two-dimensional character of the pendulum motion at large amplitudes are foresworn in this illustration as irrelevant to the *one*-dimensional vibrations that are really the subject of discussion in the text.

same dependence of excursion $X(E)$ upon excitation (Figure 3) and therefore the same ladder of levels. Fixed by the energy levels $E_0, E_1, \cdots$ is neither the left hand branch of the potential energy curve nor the right hand branch but only the separation between them. Arbitrarily specify the left hand branch by giving $r_1(E)$ as some smooth decreasing function of energy. Let this function decrease with energy no faster than the excursion increases. Then the right hand turning point, $r_2(E)$, (1) is fully determined,

$$r_2(E) = r_1(E) + X(E) \ , \tag{32}$$

(2) is a non-decreasing function of energy, and (3) provides the unique acceptable continuation of the left hand branch of the potential energy curve from the potential energy minimum to the right.

As a fourth, illustrative, commentary on the "JWKB inversion procedure" schematized in Figure 2, it is interesting to examine some of the spectrum-equivalent potentials associated with a few standard spectra. As the first example, take the case of a ladder of equally spaced levels,

$$E_n = \left(n + \frac{1}{2}\right) \hbar\omega \ . \tag{33}$$

Here the origin of the scale of energy is so chosen that E_{min} [see Equation (26)] is zero. Following the flow diagram of Figure 2, we evaluate in turn (1) the phase difference between turning points,

$$\left[n(E) + \frac{1}{2}\right]\pi = \pi E/\hbar\omega \ ; \tag{34}$$

(2) "the inclusion" (Figure 1),

$$I(E) = (2/\pi^{\frac{1}{2}})(\hbar^2/2\mu)^{\frac{1}{2}}{}_{E_{min}}D_E^{-\frac{1}{2}}\left[n(E) + \frac{1}{2}\right]\pi$$

$$= (2/\pi)(\hbar^2/2\mu)^{\frac{1}{2}}(\pi/\hbar\omega)\int_0^E (E-E')^{-\frac{1}{2}}\, E' dE'$$

$$= (2^{5/2}/3)\, E^{3/2}/(\mu\omega^2)^{\frac{1}{2}} \ ; \tag{35}$$

(3) the excursion,

$$X(E) = D_E I(E) = 2(2E/\mu\omega^2)^{\frac{1}{2}} ; \tag{36}$$

and finally, only for the sake of completeness, (4) the period,

$$T(E) = 2\hbar D_E \left[n(E) + \frac{1}{2}\right]\pi = 2\pi/\omega . \tag{37}$$

Every quantity agrees with what we would expect for a two-body system of reduced mass μ and zero angular momentum undergoing radial vibrations under the influence of an interaction potential

$$V(r) = (\mu\omega^2/2)(r - r_0)^2 . \tag{38}$$

However, as one sees from Figure 3, one reproduces the same "excursion function" $X = X(E)$ that appears in (36) and therefore the same spectrum (and period, and "inclusion") with a potential.

$$V(r) = (\mu\omega^2/2)(r - r_0)^2/(1-\sigma)^2 \quad \text{for} \quad r < r_0 ,$$

$$V(r) = (\mu\omega^2/2)(r - r_0)^2/(1+\sigma)^2 \quad \text{for} \quad r > r_0 . \tag{39}$$

Here σ is any real number greater than -1 and less than $+1$. Thus, the movement to the left of r_0 is reduced in the ratio $(1-\sigma)$ compared to that given by (20), and the movement to the right is increased in the ratio $(1+\sigma)$; but the sum of the two distances, the excursion itself, $X(E)$, is unaltered.

The complete family of potentials that reproduce the uniform ladder spectrum (16) requires for its parametrization, not a single freely disposable number σ, as in (39), but a function. This function is most conveniently expressed by giving up the idea of writing $V(r)$ as a function of r and instead writing r as a function of V; thus,

$$r = r_0 - (2V/\mu\omega^2)^{\frac{1}{2}} + f(V) \quad \text{(left branch)}$$

$$r = r_0 + (2V/\mu\omega^2)^{\frac{1}{2}} + f(V) \quad \text{(right branch)} \tag{40}$$

Here f(V) is understood to be a function of V that is continuous, and has a well defined derivative, limited by the inequality

$$|df(V)/dV| < (d/dV)(2V/\mu\omega^2)^{\frac{1}{2}} \tag{41}$$

("no overhanging cliffs" in the potential energy curve V(r)).

A similar analysis, following the pattern of Figure 2, reveals the potentials that reproduce other standard spectra* (Table I). In this table the Morse potential [Morse (1929)] is the potential used so often and so conveniently [see for example Eyring, Walter and Kimball (1944) or Slater (1963)] in the analysis of the vibrations of diatomic molecules,

$$V(r) = D[1 - e^{-a(r-r_0)}]^2 . \tag{42}$$

For mathematical convenience, this potential is normalized to zero, not at $r = \infty$, but at $r = r_0$. The value of r_0 cannot be found from the spectrum of vibrational levels (for a single ℓ-value!). Neither can one find from the spectrum the "shift" f(V) (such as is illustrated in Equation (40)) of some other potential that produces the same spectrum to the right or left relative to the Morse potential.

The "box potential" presents two special features. First, the sudden rise of the box potential from zero to infinity at the walls excludes a JWKB analysis. The effective phase of the wave function at each wall is not

---

*The constants employed in the formulas in Table I for the most part have simple standard interpretations: μ, reduced mass of the 2-body system, $\hbar = h/2\pi$ = quantum of angular momentum; ω, constant in expansion of potential about its minimum, $V(r) = E_{min} + (\mu\omega^2/2)(r - r_{min})^2 + \cdots$ or (last case in table) constant in expansion of potential about a *maximum*; D, dissociation energy measured, not from actual ground state to continuum, but from minimum of potential to continuum, L, width of "box" in box potential; Z, integral number of unit charges, e, contained in nucleus that interacts with electron in one-electron atom, $\ell\hbar$, angular momentum of the family of one-electron states under consideration; e = 2.718 ⋯; r_s and E_s, location and height of barrier summit; n_s, interpolated number of states bound in the potential well that lies behind this barrier (ordinarily not an integer); r_1, inner turning point for motion in this well with energy $E = E_s$; I_s, the inclusion for this energy; T_0, T_1, F_1^{-1}, etc., expansion coefficients.

Table I. Standard spectra and the associated inclusion, I(E), excursion, X(E), and period, T(E).

Prototype potential	Spectrum	"Inclusion"	Excursion	Period
Harmonic	$E = \left(n + \frac{1}{2}\right)\hbar\omega$	$I(E) = (2^{5/2}/3)\,E^{3/2}/(\mu\omega^2)^{\frac{1}{2}}$	$X(E) = 2(2E/\mu\omega^2)^{\frac{1}{2}}$	$T(E) = 2\pi/\omega$
Morse	$E = \left(n + \frac{1}{2}\right)\hbar\omega - \left(n + \frac{1}{2}\right)^2 (\hbar\omega)^2/4D$	$I(E) = \frac{2^{3/2} D^{3/2}}{(\mu\omega^2)^{\frac{1}{2}}}\left[\left(\frac{E}{D}\right)^{\frac{1}{2}} - \frac{1}{2}\left(1 - \frac{E}{D}\right)\ell n \frac{1 + (E/D)^{\frac{1}{2}}}{1 - (E/D)^{\frac{1}{2}}}\right]$ $= \frac{2^{5/2} E^{3/2}}{(\mu\omega^2)^{\frac{1}{2}}}\left[\frac{1}{1.3} + \frac{E}{3.5D} + \frac{E^2}{5.7D^2} + \cdots\right]$ for $E \leq D$ $= \frac{2^{3/2} D^{3/2}}{(\mu\omega^2)^{\frac{1}{2}}}$ for $E = D$	$X(E) = \frac{2^{\frac{1}{2}} D^{\frac{1}{2}}}{(\mu\omega^2)^{\frac{1}{2}}}\, \ell n \frac{1 + (E/D)^{\frac{1}{2}}}{1 - (E/D)^{\frac{1}{2}}}$ $= \frac{2^{3/2} E^{\frac{1}{2}}}{(\mu\omega^2)^{\frac{1}{2}}}\left[1 + \frac{E}{3D} + \cdots\right]$ for $E < D$ $= \infty$ for $E = D$	$T(E) = (2\pi/\omega)/(1 - E/D)^{\frac{1}{2}}$

Table I (continued)

Prototype potential	Spectrum	"Inclusion"	Excursion	Period
Box	$E = n^2 \pi^2 \hbar^2/2\mu L^2$	$I(E) = EL$	$X(E) = L$	$T(E) = (2\mu L^2/E)^{\frac{1}{2}}$

Coulomb

$$E = -\frac{\mu}{2}\left[\frac{Ze^2/\hbar}{\ell + \frac{1}{2} + n_r + \frac{1}{2}}\right]^2 \quad \text{and} \quad E_{min} = -\frac{\mu}{2}\left[\frac{Ze^2/\hbar}{\ell + \frac{1}{2}}\right]^2$$

$$I(E) = Ze^2\left[-2(1 - E/E_{min})^{\frac{1}{2}} + \ell n \frac{1 + (1-E/E_{min})^{\frac{1}{2}}}{1 - (1-E/E_{min})^{\frac{1}{2}}}\right]$$

$$\simeq (2/3)(Ze^2)(1-E/E_{min})^{3/2} \quad \text{for } E \text{ close to } E_{min}$$

$$\simeq (Ze^2)\ell n(4E_{min}/e^2 E) \quad \text{for } E \text{ close to } 0$$

$$X(E) = -(Ze^2/E)(1-E/E_{min})^{\frac{1}{2}}$$

$$T(E) = \pi(\mu Z^2 e^4/2)^{\frac{1}{2}}(-E)^{-3/2}$$

Potential well that (1) binds n_s states and (2) is separated from "outside" by a barrier which (3) behaves near summit as $V(r) \simeq E_s - (\mu\omega^2/2)(r-r_s)^2$.

$$\left[n(E) + \frac{1}{2}\right]\pi = \left(n_s + \frac{1}{2}\right)\pi - [(E_s-E)/2\hbar\omega]\ell n[e\hbar\omega/(E_s-E)] - (T_0/2\hbar)(E_s-E) - (T_1/4\hbar)(E_s-E)^2 - \cdots$$

$$I(E) = I_s - (r_s-r_1)(E_s-E) + (2/3)(2/\mu\omega^2)^{\frac{1}{2}}(E_s-E)^{3/2} + (2F_1)^{-1}(E_s-E)^2 + \cdots$$

$$X(E) = (r_s-r_1) - (2/\mu\omega^2)^{\frac{1}{2}}(E_s-E)^{\frac{1}{2}} - F_1^{-1}(E_s-E) - \cdots$$

$$T(E) = T_0 + (1/\omega)\ell n[\hbar\omega/(E_s-E)] + T_1(E_s-E) + \cdots .$$

$\pi/4$, but zero. To nevertheless employ the flow diagram of Figure 2 for this problem is only feasible when one lets the n here play the role of the $\left(n+\frac{1}{2}\right)$ of the JWKB analysis. There is a simple signal that one should make this change. The signal appears when one minimizes the *mathematical* expression for the energy, $E = E(n)$, to find E_{min}. For the box spectrum that minimum is found, not by setting $n + \frac{1}{2} = 0$ (as called for in the JWKB analysis, and to be contrasted with the lowest *physical* state, at $n = 0$ in the JWKB analysis), but by setting $n = 0$ (to be contrasted with the lowest *physical* state, at $n = 1$). Second, the excursion, $X(E) = L$, calculated along these modified lines $\left(n + \frac{1}{2} \rightarrow n\right)$ from the spectrum is independent of energy. Therefore no shift at all is possible in the potential (like $f(V)$ in Equation (40)) without producing an "overhanging" potential energy curve. Moreover, such an overhang would violate the requirement that the potential should be a single-valued function of energy. Thus one has to set $f(V) = 0$. In other words, the potential is unique apart from the indeterminateness in where the left hand wall is located.

If an energy rising as rapidly as n^2 implies walls rising straight up, then an energy rising more rapidly than n^2 would imply walls closing in as the energy increases. But such a potential makes no sense. Therefore the energy of a particle moving non-relativistically in a one-dimensional single-minimum potential can rise no faster than n^2.

The "Inverse Scattering Problem" via JWKB: "No Slideability Above the Line"

It is a short step to extend the analysis from the discrete spectrum to the continuum (Figure 4). The right hand wall of the potential well has moved off to infinity. Its place is taken by the curve of the "fiducial potential" of centrifugal force that appears on the right hand side of Equation (20). In place of the excursion appears the "incursion": the inward shift of the actual potential as compared to the fiducial potential, at a given level of energy, $V = E$; thus,

$$V_\ell(r) = E \tag{43}$$

and

$$V_{\text{fiducial}}(r + \text{incursion}) = E \ . \tag{44}$$

The dotted area in Figure 4 under the line $V = E =$ constant is made of two parts. The part below the line $E = 0$ is the inclusion as calculated by one semi-integration from the phase difference $\left[n(E) + \frac{1}{2}\right]\pi$ between turning point and turning point. The new part, above $E = 0$, arises in the same way from the phase shift. The total "inclusion" or (synonym) "inclosure" is defined now as the sum of the two areas, and is given as before (Figure 2) by a single semi-integration,

$$\begin{aligned} I(E) &= (2/\pi^{\frac{1}{2}})(\hbar^2/2\mu)^{\frac{1}{2}}_{E_0} D_E^{-\frac{1}{2}} \left\{ \left[n(E) + \frac{1}{2}\right]\pi, \ \eta(E) \right\} \\ &= (2/\pi^{\frac{1}{2}})(\hbar^2/2\mu)^{\frac{1}{2}}(1/\pi^{\frac{1}{2}}) \left\{ \int_{E_0}^{0} \frac{\left[n(E') + \frac{1}{2}\right]\pi}{(E-E')^{\frac{1}{2}}} \, dE' + \int_0^E \frac{\eta(E')}{(E-E')^{\frac{1}{2}}} \, dE' \right\} . \end{aligned} \tag{45}$$

From this the incursion, $X(E)$, follows on one differentiation:

$$X(E) = dI(E)/dE \ . \tag{46}$$

This is the solution of the "inverse scattering problem" in the semiclassical approximation.

Nothing is more satisfying about this standard result than the insight it gives into Bargmann's phase-equivalent potentials. Given the bound state energies, and given the phase shift as a function of function of energy in the continuum, one gets by the operation $D_E^{-\frac{1}{2}}$ a perfectly definite inclusion $I(E)$ and from it a perfectly definite "excursion" $X(E)$ (negative energies) or "incursion" $X(E)$ (positive energies). The only difference is this, that for negative energies, two unknown wall positions are involved, with only one condition to fix them – hence "slideability"; whereas for positive energies only one unknown wall position comes in –

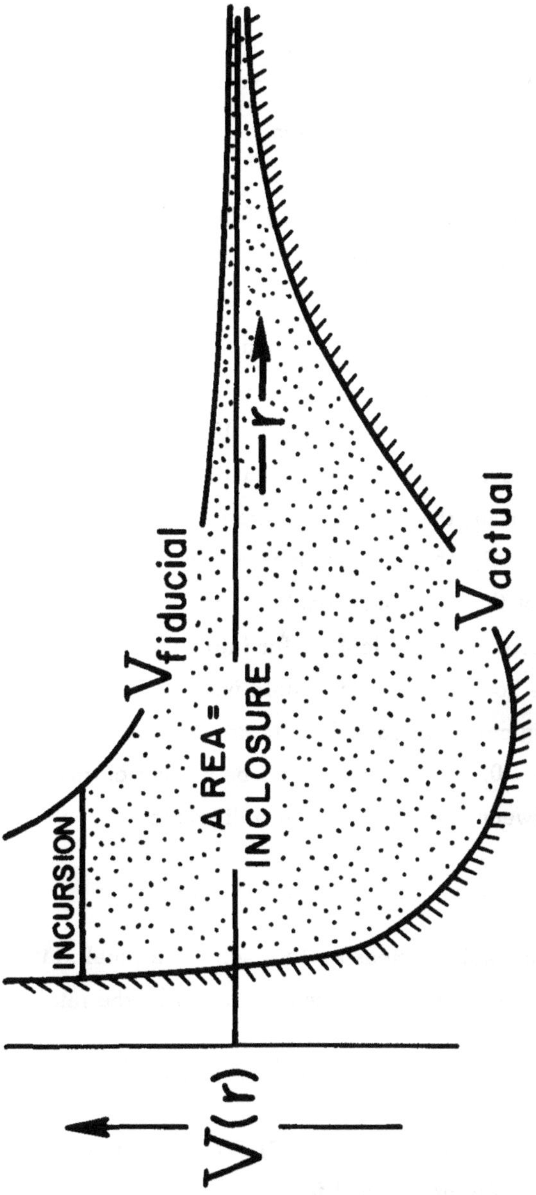

Figure 4. For a positive energy, the ''inclusion'' or ''inclosure'' I(E) (synonyms) includes not only the contribution of bound states, but also a contribution governed (through semi-integration) by the phase shift.

hence no slideability, but a perfectly well determined result. In other words, in the JWKB approximation, *phase-equivalent potentials differ "below the line" but not above it.* There is no easier way to test the semiclassical treatment for accuracy in the present context than to verify numerically how well Bargmann's (1949a, 1949b) phase equivalent potentials comport with this conclusion.

The Difficulty of Resolving Slideability "Below the Line"

"Slideability below the line but none above the line": how is it to be removed? Or can it be removed? Marchenko's formula disposes of the slideability of each bound state by demanding for each a specified value of the constant a_n in the asymptotic expression $\sim a_n \exp(-\kappa_n r)$ for the radial part of the wave function. If one could think of a marksman firing at the system with a suitable rifle, and scoring a point only when at the moment of hit the two centers are further separated than some substantial distance R, then one might have a way to determine this factor a_n in the tail of the wave function.

As an alternative to looking for information about the right hand turning point of the bound states, one could imagine getting information on the left hand turning point via something to do with the strength of the wave function near $r = 0$. An example would be a reaction induced by a close encounter between the two nuclei of a diatomic molecule; for instance, the reaction

$$^{6}\mathrm{Li} + \mathrm{D} \rightarrow 2\mathrm{He} \quad . \tag{47}$$

Imagine being able to measure for each vibrational state $n = 0, 1, 2, \cdots$ of LiD of the specified angular momentum, $\ell\hbar$, the rate,

$$A_n(\sec^{-1}) = \nu_n N e^{-2P} \quad , \tag{48}$$

of the "spontaneous" reaction (47). One knows from thermonuclear reaction rates the factor N, and from the spacing of vibrational states the frequency ν_n of impact on the potential barrier against the present inward motion. Therefore, one would have a way to get the Gamow penetration factor e^{-2P}. From it, information would at once follow on the distance from $r = 0$ to $r = TP_1$.

The black bird will not regard either program to get the missing information as very feasible – either the marksman or the transmutations. This is where he will part company with the crimson bird. It continues its loyal attachment to the "ℓ = constant" program. He goes off to other ways to find $V(r)$.

The Rydberg-Klein-Rees Method

Determine the inclusion I, not only as a function of energy, E, as heretofore, but also as a function of angular momentum or as a function of the parameter

$$K = \left(\ell + \frac{1}{2}\right)^2 . \tag{49}$$

Then, according to the now standard Rydberg-Klein-Rees method of analyzing data on atom-atom interactions,

$$\partial I(E, K)/\partial E = \int_{r_1}^{r_2} dr = r_2 - r_1 , \tag{50}$$

as before; but also

$$-(2\mu/\hbar^2)\,\partial I(E, K)/\partial K = \int_{r_1}^{r_2} dr/r^2 = (1/r_1) - (1/r_2) . \tag{51}$$

Equations (50) and (51) together determine both turning points, allowing one to construct the potential energy $V(r)$ in its entirety throughout the region of interaction. Summarizing much experience with applications in the field of molecular physics, Mason and Monchick (1967) remark, "The high accuracy of the RKR method, in spite of its being a first-order WKB approximation, can be rationalized. Near the minimum, it is known that the first-order WKB energy levels are exact. Near the dissociation limit, the motion becomes more clearly classical and first-order WKB results should also be accurate. The RKR method stands alone among those discussed in this chapter as far as ease and accuracy of reconstructing a potential curve are concerned. From the standpoint of a mathematician, it is also the most satisfying because the reconstruction is almost unique." Here these reviewers are referring to what one can and does get out of the analysis of bound states. However, they go on to discuss scattering; and their review is so brief and to the point that it can be quoted here:

Semiclassical Analysis of Scattering in a Broader Context

"A very elegant body of theory, developed primarily for nuclear scattering, has grown up for methods of abstracting [phase shifts from the cross section for differential scattering] and of subsequently reconstructing the potential. If the energies and eigenfunctions of the bound states are known, this potential may even be determined unambiguously. These methods do not seem to have been applied to atomic and molecular scattering, except in the classical limit. One reason for this is the very large number of phase shifts that contribute appreciably to [the cross section] in atomic and molecular scattering; another reason is that extensive experimental data on cross sections as functions of energy and scattering angle have only recently become available.

"When many partial waves contribute to the observed scattering, several mathematical approximations become permissible, which lead to the so-called semiclassical description of scattering. Under certain limiting conditions, the semiclassical description of wave scattering becomes identical to the classical description of particle scattering. Since the mathematical description of wave scattering is the same whether photons, electrons, atoms, or baseballs are being considered, there are many analogies between particle scattering and optical phenomena, some of which were first pointed out by Ford and Wheeler. The classical limit of particle scattering corresponds to the ray tracing of geometric optics; a classical description is therefore inadequate whenever diffraction or interference effects are important. There are three distinct parts to the semiclassical approximation for scattering. The first is the use of the JWKB approximation for the η_ℓ. The second is the replacement of summations over $\ell, \cdots$ by integrations, and the replacement of differences by derivatives, as

$$\eta_{\ell+1} - \eta_\ell \approx \frac{d\eta_\ell}{d\ell} \qquad \eta_{\ell+2} - \eta_\ell \approx 2\frac{d\eta_\ell}{d\ell} \cdots . \tag{47}$$

The third is the use of asymptotic formulas for $P_\ell(\cos\theta)$. A fourth approximation sometimes used is the evaluation of integrals over ℓ by the method of stationary phase.

" ... the measured cross sections at thermal energies usually show different features than those at high energies, because of the different de Broglie wavelengths and because the depth of the potential well is usually comparable to thermal energies. One result is that true total scattering cross sections can be measured because the aperture angle θ_0 can easily be made small compared to the width of the forward scattering peak, which is comparatively broad at thermal energies. Another result is that interference phenomena and other optical analogies are more commonly observed.

"The mathematical theory of most of the interesting optical analogies in particle scattering has been given by Ford and Wheeler. These analogies correspond to the semiclassical limit of the quantal theory of particle scattering, in which many partial waves must be considered.

" ... three optical analogies ... are of current interest in beam scattering studies: rainbows and haloes, shadows (total cross sections), and glories and glorified shadows." [Reference should be made here to a correction of $(\pi/2)$ required at one point in the treatment by Ford and Wheeler of the Airy function near the rainbow angle: Berry and Mount (1972).] Mason and Monchick continue with a description of these interesting effects and a summary of how one today uses them to determine molecular interactions. [See Hundhausen and Pauly (1965), especially p. 330, Figure 15b, for rainbow scattering as it shows up observationally.]

Evidently, semiclassical or JWKB methods are not limited in their usefulness to going from phase shift to potential. In the disentangling of the scattering to find the phase shifts, they also prove of value. One example is rainbow angle scattering. There, as many as three different branches of the classical deflection function (angle as a function of impact parameter) contribute to the scattering observed at a given angle. Slight changes in the angle of observation change the relative phases of the several contributing probability amplitudes in such a way that by a kind of Fourier analysis, one can unravel the two or three components individually. Another example is resonances associated with penetration through a predissociation barrier, which serve to sort out and show up the contribution of one ℓ-value against the background of all the other ℓ-values. The number of contributing ℓ-values may run to scores or hundreds in fractional-eV to few-eV atom-atom and ion-atom scattering.

Nucleus-Nucleus Interaction and the Semiclassical Description of Scattering

The scattering of one atomic nucleus by another in the region of energies of several hundred MeV allows an analysis similar to that nowadays employed in molecular physics. Again the number of partial waves that contribute significantly to the scattering may run into the hundreds. Again the circumstance provides heavy motivation to employ the semiclassical analysis. Again the assumption is made that the potential is independent of velocity. However, there is a complication for nucleus-nucleus interactions which is not normally envisaged for atom-atom interactions. The collision is often inelastic. Fewer particles emerge from the region of reaction with the original energy than entered it. On this account, one has to regard the "optical model potential" of the nucleus-nucleus interaction as a complex quantity, of which the imaginary part governs the "damping" of the wave in the primary channel. Yet to be developed is any systematic means to determine both real and imaginary parts of the potential (as functions of the distance between the two centers) out of a knowledge of real and imaginary parts of the phase shift (as functions of angular momentum and energy). Finding these phase shifts out of the observed scattering is a second problem awaiting complete resolution. At the moment, one finds it easier to take reasonable trial optical model potentials, calculate from each the expected inelastic events and elastic scattering and compare with observation. No highlights of the scattering cross section turn out to furnish more useful guidance in the choice of parameters in the optical model potential than glories and rainbow angles – features that can hardly even be talked about except within the framework of the semiclassical description of scattering. For a few reviews of nucleus-nucleus encounters in the several hundred MeV domain, reference is made to Ghiorso, Diamond and Conzett (1963), Thomas (1968), Hodgson (1971), and Jackson (1974).

The Madelung Rule for the Filling of Atomic Levels, the Necklace Orbit, and the Demkov-Ostrovski Potential

Has anything been done via the semiclassical approach to "inverse scattering theory" to learn about the potential experienced by an electron in an atom? At first sight the topic appears ridiculous. Does one not have by the method of the self-consistent atomic field far better information about the effective potential experienced by an electron inside an atom [see for example Hartree (1957) and Herman and Skillman (1963)] than one could acquire by any number of scattering experiments? One's reaction to the question changes when he asks, not for the most precise way to say, but for the simplest way say, what he knows about the atomic potential. In this connection, interest naturally focuses, not on the potential close to the nucleus, where it obviously has the Coulomb value, $V(r) = -Ze^2/r$, nor outside the atom, where polarization and other complications play a part, but in the part of the atom where the great bulk of the electrons reside, the "main region." How can one carry out an experiment on the scattering of an electron in the "main region"? And what will such an experiment reveal? Happily one already indirectly knows something about the answer. The information in question does not derive from an electron that comes in from infinity, interacts in the "main region" and goes off to infinity. Rather it comes from the motion of an electron that has stayed inside the main region almost the whole time and one which is endowed with the right energy to be at the top of the filled electron states. Rather than asking for the change in azimuth experienced by this electron in the course of its run from $r = \infty$ to $r = r_{min}$ and back to $r = \infty$, one asks for the change in azimuth in the orbital loop from $r = r_{max}$ into $r = r_{min}$ and back out to $r = r_{max}$. To a certain rough approximation, this deflection has the value 4π for all electrons through the "main region"! Where does this result come from? And what does it mean?

The Madelung rule has long been known: that the electrons at the top of the filled distribution in an atom of any substantial atomic number have in common, not the same angular quantum number, ℓ, not the same radial quantum number, n_r, not even the same total quantum number, $n = n_r + \ell + 1$, but to a good approximation the same value of the integer,

$$N = n_r + 2\ell + 1 . \tag{52}$$

[See Wheeler (1971) for references to an appreciable fraction of the very extensive literature on this topic; see also the book on this topic by Klechkovski (1968); see also Demkov, Ostrovski and Berezina (1971) for two diagrams that dramatically show how much better the order of level filling in atoms correlates with N than with n.] In other words, for

$$E(n_r, \ell) = 0 \quad \text{(top of the filled distribution)} \tag{53}$$

there is in this approximation a zero change in energy (the electron *stays* at the top of the filled distribution),

$$(\partial E/\partial n_r)\, dn_r + (\partial E/\partial \ell)\, d\ell = 0 , \tag{54}$$

for any change in the individual quantum numbers that leaves unaltered the new number, N:

$$dN = dn_r + 2d\ell = 0 . \tag{55}$$

It follows from comparison of (6) and (7) that the classical frequency of the θ component of the motion for an electron at the top of the distribution,

$$\nu_\theta = (1/h)(\partial E/\partial \ell) \text{ (evaluated at } E = 0) \tag{56}$$

is twice as great as the classical frequency of excursion in the r direction,

$$\nu_r = (1/h)(\partial E/\partial n_r) \text{ (evaluated at } E = 0) . \tag{57}$$

In other words, as r goes from r_{max} to r_{min} and back to r_{max}, and as the "phase" associated with this radial vibration therefore increases by 2π, the phase associated with the θ motion, namely θ itself, increases by 4π. The orbit, far from having the shape popularized throughout the world, has the "necklace" form of Figure 5.

What does this necklace orbit imply for the atomic potential? Or what does it imply that ℓ differs from orbit to orbit for the orbits "at" $E = 0$, and n_r also differs, but not so the sum N? Write

$$N + \frac{1}{2} = 2\left(\ell + \frac{1}{2}\right) + \left(n_r + \frac{1}{2}\right)$$

$$= 2\left(\ell + \frac{1}{2}\right) + (1/\pi)\int_{TP_1}^{TP_2} \left[f(r) - \left(\ell + \frac{1}{2}\right)^2\right]^{\frac{1}{2}} (dr/r) , \tag{58}$$

and exploit the idealization that this is independent of ℓ. Let attention focus on finding the unknown function,

$$f(r) = -(2\mu/\hbar^2)\, r^2\, V(r) , \tag{59}$$

as a function of $\ell n\, r$. The analysis at this point follows the semiclassical method as used for the "inverse problem" at fixed energy by several investigators, including Rydberg (1931, 1933), Klein (1932), Hoyt (1939), Rees (1947), Firsov (1953), Wheeler (1955), Sabatier (1966, 1972, 1974a,

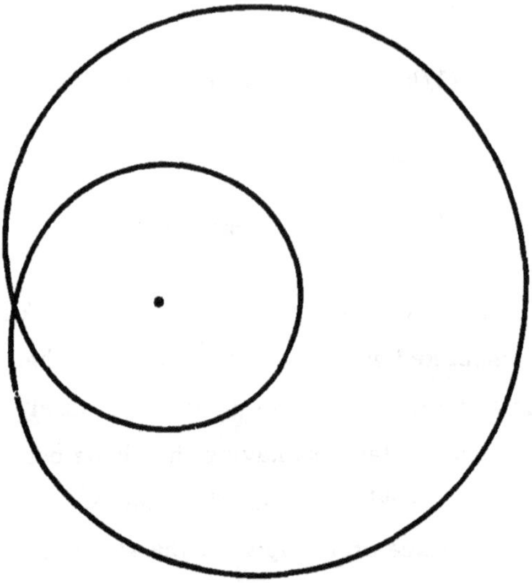

Figure 5. The necklace orbit. This curve is more appropriate than the familiar ellipse as a semiclassical description of the motion of an electron at the top of the sea of filled atomic states.

1974b, 1974c), Miller (1969, 1971), Sabatier and Quyen van Phu (1971), and others, and as reviewed by Bernstein (1966), Mason and Monchick (1967), Bernstein and Muckerman (1967), Berry and Mount (1972) and Buck (1974). The concepts illustrated in Figures 1 and 4, and the methods of semi-integration and differentiation indicated in the flow diagram of Figure 2, to advance from data to "inclusion," and from "inclusion" to "excursion," go through with only minor changes, as indicated in the following Table II.

It is natural to think of $f(r)$ as going to zero at small r, and also at large r, and having in between a maximum value. That maximum determines the largest ℓ-value that an occupied state can have:

$$f_{max}(r) = \left(\ell_{max} + \frac{1}{2}\right)^2 . \tag{60}$$

Table II. The concepts appropriate for solving the "inverse problem" in the context of semiclassical theory (1) at fixed ℓ-value and (2) at fixed energy, compared and contrasted.

Case	Fixed ℓ	Fixed E
The parameter that catalogs the observational data	E	ℓ or $\left(\ell+\frac{1}{2}\right)^2$
The nature of this data	$n_r(E)$ or $\eta(E)$ or both	$n_r(\ell)$ or η_ℓ or both
The unknown quantity	V(r)	$f(r) = -(2\mu/\hbar^2)\, r^2 V(r)$
The variable against which it is envisaged to be plotted	r	ℓn r
"Inclusion" is area between curve so plotted and a horizontal line of	constant E	constant $\left(\ell+\frac{1}{2}\right)^2$
Derivative of inclusion with respect to this parameter gives "excursion" in	r	ℓn r
The inclusion is found from the data by semi-integration with respect to	dE′	$d\left(\ell'+\frac{1}{2}\right)^2$
The kernel in the integration is	$(E-E')^{-\frac{1}{2}}$	$\left[\left(\ell'+\frac{1}{2}\right)^2-\left(\ell+\frac{1}{2}\right)^2\right]^{-\frac{1}{2}}$
The limit in the integration is	$E_{min} = V_{min}$	$\left(\ell_{max}+\frac{1}{2}\right)^2 = f_{max}$

Take some fixed smaller value of ℓ -- *call* it ℓ -- and some "running" value, ℓ', intermediate between ℓ and ℓ_{max}. Evaluate (58) for $\ell = \ell'$, divide both sides of (58) by

$$\left[\left(\ell'+\frac{1}{2}\right)^2-\left(\ell+\frac{1}{2}\right)^2\right]^{\frac{1}{2}} , \tag{61}$$

(employing here the kernel appropriate in the present instance for semi-integration), and integrate with respect to $\left(\ell'+\frac{1}{2}\right)^2$. After multiplying through by 2, for convenience, find

$$\frac{2}{\pi}\int_{(\ell+\frac{1}{2})^2}^{f(r)} \frac{\left[f(r)-\left(\ell'+\frac{1}{2}\right)^2\right]^{\frac{1}{2}}}{\left[\left(\ell'+\frac{1}{2}\right)^2-\left(\ell+\frac{1}{2}\right)^2\right]^{\frac{1}{2}}}\, d\left(\ell'+\frac{1}{2}\right)^2 = \left[f(r)-\left(\ell+\frac{1}{2}\right)^2\right] . \tag{62}$$

Integrating this with respect to $\ell n\, r$ from turning point to turning point gives the new "inclusion" of Table II, an inclusion expressible through a reversal in the order of integration as

$$I\left[\left(\ell+\frac{1}{2}\right)^2\right] = \int_{(\ell+\frac{1}{2})^2}^{f_{max}} X\left[\left(\ell'+\frac{1}{2}\right)^2\right] d\left(\ell'+\frac{1}{2}\right)^2 , \tag{63}$$

where X is the excursion in $\ell n\, r$. This inclusion, derived from the *last* term in (58) by the indicated operations [Equations (61) to (62)], is now to be computed by applying the same operations to all the other terms in (58). Next, take the resulting function I of the independent quantity $\left(\ell+\frac{1}{2}\right)^2$, differentiate it with respect to $\left(\ell+\frac{1}{2}\right)^2$, reverse the sign of the result, and obtain the excursion in $\ell n\, r$,

$$X\left[\left(\ell+\frac{1}{2}\right)\right]^2 = 4\,\ell n\left\{\frac{\left(\ell_{max}+\frac{1}{2}\right)}{\left(\ell+\frac{1}{2}\right)} - \left[\frac{\left(\ell_{max}+\frac{1}{2}\right)^2}{\left(\ell+\frac{1}{2}\right)^2}-1\right]^{\frac{1}{2}}\right\} .$$

In this expression one has now to replace $\left(\ell+\frac{1}{2}\right)^2$ by f(r). Thus one ends up with the excursion in $\ell n\, r$ for every value of f(r) up to f_{max}.

Two conclusions stand out at once. First, from the given data, one will never be able to find out more about the function f than the amount of excursion in $\ell n\, r$ between the two turning points at which $f(r)$ takes on the same arbitrarily specified value $\left(\ell + \frac{1}{2}\right)^2$. Second, granted any function $f(r)$ that satisfies this excursion condition, one can obtain another from it of the same "shape" (curve of f versus $\ell n\, r$) by increasing $\ell n\, r$ by a constant; that is, by simple scaling.

As a simple analytical solution of this excursion condition, Demkov and Ostrovski give the potential

$$V(r) = \frac{-\,\text{constant}}{r(r+R)^2} \,. \tag{65}$$

It is easy to obtain one relation on the two constants in this potential from the requirement that it go over into the Coulomb potential at small r; and another from the requirement that the total of the electrons in all filled states together should be just enough to balance the nuclear charge.

It is not necessary to mention that this potential gives a good fit to the Fermi-Thomas potential throughout the main region, nor that its simple analytic character permits one by JWKB methods to find out much about bound states with no numerical work.

The Demkov-Ostrovski potential has a remarkable feature. Let the strength of the potential be varied by slowly decreasing the value of the quantity N in (52) from some initially large value such as 10. In this transformation, the energy of each bound state is slowly increased until it reaches the ionization limit and is expelled into the continuum. What is more, the model atom expels the uppermost level for every bound ℓ-value at the same instant. This happens for $N = 9, 8, 7 \cdots$. In other words, the Demkov-Ostrovski atom model for any integer value of N shows a zero energy resonance for every ℓ-value up to and including $\ell_{max} = 4, 3, 3, 2, 2, \cdot$

Many are the examples of the dramatic effect it has on the interaction cross section for a compound system to have a resonance at zero energy, and none more so than the $2.7 \times 10^6 \times 10^{-24}\ \text{cm}^2$ cross section presented by ^{135}Xe for the capture of a thermal neutron [for the story of the discovery of which reference may be made, for example, to Babcock (1964)].

However, it is not clear that any model system in physics except the Demkov-Ostrovski atom model even comes close to having resonances for several different ℓ-values simultaneously at zero energy. The analysis of Equations (58), (63) and (64) can be viewed as one means to determine the potential that will have such a special property; and Firsov's (1953) analysis, another. No wonder that Demkov, Ostrovski and Berezina (1971) gives such an interaction the name of "focusing potential."

The Superdirectional Antenna

If such a remarkable feature as focusing attaches to a law of force that puts the resonances for all ℓ-values at zero energy, what remarkable features might be expected if one could, while keeping each resonance sharp, push the resonances for all ℓ-values up into the spectrum of energy (in the case of particles) or the spectrum of frequency (in the case of radiation) into coincidence at exactly the point where one wants the system to show a powerful cross section? One has only to ask this question to be transported from the world of the atom to the world of electrical engineering and the so-called "superdirectional" or "supergain" antenna. There is hardly a better way to introduce this idea than the words of the originators themselves, Schelkunoff and Friis (1952) [see also Friis (1925); Schelkunoff (1942, 1943); Bouwkamp and de Bruijn (1946); Riblet (1948); Yaru (1951); Stearns (1961); and Casimir (1966)]: "It is not difficult to prove that theoretically there is no limit to the directivity of an arbitrarily small antenna ... . [The] more superdirective the antenna, the narrower the bandwidth. Also, a much greater precision will be required in the adjustments of the amplitudes and phases of the various elements ... . [There] is no limit to [the effective interception area of the antenna] *in absence of heat loss*. This is a highly theoretical property since all practical antennas have internal resistance. Nevertheless, this is a property requiring some explanation, for it implies that a perfectly conducting antenna of infinitesimal dimensions is capable of intercepting from a plane wave the amount of power passing through a very large area ... .

The answer is to be found in the combination of resonance and low radiation resistance. These two factors enable the antenna to create a strong reactive field extending to large distances from the antenna which redirects the power passing through a large area of the incoming plane wave and forces it to flow towards the antenna.''

Theoretically attractive, and ideologically linked to the Demkov-Ostrovski atom model, the supergain antenna nevertheless has its problems, as described especially clearly in these words of Casimir (1966): ''Now in actual practice ... your currents get so high and you have so many compensating currents in different directions, that ohmic losses get much too high. Then, of course, you might start dreaming about superconductivity; but even then you get into difficulties, because very soon the tolerances, the geometrical tolerances which such a system would have to satisfy are such that they get below atomic dimensions. All the same it is possible to supergain a combination of a dipole and a quadrupole and I think that is about the most people have ever been able to do. Even adding an octupole would ask for such terrific accuracy that it is not done in practice. So, although first you can tell your radio-astronomer friends they are just fools building large mirrors, because with a small mirror or a small system they could have all the angular definition and all the intercepting area they require, the practical result is that although this is an interesting theorem, you cannot do very much with it.''

With this discussion of a superdirective antenna, have we not come far from anything with any flavor of the semiclassical approximation? Not at all. The existence of an ''inner region'' connected to the ''outside'' through penetration of a barrier or passage through a wave guide channel under appropriate conditions lends itself to analysis by the JWKB method as well here as anywhere.

''Electron Leakage''

It would be a mistake to conclude that the analysis of the Madelung rule exhausts the applications of semiclassical scattering theory to the interaction of electrons with atoms. As one more of the many other applications, and spheres of application [in which connection see for example McDowell (1964), Mott and Massey (1965), and Flaks (1967)], it might be enough to consider the interaction of a low energy (few eV or less) electron with an atom of atomic number well up in the periodic table. In certain regions of Z the effective potential has, not one minimum, but two, as pointed out long ago by Fermi (1928) and Goeppert-Mayer (1941) among others, and evidenced especially clearly in the results of Latter (1955). Moreover, the potential barrier separating the inner and outer minimum lies in the region of positive energy. It can only be surmounted by an electron of still higher energy. An electron of lesser but still positive energy has the option to be to the left of the barrier (bound) or to the right of the barrier (free). An electron in the bound region can therefore undergo the phenomenon of "electron leakage" [Wheeler (1971)] in complete analogy to the process by which an α particle escapes from an atomic nucleus in α-decay. The location of the bound states, the probability of penetration through the barrier, and the phase shift of an electron wave coming from outside and scattered by the barrier — as influenced by coupling with the bound states — all let themselves be analyzed to good advantage by the semiclassical method.

Cases where the Potential Depends on Velocity: Their Prevalence and Catastrophic Consequences for "Inverse Scattering Theory"

One cannot put the red bird of the ℓ = constant exact theory of the "inverse scattering problem" and the black bird of semiclassical theory into a larger perspective without asking, is it always justified to describe the interaction in terms of a potential, V(r), independent of velocity? And what does one do when this description is not justified?

In no context is one the readier to forget the velocity dependence than that of atom-atom collisions. It is the whole point of the Born-Oppenheimer (1927) approximation that it makes sense to treat the electronic motion, and calculate the interatomic potential, as if the nuclei were at rest, despite the fact that in actuality they are moving. How good this approximation is shows best in the instance where one might think it would be worst. The velocity will be greatest and departures from "adiabaticity" of the kind one first fears will be strongest for the system with the lightest atoms, H_2 and H_2^+. The calculated correction [Bunker (1973); see also Bunker (1968)], for nonadiabaticity, in the lowest state of $H_2^+(n_r=0, \ell=0)$ is about 0.1cm^{-1} as compared to an energy of dissociation of 21379.2cm^{-1} (2.65eV).

Much more important than such a minor effect in limiting the concept of a velocity independent potential $V(r)$ is "crossing" of molecular potential energy curves. The separation between the two potential energy curves changes near the point of crossover at a rate

$$L = d(\Delta E)/dr \ . \tag{66}$$

The inevitable coupling, H_{ab}, between the two electronic states in question, ψ_a and ψ_b, changes the separation from linear to hyperbolic,

$$\Delta E = [L^2(r-r_0)^2 + (2H_{ab})^2]^{\frac{1}{2}} \ . \tag{67}$$

The system follows the lower potential energy curve when the separation changes slowly with time. When it changes rapidly, the system "jumps the gap" between the lower hyperbola and the upper hyperbola in such a way as to remain as nearly as possible on its original "straight line course" (straight line over the small scale of r-values relevant for crossover, as distinguished from the much larger range of r-values relevant for the whole course of the collision process, over which the potential is of course far from linear). The probability of crossover [Zener (1932)] is governed by the dimensionless "interaction parameter" G [Hill and Wheeler (1953, Figure 34)],

$$G^2 = 2|H_{ab}|^2/(\hbar \dot{r} L) \tag{68}$$

and has the value

$$(\text{probability}) = e^{-\pi G^2} \tag{69}$$

(small for small $\dot{r} = dr/dt$; close to unity for large $\dot{r}$). In a stationary state, the partial waves corresponding to the two possibilities are recombined, of course, in each cycle of vibration. The consequence, one has to expect, is an effective potential dependent both on r and on $\dot{r}$, in any situation in which one or more level crossings are encountered in the zone of motion.

The concept of a potential affected by level crossings, and this to an extent dependent on the speed, is by no means restricted to molecular physics. The same ideas apply to nuclear physics [Hill and Wheeler (1953)]. The place of the nuclear separation, r, of the molecule is taken by the dimensionless parameter α that measures the collective deformation of the nucleus. Instead of speaking of the states of the individual electrons in the molecule and the total electronic state to which they couple, one speaks of the quantum states of the individual nucleons and the overall quantum state of the nucleonic motion to which these much heavier particles couple. Otherwise the analysis goes through unchanged and Equations (68) and (69) continue to apply (or the appropriate variant thereof when due account is taken of the coupling between partial waves that characterizes any stationary state). Whether there will or will not be any level crossings depends very much on the spin of the compound nucleus. When an even-even heavy nucleus undergoes a deformation leading toward fission, and in the process of this deformation passes over or through the potential energy barrier against fission, it encounters zero or few level crossings as in the case of ^{238}U. When a nucleus of odd mass (such as ^{235}U) goes over the barrier, it encounters a number of crossovers. In consequence, what would otherwise be a smooth potential energy barrier is endowed with cusps equal in number to the number of crossovers encountered. These cusps raise the effective height of the barrier by an amount known as the "specialization energy" [Wheeler (1955a)] and reduce the rate of fission compared to what it would otherwise be. They have other effects. One of them is far from having been explored in full: an alteration in the effective mass or inertial parameter associated with the deformation coordinate, α.

From this discussion, it is clear that when one tries to treat the many-body system, whether a molecule or a nucleus, by the concepts of one-body physics, he ordinarily has to be open to the possibility that the effective potential (and in the nuclear case even the effective mass parameter) can and will depend upon velocity. This is not a new conclusion. Analysis of the interaction between nuclei by the method of "resonating group structure" [Wheeler (1937)] long ago led to a systematic procedure to calculate a "kernel" to describe the velocity-dependent part of the interaction potential and the velocity-dependent part of the inertia factor.

It is enough to limit attention to a velocity-dependent potential (forgetting any dependence of inertia on velocity) in order to see the consequences that follow for scattering theory and for the inverse problem of scattering theory: "If we assume an interaction, $V_{12}\cdot$, depending on momentum and separation in a general way, $V(X, P)$, we will find it convenient to represent it not as a differential operator by replacing P by $-i\hbar\partial/\partial X$, but as an integral operator [Dirac (1930)], according to the equation

$$V_{12}\cdot\psi(x) = \int J(X, \xi)\psi(\xi)\,d\xi \ , \tag{70}$$

where

$$J(X, \xi) = h^{-3}\int V(X, P)\exp\{iP(X-\xi)/\hbar\}\,dP \ . \tag{71}$$

As is well known, the Majorana force in this representation is simply $J(X, \xi) = V(X)\delta(X+\xi)$ and the Wigner force $V(X)\delta(X-\xi)$. ... To be in accord with the law of conservation of energy, $J(X, \xi)$ must be self-adjoint: $J(\xi, X) = J^*(X, \xi)$. ... Referred to a frame of reference in which the center of gravity of the two [centers] is at rest, the wave equation is

$$(\hbar^2/2\mu)\nabla^2\psi + E\psi - \int J(X,\xi)\,\psi(\xi)\,d\xi = 0 \,.$$

"This equation may be separated in polar coordinates, r, θ, ϕ, by writing ψ as a spherical harmonic $Y_L(\theta,\phi)$ of order L ($L\hbar$ being the angular momentum of the system) times an undetermined function $f_L(r)/r$, and by analyzing the total interaction $J(X,\xi) = J(r,\rho,\theta_{12})$ into parts referring to the interaction of two [centers] of definite angular momentum, thus:

$$J(r,\rho,\theta_{12}) = \sum_L (2L+1)\,P_L(\cos\theta_{12})\,J_L(r,\rho)/4\pi r\rho \,. \tag{72}$$

The result is that $f_L(r)$ satisfies the equation

$$(\hbar^2/2\mu)\,f''(r) + [E - L(L+1)\hbar^2/2\mu r^2]\,f(r) = \int J_L(r,\rho)\,f(\rho)\,d\rho \,. \tag{73}$$

One fact at once emerges of interest in connection with the analysis of scattering experiments: From the behavior of two [centers] with angular momentum $L\hbar$ we cannot in general, without further information, draw conclusions as to the behavior of the same two [centers] moving with some other angular momentum." [Wheeler (1936).]

Several problems of nuclear physics have been analyzed by now by the method of resonating group structure. In this work, starting with an assumed law of interaction between nucleon and nucleon, with or without adjustable parameters, one determines the interaction kernel $J(X,\xi)$ and from that, employing Equation (73), calculates scattering or bound states or both for comparison with observation. [For one cross section through the extensive work in this field of "resonating group structure" or the "cluster approximation," see the review report of Tombrello (1968) comparing this and other methods for treating the few-nucleon problem and reports presented at the same international conference.] But is there a way to go in the reverse direction ("inverse scattering theory")? That is the question we put to the crimson bird and the black bird alike. Both answer no! For this problem the semiclassical analysis no longer has its previous advantage over the Gel'fand-Levitan-Marchenko treatment. Previously it took into its grip the information about phase shifts as a function of energy for many ℓ-values simultaneously. There is no point now in bringing all that information together. For a velocity-dependent interaction, each ℓ-value presents a problem unto itself [$j_0(r,\rho)$ independent of $j_1(r,\rho)$ independent of $j_2(r,\rho)$ ...]. Moreover, for any one ℓ-value, where the crimson bird and the black bird plainly have identical rations of information (phase shift as a function of a *single* variable, $\eta_\ell = \eta_\ell(E)$, plus appropriate information about bound states), it is equally plainly beyond the level of the food supplied to either bird to do the task demanded: determine an unknown function of *two* variables, the ℓ-th interaction kernel, $j_\ell(r,\rho)$. Nor would it help to know another function of a single variable, such as the multiplicative factor $c_\ell(E)$ in the small-distance behavior of the wave function,

$$f_\ell(r, E) \sim c_\ell(E)(kr)^{\ell+1} , \qquad (74)$$

as found for example by measuring the chance for the two centers to come close together. In brief, information about the asymptotic properties of the wave function is not enough to determine the wave function itself everywhere. And only information about that wave function, or about the radial factor $f_\ell(r, E)$ (for a given ℓ) for all r and all E, or something equivalent to it, would seem adequate for working out the kernel $j_\ell(r,\rho)$. That information we do not have.

In spite of the considerations mentioned here, why not go ahead anyway and give the crimson bird its standard food: information about phase shifts and bound states for some selected ℓ-value? Very well. *Let* the bird do its standard act. Let it deliver up a potential $V_\ell(r)$. But no longer will we respond with our customary gratitude. Instead, we will say, "Don't you see, you have given us only one solution of our problem,

$$j_\ell(r,\rho) = V_\ell(r)\delta(r-\rho) \ . \tag{75}$$

What about the infinitude of other solutions? And don't you realize that even if you displayed to us all the others, you would still leave us as far as ever from any clue as to which is the right one?" – not a very appreciative response.

In conclusion, the "inverse scattering problem" comes to an unhappy end when the force between the two centers depends significantly on velocity. Yet such a dependence on velocity is inescapable in the interaction between two complex nuclei, and is not rare in the interaction between two atoms. Consideration of such problems forced the introduction of the scattering matrix in the first place [Wheeler (1937)]. It summarizes what one can observe after the interacting centers have separated. It was not conceived as, and never can serve as, a substitute for detailed information about the coupling within the zone of interaction.

The Black Bird and the Crimson Bird: Instability

This chapter up to now has perhaps served to recall something of the scope of that black bird, the semiclassical theory of scattering. What now of the other side of our subject? By studying the zoology of the black bird, what questions are we led to ask and what conjectures are we led to make about the structure of its close relative, the crimson bird of the Gel'fand-Levitan-Marchenko exact "inverse scattering theory"? Two issues have already received attention: (1) Why is it necessary to know for each bound state, in addition to the energy E_n, the constant a_n in (12) that measures the amplitude of the exponentially dying normalized bound-state wave function far outside the zone of binding? Our answer was brief. This or some equivalent information is required to fix the otherwise arbitrary "shift" in the potential minimum to larger or smaller r-values. (2) What does one do when one does not have the complete information demanded by the exact procedure of the crimson bird which makes no concession for incompleteness of data? "Fill in arbitrarily but reasonably the data for energies higher than those reached in the observations," the black bird replies. "I can tell you at once to what extent the resulting potential curve is sensitive to this choice of made-up data, and why." Now we come to a third question: (3) Is the crimson bird stable or unstable? Are there conditions in which the extremely small change in the data that the bird takes in can produce a very great alteration in the potential, $V(r)$, that the bird puts out? The answer is yes. Nowhere does this instability show more clearly than in the analysis of the "inverse energy level problem" for the case of a potential with a double minimum.

The Double-Minimum Potential and the Unravelling of Overlapping Spectra

How does the semiclassical analysis move forward when the potential $V_\ell(r)$ has two minima, separated by a potential hill? So long as the energies under consideration are significantly below the summit of this barrier, leakage through the barrier can be neglected in a first account of the situation. Then the one level spectrum falls apart into two distinct spectra. No one who has ever disentangled on a photographic plate two overlapping band spectra can be in any doubt what this operation is or how it proceeds. The criterion of success is a smooth progression of energy with ordinal number in each of the two spectra individually. Inverted, these progressions yield two curves for phase change between turning point and turning point,

$$\left[n'(E) + \frac{1}{2}\right]\pi \tag{76}$$

and

$$\left[n''(E) + \frac{1}{2}\right]\pi \ . \tag{77}$$

Extrapolated to zero ordinate, these two curves give the energies, E'_{min} and E''_{min}, of the two potential minima. From the information so far, the black bird's analysis proceeds as before (Figure 2) to find in turn the "inclosures" $I'(E)$ and $I''(E)$ by semi-integration,

$$I'(E) = [(2/\pi^{\frac{1}{2}})(\hbar^2/2\mu)^{\frac{1}{2}}]\pi^{-\frac{1}{2}} \int_{E'_{min}}^{E} (E-\tilde{E})^{-\frac{1}{2}} \left[n'(\tilde{E}) + \frac{1}{2}\right] \pi \, d\tilde{E} \qquad (78)$$

and from them by differentiation the "excursions" $X'(E) = dI'(E)/dE$ and $X'' = dI''/dE$ in the two separate minima. It is not known which is the "inner" minimum and which is the "outer" one, until independent physical evidence or the further working out of the semiclassical analysis itself, at higher energies, close to the point of dissociation, settles this point. For simplicity of discussion, this binary bit of information is assumed already to be known. Accordingly, $X'(E)$ is used to refer to the excursion in the inner minimum, between turning points $r = TP_1(E)$ and $r = TP_2(E) = TP_1(E) + X'(E)$; and corresponding labels, TP_3 and $TP_4 = TP_3 + X''$, for the turning points in the outer potential well.

Barrier Penetrability Revealed by Level Perturbations

As the next step the semiclassical analysis recognizes the effect, previously neglected, of leakage through the barrier. This effect is relevant when the energy E is greater than both barrier bottoms, E'_{min} and E''_{min}. The magnitude of the effect is measured by the same kind of quantity that measures the probability per second of penetration through the nuclear potential barrier in the phenomenon of α-decay:

$\Gamma/\hbar$, probability per sec of passage through the barrier;

Γ, the "width for barrier penetration"; more precisely, the full width (expressed in energy units) of the decaying state at half maximum under conditions, as in α-decay, where everything that penetrates through the barrier from left to right is removed on the right (a condition not fulfilled here because the wave travelling to the right is reflected back to the left in the outer potential well); given by

$$\Gamma_{n'}/\hbar = \nu_{n'} e^{-2P} , \qquad (79)$$

where

$$P(E_{n'}) = \int_{TP_2}^{TP_3} \kappa(r, E_{n'})\,dr \qquad (80)$$

is the standard Gamow barrier-penetration integral, and

$$\nu_{n'} = h^{-1}(dE/dn') = (1/2\hbar)\,\frac{dE}{d\left[\left(n' + \frac{1}{2}\right)\pi\right]}$$

$$= (1/2\hbar)\,[d(\text{phase change from } TP_1 \text{ to } TP_2)/dE]^{-1} \qquad (81)$$

is the number of impacts per second on the barrier when the system is oscillating in the n'-th quantum state in the inner potential well.

The wave reflected back from the outer wall of the outer well alters the boundary condition on the wave function at the outer turning point in the inner well. In consequence, the energy of the n'-th quantum state is changed from the value otherwise to be expected, $E_{n'}^{(0)}$ (smooth curve relating $E_{n'}^{(0)}$ to n') to the actual value, $E_{n'}$. The alteration in energy, as found by estimating the difference between the actual points and a smooth curve, is typically of order Γ, and varies in what is at first sight an irregular way from level to level. A closer study of these level perturbations permits the "black bird" to evaluate the barrier's penetrability for a whole sequence of energies from bottom to top, and thus to find the width of the barrier, $X_{barrier}(E)$.

Briefly stated, the perturbation in energy is connected with the "width for barrier penetration," Γ, by the formula

$$E_{n'} - E_{n'}^{(0)} = -\frac{1}{2}\,\Gamma_{n'}\tan\left[\left(n'' + \frac{1}{2}\right)\pi\right] . \qquad (81)$$

Here $\left(n'' + \frac{1}{2}\right)\pi$ is the phase change from TP_3 to TP_4 in the outer potential well, evaluated at energy $E_{n'}$. It can be read off from the smooth plot supposed already in hand for the "outer" spectrum. Say the energy $E_{3'}$ in the "inner" spectrum falls at a point on this curve for the outer spectrum where n'' interpolates to have the value $n'' = 7.50$. Then the tangent on the right hand side of (81) happens to have the value zero. There is no perturbation in the location of $E_{3'}$. One can translate this result into another language by saying that $E_{3'}$ is "pushed up" in this case by "repulsion" from the lower levels $E_{7''}$, $E_{6''}$, $E_{5''}$, $\cdots$ just as much as it is "pushed down" by repulsion from the higher levels $E_{8''}$, $E_{9''}$, $\cdots$. Now imagine an adiabatic lowering made in the whole of the outer potential well, so that all of the $E_{n''}$ are slowly dragged downward. This drop on the right brings $E_{8''}$ closer to $E_{3'}$ and augments the "repulsive force" driving $E_{3'}$ down. At the same time, the lowering of $E_{7''}$ takes it further beneath $E_{3'}$ and attenuates the "repulsive force" that drives $E_{3'}$ up. The net consequence is a lowering, a small lowering of order Γ, in $E_{3'}$.

There is no need to enter into any such theology of repulsive forces in order to analyze the level perturbation via Equation (81). Nor does one have to ask whether the "neutral point" is closer in *energy* to $E_{7''}$ or to $E_{8''}$. The central concept in the analysis of the perturbation is not energy, but phase, as evidenced in the argument $\left[n''(E_{n'}) + \frac{1}{2}\right]\pi$ of the tangent function in (81).

From here on, the analysis is a simple problem in curve-fitting: find smooth curves, with a minimal number of parameters, for $E_{n'}^{(0)}$ and $\Gamma_{n'}$ as functions of n' that will reproduce [via perturbation Equation (81)] the observed levels $E_{n'}$. This done for the inner spectrum, the process of analysis into smooth curve plus perturbations is repeated for the outer spectrum. It hardly makes sense to iterate the analysis to produce improved curves. It does make sense to demand – and to test that the demand is satisfied – that the analyses of the perturbations of the inner and outer spectra yield curves for the dependence of barrier penetration probability on energy that are identical within the expected margin of accuracy:

$$e^{-2P} = 2\Gamma_{n'}\, d\left[\left(n' + \frac{1}{2}\right)\pi\right]/dE \tag{82}$$

and

$$e^{-2P} = 2\Gamma_{n''}\, d\left[\left(n'' + \frac{1}{2}\right)\pi\right]/dE \ . \tag{83}$$

Inverting the Barrier Penetration Integral to Find the Distance Through the Barrier

Whatever the energy at which the empirically determined Gamow penetration integral extrapolates to zero, that is the empirically determined height, E_{max}, of the barrier:

$$\text{``Limit''}_{E \to E_{max}} P(E) = 0 \ . \tag{84}$$

This quantity known, and the function $P(E)$ known, the "black bird" has all that it needs to find the "inclusion," $I_{barrier}(E)$, under the barrier (Figure 6) by semi-integration,

$$I_{barrier}(E) = [(2/\pi^{\frac{1}{2}})(\hbar^2/2\mu)^{\frac{1}{2}}]\,\pi^{-\frac{1}{2}} \int_E^{E_{max}} (\tilde{E}-E)^{-\frac{1}{2}} P(\tilde{E})\, d\tilde{E} \ . \tag{85}$$

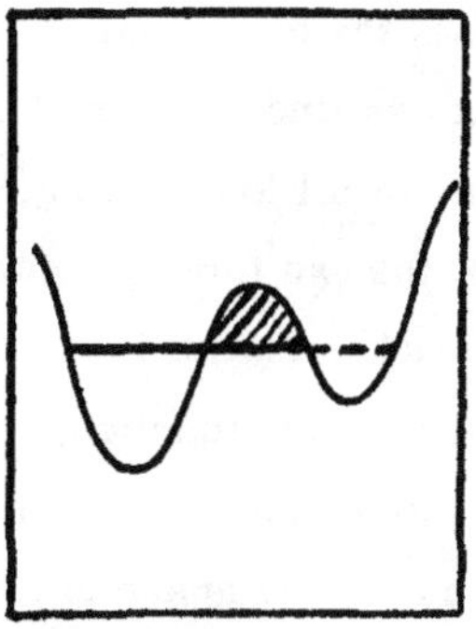

Figure 6. The "inclusion" (shaded area) under the potential barrier, differentiated once with respect to energy, yields the distance through the barrier.

One is naturally led to this equation by "turning upside down" the previous equation for the inclusion [say Equation (35)]. Where there the integration ran from E_{min} to E, and the kernel of the semi-integration was $(E-\tilde{E})^{-\frac{1}{2}}$, here the integration goes from E to E_{max} and the kernel is $(\tilde{E}-E)^{-\frac{1}{2}}$. It is only necessary to insert into this integral the expression for the Gamow penetration integral,

$$P(E) = \int_{TP_2}^{TP_3} \kappa(r, E)\, dr = (2\mu/\hbar^2)^{\frac{1}{2}} \int_{TP_2}^{TP_3} [V(r) - E]^{\frac{1}{2}}\, dr \qquad (86)$$

to verify that the new inclusion is indeed

$$I_{barrier}(E) = \int_{TP_2}^{TP_3} [V(r) - E]\, dr \qquad (87)$$

$$= \int_{E}^{E_{max}} X_{barrier}(\tilde{E})\, d\tilde{E}\ , \qquad (88)$$

as expected. Differentiation immediately gives the distance through the barrier,

$$X_{barrier}(E) = -dI_{barrier}/dE \qquad (89)$$

as desired. With this done, the black bird has found out from the spectrum of bound states all that it can find out from that information alone: It has determined the separations of all four turning points below the barrier summit, and the separation of the two turning points that remain above the summit. The double-minimum potential is in that way completely specified up to a single unknown, the "slide function," exactly as in (40).

Probably wise to skip in a first reading, but necessary in the end for completeness, is the more detailed analysis required when the simple perturbation formula,

$$E_{n'} - E_{n'}^{(0)} = -\frac{1}{2}\Gamma_{n'} \tan\left[\left(n'' + \frac{1}{2}\right)\pi\right], \tag{90}$$

fails to apply. That situation can develop on occasion well below the top of the barrier and always does develop close to the top of the barrier. It comes about when the perturbation calculated from the naïve application of (90) is, or when Γ itself is, an appreciable fraction of the level spacing.

The Regime of Level "Crossover"

The "transition width" Γ, given by

$$\Gamma = \hbar\nu e^{-2P} = \frac{(\text{level spacing})}{2\pi} e^{-2P} , \tag{91}$$

is very small compared to the spacing well below the top of the barrier. However, the factor, $\tan\left[\left(n'' + \frac{1}{2}\right)\pi\right]$, that multiplies Γ will become arbitrarily large when the perturbing level in the outer well ($n'' = 8$ in the earlier example) comes arbitrarily close to the unperturbed position of the inner-well level under consideration; thus,

$$\begin{aligned} \tan\left[n''(E_{n'}^{(0)}) + \frac{1}{2}\right]\pi &\simeq \frac{-1}{[n''(E_{n'}^{(0)}) - n''(E_{n''})]\pi} \\ &= -\frac{1}{\pi}\frac{dE}{dn''}\frac{1}{E_{n'}^{(0)} - E_{n''}} \\ &= -\frac{2\hbar\nu''}{E_{n'}^{(0)} - E_{n''}} . \end{aligned} \tag{92}$$

In the same approximation, the perturbation formula (81) becomes

$$E_{n'} - E_{n'}^{(0)} = -\frac{(\hbar\nu')(\hbar\nu'')e^{-2P}}{E_{n''} - E_{n'}^{(0)}} . \tag{93}$$

But what is to be understood here by $E_{n''}$: the actual position of that level in the outside well? or the position as modified by the perturbative influence of $E_{n'}$? The difference between those two energies will be given in the same approximation by a similar formula,

$$E_{n''} - E_{n''}^{(0)} = \frac{-(\hbar\nu')(\hbar\nu'')e^{-2P}}{E_{n'} - E_{n''}^{(0)}} . \tag{94}$$

In actuality, both (93) and (94), when pushed to the extreme, are incompatible with "the principle that action equals reaction." They break down when tne positions of the two unperturbed levels in question, $E_{n'}^{(0)}$ and $E_{n''}^{(0)}$, have a separation comparable to or less than the quantity Δ defined by the equation

$$\Delta = [(\hbar\nu')(\hbar\nu'')]^{\frac{1}{2}} e^{-P} . \tag{95}$$

When the level separation is of this order or less, the positions of the perturbed levels are to be calculated from the formulas

$$E_a = \left[\tfrac{1}{2} E_{n'}^{(0)} + \tfrac{1}{2} E_{n''}^{(0)}\right] + \left\{\left[\tfrac{1}{2} E_{n'}^{(0)} - \tfrac{1}{2} E_{n''}^{(0)}\right]^2 + \Delta^2\right\}^{\frac{1}{2}} \tag{96}$$

and

$$E_b = \left[\tfrac{1}{2} E_{n'}^{(0)} + \tfrac{1}{2} E_{n''}^{(0)}\right] - \left\{\left[\tfrac{1}{2} E_{n'}^{(0)} - \tfrac{1}{2} E_{n''}^{(0)}\right]^2 + \Delta^2\right\}^{\frac{1}{2}} . \tag{97}$$

Here the levels are given the neutral designations E_a and E_b because the states that they describe are mixtures of states in the inner well and the outer well, as is seen from the following transition:

$$E_a = E_{n'}^{(0)} + \frac{\Delta^2}{E_{n'}^{(0)} - E_{n''}^{(0)}} \quad \text{for} \quad E_{n'}^{(0)} - E_{n''}^{(0)} >> \Delta ;$$

$$E_a = \tfrac{1}{2} E_{n'}^{(0)} + \tfrac{1}{2} E_{n''}^{(0)} + \Delta \quad \text{for} \quad E_{n'}^{(0)} = E_{n''}^{(0)} ;$$

$$E_a = E_{n''}^{(0)} + \frac{\Delta^2}{E_{n''}^{(0)} - E_{n'}^{(0)}} \quad \text{for} \quad E_{n''}^{(0)} - E_{n'}^{(0)} >> \Delta . \tag{98}$$

The separation between E_a and E_b is never less than 2Δ. This "level crossover without level crossover" reminds one of the familiar question, "What happens when an irresistible force meets an immovable object?" and of the answer, "It goes through without leaving a hole behind it." The following Table III boils down the features of this crossover regime and of the other regimes mentioned or to be mentioned.

Table III. Perturbation regimes. Bound state n' in inner well is perturbed by the amount $\Delta E = E_{n'} - E_{n'}^{(0)}$ through interaction with bound states in the outer potential well. Relevant magnitudes are (1) H, the height of the barrier summit relative to the energy level under consideration; (2), $D' = h\nu'$ = level spacing in the inner well, and $\hbar\nu' = D'/2\pi$; (3), "the coupling," Δ = (geometric mean of the level spacings in the two wells $/2\pi)e^{-P}$; (4), "the width for barrier transmission," Γ' = (level spacing $/2\pi)e^{-2P}$; and (5), the circular frequency, ω, with which the system would oscillate if the barrier were turned upside down [defined by $V(r) \simeq E_{max} - \frac{1}{2}\mu\omega^2(r - r_{max})^2$ near the summit].

Distance below barrier	$H >> (\hbar\omega)/2\pi$	$H >> (\hbar\omega)/2\pi$	$H >> (\hbar\omega)/2\pi$	$H \sim (\hbar\omega)/2\pi$
P in e^{-2P} substantially larger than unity?	Yes	Yes	Yes	No
Γ' and Δ significant in comparison with $D'/2\pi$?	No	No	No	Yes
Separation of unperturbed $E_{n'}^{(0)}$ and $E_{n''}^{(0)}$	$\sim D'$ and $>> \Delta$	$<< D'$ and $>> \Delta$	$<< D'$ and $\sim \Delta$	$\sim D'/2\pi$ and $\sim \Delta$ and $\sim \Gamma'$
Formula to use for perturbation, ΔE?	"tangent formula" (90)	"energy denominator formula" (93)	"root of quadratic equation" (96)	analysis in terms of parabolic cylinder functions
Order of magnitude of ΔE?	$<<< D'$ $<< \Delta$ $\sim \Gamma'$	$<<< D'$ $<< \Delta$ $\sim \Gamma'$	$<< D'$ $\sim \Delta$ $>> \Gamma'$	$\sim D'/2\pi$ $\sim \Delta$ $\sim \Gamma'$

The purpose of the whole analysis is still to extract information about the potential energy curve. For this purpose, levels that interact closely (separation $\sim \Delta$) have to be given special treatment. For them the perturbation $(\sim \Delta)$ is greater than the perturbations considered so for $(\sim \Gamma)$ by a factor of the order of e^P. In constructing the curves (minimal number of parameters!) that give the best estimates of the course of the two spectra, $E_{1'}^{(0)}, E_{2'}^{(0)}, E_{3'}^{(0)}, \cdots$ and $E_{1''}^{(0)}, E_{2''}^{(0)}, \cdots$ as they would have gone in the absence of interaction, it is advisable to use neither E_a nor E_b individually. More relevant is the combination

$$E_a + E_b = E_{n'}^{(0)} + E_{n''}^{(0)} . \tag{99}$$

This one bit of information can only then be properly exploited when the data matching is done by fitting the two curves $E(n')$ and $E(n'')$ simultaneously. When these curves have been constructed, the value of the "coupling" Δ at the relevant point in the spectrum follows at once from the formula

$$4\Delta^2 = (E_a - E_b)^2 - (E_{n'}^{(0)} - E_n^{(0)})^2 . \tag{100}$$

No better opportunity offers itself to get a good fix on the value of Δ than such an instance of two closely interacting levels. From Δ the value of the desired barrier penetration integral, P, can be obtained:

$$e^{-P} = \Delta/[(\hbar\nu')(\hbar\nu'')]^{\frac{1}{2}} . \tag{101}$$

It is not necessary to go through the detailed derivation of the foregoing formulas to say that they rest on the standard procedures for fitting together at a turning point the JWKB formulas for the oscillatory regime and the exponential regime, formulas available for example in Fröman and Fröman (1965), in Kemble (1937), and in Pauli (1933). The perturbation in energy,

$$\Delta E = E_{n'} - E_{n'}^{(0)} , \tag{102}$$

raises the phase change between TP_1 and TP_2

$$\text{from } \left(n' + \frac{1}{2}\right)\pi$$

$$\text{to } \left(n' + \frac{1}{2}\right)\pi + \delta' , \tag{103}$$

an increase of

$$\delta' = \Delta E \, d\left[\left(n' + \frac{1}{2}\right)\pi\right]/dE = \frac{1}{2}\,\Delta E/(\hbar\nu') \; . \tag{104}$$

This change is very small, but it sucks towards the inner well the node of the wave function that lies outside of, and to the right of, that well. The wave function at TP_2 approaches closer to zero. This decrease in value is symbolized by the drop from $\pm\sin(\pi/4)$ to $\pm\sin[(\pi/4)-\delta]$ in the trigonometric part of the JWKB formula at TP_2. Here are the JWKB expressions for the wave function in the successive domains:

$$(1/2\kappa^{\frac{1}{2}})\exp\left(-\int_r^{TP_1}\kappa\,dr\right) \; ;$$

$$(1/k^{\frac{1}{2}})\sin\left[(\pi/4) + \int_{TP_1}^{r} k\,dr\right]$$

$$= (-1)^{n'}(1/k^{\frac{1}{2}})\sin\left[\int_r^{TP_2} k\,dr + (\pi/4) - \delta'\right] \; ;$$

$$(-1)^{n'}(1/\kappa^{\frac{1}{2}})\left\{\frac{1}{2}\cos\delta'\exp\left(-\int_{TP_2}^{r}\kappa\,dr\right) - \sin\delta'\exp\left(\int_{TP_2}^{r}\kappa\,dr\right)\right\} ;$$

$$(-1)^{n'}(1/k^{\frac{1}{2}})\left\{\cos\delta'\,e^{-P}\cos\left[\int_{TP_3}^{r} k\,dr + (\pi/4)\right]\right.$$

$$\left. - 4\sin\delta'\,e^{P}\sin\left[\int_{TP_3}^{r} k\,dr + (\pi/4)\right]\right\}$$

$$= (A/k^{\frac{1}{2}})\left\{\cos\left[\left(n'' + \frac{1}{2}\right)\pi\right]\cos\left[\int_{TP_3}^{r} k\,dr + (\pi/4)\right]\right.$$

$$\left. + \sin\left[\left(n'' + \frac{1}{2}\right)\pi\right]\sin\left[\int_{TP_3}^{r} k\,dr + (\pi/4)\right]\right\} ;$$

and

$$(\text{constant}/\kappa^{\frac{1}{2}}) \exp\left(-\int_{TP_4}^{r} \kappa \, dr\right) . \qquad (105)$$

Here $\left(n'' + \frac{1}{2}\right)\pi$ is an abbreviation for the phase difference,

$$\left[n''(E) + \frac{1}{2}\right]\pi \equiv \int_{TP_3}^{TP_4} k \, dr . \qquad (106)$$

The necessary line of reasoning is described nowhere more clearly – though in a different context – than in Breit's treatment of radioactive disintegration [Breit (1932); expounded more fully in Breit and Yost (1935)]. One is led there and here to introduce the "width for barrier penetration,"

$$\Gamma_{n'} = (1/2\pi)(dE/dn')e^{-2P} , \qquad (107)$$

to simplify the discussion, the results of which are summarized above.

The Regime of Spectrum Squeeze

New effects come in near the top of the barrier. As the energy E approaches the barrier summit, reaches it, and rises above it, the two separate spectra merge into one. Nothing better describes one's first impression of this event than the chaos of Sunday afternoon traffic, when two streams of cars have to squeeze to one, coming though they do from different sources, with different spacings, and different phases at the point of join. Out of this irregularity, regularity nevertheless lets itself be distilled. The necessary considerations will not be spelled out here, for they have already been given in a slightly different but closely related context [quantum effects near a barrier maximum: Ford, Hill, Wakano, Wheeler (1959); especially important in the phenomenon of orbiting: Ford and Wheeler (1959); also well illustrated in the relevant degree of freedom of the asymmetric top: Kramers and Ittmann (1929)]. It is more to the point to state clearly the information has to be extracted from, and can be extracted from, this "spectrum squeeze":

(1) The curve of $\left[n'(E) + \frac{1}{2}\right]\pi$ as a function of E as it *would* have been in the absence of barrier leakage and other quantum effects (over and above the canonical JWKB phase of $\pi/4$).

(2) The corresponding curve for $\left[n''(E) + \frac{1}{2}\right]\pi$.

(3) The curve for the idealized penetration integral,

$$P(E) = \int_{TP_2}^{TP_3} \kappa(r, E)\, dr \tag{108}$$

also as a function of E, also right up to the summit of the barrier, and also with all other quantum effects boiled out of it.

(4) The curve for $\left[n_{merged}(E) + \frac{1}{2}\right]\pi$ as a function of E for a little way above the summit of the barrier, again as it *would* have been in the absence of barrier-induced corrections to the JWKB analysis. These corrections fall off exponentially with increase of energy above the barrier summit. The characteristic distance of fall off is of the order of the energy $\hbar\omega$, where ω is the classical circular frequency of oscillation in the overturned potential barrier.

Above this point it is normally reasonable to expect to be able to read off the phase difference between TP_1 and TP_4, $\left[n_{merged}(E) + \frac{1}{2}\right]\pi$, directly from the count of levels as they are encountered one by one running up the spectrum.

Because of the sophistication of the quantum corrections near the barrier summit, it is reasonable to forget them in appropriate situations for a first analysis, and determine items (1, 2, 3, 4) at and near the peak by elementary extrapolation from lower and higher energies. It is one the tests of such an extrapolation that at E_{max} itself the phase difference between TP_1 and TP_4 must have the same value whether the summit is approached from above or below:

$$\left[n_{merged}(E_{max}) + \frac{1}{2}\right]\pi = \left[n'(E_{max}) + \frac{1}{2}\right]\pi + \left[n''(E_{max}) + \frac{1}{2}\right]\pi \,. \quad (109)$$

Above the Barrier

If a small change in level position below $E = 0$ has a big consequence for the calculated barrier thickness, a minor change in "level width" above $E = 0$ gives a comparable effect; and we sustain our impression of the instability of the whole inversion process. We are considering a situation where a barrier projects up into the continuum, and, as throughout, we limit attention to situations where the JWKB analysis is appropriate (smooth potential; wave length short compared to the relevant dimensions of the zone of interaction). The phase shift varies smoothly with energy, as if the incoming particle felt only, and were scattered only, by the outer turning point, except at "resonances," (actually more like steps than resonances) where the phase shift suddenly jumps by π. The width, Γ, of the "resonance" as determined by this kind of observation from outside is known to be identical, up to the factor $\hbar$, with the width as deduced from the rate of radioactive decay, $A(\sec^{-1})$, of systems located initially in the same state inside the barrier and decaying to the outside; thus,

$$\Gamma(\text{erg}) = \hbar A(\sec^{-1}) \,. \quad (110)$$

From the Γ values for the resonances, we deduce the penetration integral, $P(E)$, as before. Then semi-integration yields the "inclusion" under the barrier, and differentiation of that quantity gives the barrier thickness, $X_{barrier}(E)$. One finds all the turning points "above the line $(E = 0)$." There is no slideability. But the width of the region in which the phase suddenly rises by the value π has to be determined with reasonable percentage accuracy if the thickness and other details of the barrier are to be found with reasonable percentage accuracy and *if* no call is to be made on information from other ℓ-values. That requirement can be difficult to meet if the width in question is very small.

The Physics Hidden in the Marchenko Equation, Waiting to be Revealed

The method of the integral equation (105.15) [just after (12)] to obtain the potential, when there is a potential, from data on scattering and bound states has the appeal of uniformity. It does not ask when approximation methods are appropriate and when they are not. Still less does it speak about "regimes": the regime above the barrier versus the regime below the barrier; the regime of easy leakage through the barrier versus the regime of difficult leakage through the barrier; or the regime of many bound states versus the regime of few bound states. All these are questions to which that method of analysis is indifferent. Simply drop in the data and turn the crank: that seems to be the promise. Nothing could come closer to satisfying that student who said about a course in nuclear models, "Nuclear models? I don't want to have to understand nuclear models. I don't want to have to think. All I want to do is calculate."

In contrast to this simplistic approach that the crimson bird offers to every problem, the black bird, dependent on its semiclassical methods, even has to turn its back on problems where the effective wave length is large compared to the size of the zone of interaction. And when it can deal with a problem, it does so far from simplistically. Instead, it regards the problem as a rich zoological park, crowded with the most varied animals, each of which requires its own special treatment: interwoven spectra to be disentangled; level interaction; barrier penetration; spectrum squeeze; phase shift to be decomposed into continuous curve plus jumps; level widths and radioactive decay rates; "phase difference" and "penetration integral" as function of energy; "inclusion"; and finally the "excursion" of the potential, its "slideability," and the possibilities to resolve this slideability by additional physical information. Moreover, some of the most inconspicuous of these effects are the most important in fixing the shape of the potential when that potential contains a barrier. A very minor change in the relative position of two nearby bound levels can have a very big effect on the calculated barrier width; so can a very minor change in the width of the energy region in which the phase suddenly jumps by π.

In addition, a level spectrum that thins out too quickly with rising energy signifies to the watchful black bird a physical impossibility, a potential with overhanging walls.

All of the effects mentioned have the most direct possible physical interpretation – and every one of them the uniformitarian machinery of the integral equation sweeps under the rug. Of course truth will out. One would only have to program a computer for solving the integral equation to recognize terrifying instabilities "inexplicably" scattered here and there to very minor changes in certain parts of the input data. Moreover, if one turned from machine analysis of that integral equation to regime analysis, as one is in the habit of doing with every other equation of physics, one would rediscover the hard way every one of the regimes that has come to light so naturally in the semiclassical analysis.

The integral equation that Marchenko got by the Gel'fand-Levitan method is a great and beautiful achievement. It is a measure of its greatness and its beauty that it has telescoped away and hidden inside it so much physics that has never been unfolded and brought to light. The black bird of the semiclassical analysis now at the end bows low in respect to the crimson bird of exact inversion theory as he takes his leave, "Nobody realizes better than I how much you still have hidden away, waiting to reveal to the world." Bargmann has set an example for all who hope someday to see these truths of scattering theory.

Acknowledgment

Appreciation is expressed to V. Bargmann, H. G. B. Casimir, Yu. N. Demkov, F. J. Dyson, Ugo Fano, N. V. Fedorenko, N. Fröman, P. O. Fröman, V. I. Goldanski, S. A. Goudsmit, J. R. Pierce, R. T. Powers, R. D. Puff, B. Ross and P. C. Sabatier for discussion or correspondence about some of the topics taken up in this report; and to the Physics Department of the University of Washington at Seattle for hospitality during part of the writing of this report.

JOSEPH HENRY LABORATORIES
PRINCETON UNIVERSITY
PRINCETON, NEW JERSEY

REFERENCES

Titles are given to assist in identifying contents.

Agranovich, Z. S., and V. A. Marchenko, 1963, *The Inverse Problem of Scattering Theory*, trans. from Russian by B. D. Seckler, Gordon and Breach, New York.

Babcock, D. F., 1964, "The discovery of xenon - 135 as a reactor poison," *Nuc. News 7, No. 9,* 38-42 (September).

Bargmann, V., 1949a, "Remarks on the determination of a central field of force from the elastic scattering phase shifts," *Phys. Rev. 75,* 301-303.

__________, 1949b, "On the connection between phase shifts and potentials," *Rev. Mod. Phys. 21*, 488-493.

Bernstein, R. B., 1960, "Quantum mechanical (phase shift) analysis of differential elastic scattering of molecular beams," *J. Chem. Phys. 33,* 795-804.

__________, 1962, "Semiclassical equivalence relationship applied to the calculation of molecular-beam scattering phase shifts," *J. Chem. Phys. 36,* 1403-1404.

__________, 1966, "Quantum effects in elastic molecular scattering," pp. 75-134 in Ross, J., ed., 1966.

Bernstein, R. B., and J. T. Muckerman, 1967, "Determination of intermolecular forces via low-energy molecular beam scattering," pp. 389-486 in Hirschfelder, ed., 1967.

Berry, M. V., and K. E. Mount, 1972, "Semiclassical approximations in wave mechanics," *Rep. Prog. Phys. 35,* 315-397.

Bohr, N., 1913, "On the constitution of atoms and molecules," *Phil. Mag. 26*, 1-25.

Born, M., and J. R. Oppenheimer, 1927, "Zur Quantentheorie der Molekeln,' *Ann. der Physik. 84,* 457-484.

Bouwkamp, C. J., and N. G. de Bruijn, 1946, "The problem of optimum antenna current distribution," *Philips Res. Rep. 1.*, 135-158.

Breit, G., 1932, "A remark on Gamow's treatment of radioactive disintegration," *Phys. Rev., 40,* 127.

Breit, G., and F. L. Yost, 1935, "Radiative capture of protons by carbon," *Phys. Rev. 48*, 203-210.

Buck, U., 1974, "Inversion of molecular scattering data," *Rev. Mod. Phys. 46*, 369-389.

Bunker, P. R., 1968, "The electronic isotope shift in diatomic molecules and the partial breakdown of the Born-Oppenheimer approximation," J. Molec. Spectros., *28*, 422-443.

Bunker, P. R., 1973, "Non-adiabatic effects on the vibrational intervals of H_2^+," *J. Molec. Spectros.*, *46*, 504-505.

Casimir,* H.G.B., 1966, "Remarks on the Hanbury-Brown and Twiss experiment, and on super-gain antennas," pp. 37-40 in *Pubblicazioni del Comitato Nazionale per le Manifestazioni Celebrative, IV Centenario della Nascita di Galileo Galilei, 1564-1964, Volume II, Atti dei Convegni, Tomo 6, Atti del Convegno sulla Filosofia Naturale Oggi, Pisa, 17-21 Settembre 1964*, G. Barbèra Editore, Firenze (*also p. 422, end).

Child, M. S., 1974, *Molecular Collision Theory*, Academic Press, New York.

Demkov, Yu. N., and V. N. Ostrovski, 1972, "n + ℓ filling rule in the periodic system and focusing potentials," *Soviet Physics JETP 35*, 66-69.

Demkov, Yu. N., V. N. Ostrovski, and N. B. Berezina, 1971, "Uniqueness of the Firsov inversion method and focusing potentials," *Soviet Phys. JETP 33*, 867-870.

Dirac, P. A. M., 1930, "Note on exchange phenomena in the Thomas atom," *26*, 376-385; see equations (14, 15) on pp. 382-383.

Doll, J. D., and W. H. Miller, 1972, "Classical-limit quantization of non-separable systems: multidimensional WKB perturbation theory," *J. Chem. Phys.*, *57*, 4428-4434.

Dyson, F. J., 1976, "Old and new approaches to the inverse scattering problem," chapter 7 in the present volume.

Eyring, H., J. Walter and G. E. Kimball, 1944, *Quantum Chemistry*, Wiley, New York.

Faddeyev, L. D., 1962, "The inverse problem in the quantum theory of scattering," *J. Math. Phys. 4*, 72-104.

Falkenhagen, H., ed., 1928, *Quantentheorie und Chemie, Leipziger Vorträge*, Hirzel, Leipzig [cited under Fermi (1928)].

Fermi, E., 1928, "Über die Anwendung der statistischen Methode auf die Probleme des Atombaues," in Falkenhagen, ed., 1928, pp. 95-111; reprinted in Segré, ed., 1962, pp. 291-304.

Firsov, O. B., 1953, "Determination of the forces between atoms with the help of differential effective cross section of the elastic scattering," *Zhur. Eksp. theor. Fiz.* 3, 279-283.

Flaks, I. P., ed., 1967, *V International Conference on the Physics of Electronic and Atomic Collisions, Leningrad, July 17-23, 1967, Abstracts of Papers*, Nauka, Leningrad.

Flügge, S., and G. Vollmer, 1971, "Determination of potentials by WKB-inversion of term formulae," *Zeits. f. Physik*, *248*, 1-6.

Ford, K. W., and J. A. Wheeler, 1959, "Semiclassical description of scattering," *Ann. Phys. 7*, 259-286.

Freis, H. T., 1925, "A new directional antenna system," *IRE Proc. 13*, 685-767.

Fröman, N., 1966a, "Detailed analysis of some properties of the JWKB-approximation," *Arkiv för Fysik, 31*, 391-408.

———, 1966b, "Outline of a general theory for higher order approximations of the JWKB type," *Arkiv for Fysik, 32*, 541-548.

———, 1970, "Connection formulas for certain higher order phase-integral approximations," *Annals of Physics 61*, 451-463.

———, 1974, "A simple formula for calculating quantal expectation values without the use of wave functions," *Phys. Lett. 48A*, 137-139.

Fröman, N., and P. O. Fröman, 1965, *JWKB Approximation*, North-Holland, Amsterdam.

———, 1970, "Transmission through a real potential barrier treated by means of certain phase-integral approximations," *Nucl. Phys. A 147*, 606-626.

———, 1974a, "A direct method of modifying certain phase-integral approximations of arbitrary order," *Annals of Physics, 83*, 103-107.

———, 1974b, "On modification of phase integral approximations of arbitrary order," *Nuovo Cimento, 20B*, 121-131.

Fröman, N., P. O. Fröman, U. Myhrman, and R. Paulsson, 1972, "On the quantal treatment of the double-well potential problem by means of certain phase-integral approximations," *Annals of Physics*, 72, 314-323.

Fröman, P. O., 1974, "On the normalization of wave functions for bound states in a single-well potential when certain phase-integral approximations of arbitrary order are used," *Annals of Physics, 88*, 621-630.

Geiger, H., and K. Scheel, eds., 1933, *Handbuch der Physik*, 2nd ed., *Quantentheorie*, Springer, Berlin [cited under Pauli (1933)].

Gel'fand, I. M., and B. M. Levitan, 1951, "On the determination of a differential equation from its spectral function," *Izv. Akad. Nauk SSSR, Ser. Mat. 15*, 309-360; English trans. in *Am. Math. Soc. Translations* (2), *1*, 253-304 (1953).

Ghiorso, A., R. M. Diamond and H. E. Conzett, eds., 1963, *Proceedings of the Third* [Asilomar] *Conference on Reactions between Complex Nuclei*, University of California Press, Berkeley and Los Angeles, California.

Goeppert-Mayer, M., 1941, "Rare-earth and transuranic elements," *Phys. Rev. 60*, 184-187.

Hartree, D. R., 1957, *The Calculation of Atomic Structures*, Wiley, New York.

Heading, J., 1962, *An Introduction to Phase Integral Methods*, Methuen, London.

Herman, F., and S. Skillman, 1963, *Atomic Structure Calculations*, Prentice-Hall, Englewood Cliffs, New Jersey.

Hill, D. L., and J. A. Wheeler, 1953, "Nuclear constitution and the interpretation of fission phenomena," Phys. Rev. *89*, 1102-1145.

Hirschfelder, J. O., ed., 1967, *Intermolecular Forces*, Vol. 12 of *Advances in Chemical Physics*, Wiley-Interscience, New York [cited under Bernstein and Muckerman and under Mason and Monchick].

Hodgson, P. E., 1971, "The nuclear optical model," *Reports on Prog. in Phys., 34,* 765-819 (and p. 1249).

Hoyt, F. C., 1939, "The determination of force fields from scattering in the classical theory," *Phys. Rev. 55,* 664-665.

Hundhausen, E., and H. Pauly, 1965, Der Regenbogeneffekt und Interferenzerscheinungen bei molekularen Stössen, *Zeits. f. Phys. 187,* 305-337.

Jackson, D. F., 1974, "Nuclear sizes and the optical model," *Reports on Prog. in Phys. 37,* 55-146.

Jean, C., and P. C. Sabatier, 1973, "On an inverse scattering problem of quantum theory," Il Nuovo Cimento 18, 105-111.

Karlsson, F., 1975, "Transmission through a real Eckart-Epstein potential barrier treated by means of certain phase-integral approximations," *Phys. Rev. D11,* 2120-2123.

Kemble, E. C., 1937, *The Fundamental Principles of Quantum Mechanics, with Applications,* McGraw Hill, New York; reissue, New York, 1958.

Klein, O., 1932, Zur Berechnung von Potentialkurven für zweiatomige Moleküle mit Hilfe von Spektraltermen, *Zeits. f. Phys. 76,* 226-235.

Knoll, J., and E. Schaeffer, 1975, "Semi-classical scattering theory with complex trajectories. I-Elastic Waves," issued as a preprint from Service de Physique Theorique, Centre d'Etudes Nucleaires de Saclay, Gif-sur-Yvette, France. 101pp.

Krainov, V. P., 1971, "WKB method for a strong Coulomb field," *Zhur. Eksp. Teor. Fiz. Pis.*: MA 359-362. English translation in *Soviet Phys. JETP Lett., 13,* 255-257 (1971).

Kramers, H. A., 1926, "Wellenmechanik und halbzahlige Quantisierung," *Zeits. f. Physik, 29,* 828-840.

Kramers, H. A., and G. P. Ittmann, 1929, "Zur Quantelung des asymmetrischen Kreisels. II," *Zeits. f. Physik, 58,* 217-231.

Landau, L. D., and E. M. Lifshitz, 1958a, *Mekhanika,* Moscow; English version, *Mechanics,* translated by J. B. Sykes and J. S. Bell, 2nd ed., Pergamon, New York, 1969.

________________, 1958b, *Quantum Mechanics: Non-Relativistic Theory,* trans. by J. B. Sykes and J. S. Bell, Addison-Wesley, Reading, Massachusetts, pp. 398-399.

Langer, R. E., 1932, "On the asymptotic solutions of differential equations, with an application to the Bessel functions of large complex order," *Transact. Amer. Math. Soc., 34,* 447-480.

__________, 1934, "The asymptotic solutions of ordinary linear differential equations of the second order, with special reference to the Stokes phenomenon," *Bull. Amer. Math. Soc. 40,* 545-582.

__________, 1937, "On the connection formulas and the solutions of the wave equation," *Phys. Rev., 51,* 669-676.

Latter, R., 1955, "Atomic energy levels for the Thomas-Fermi and Thomas-Fermi-Dirac Potential," *Phys. Rev. 99,* 510-519.

Lu, P., and S. S. Wald, 1972, "Modified WKB approximation applied to the solution of the repulsive singular potential,"

Marchenko, V. A., 1955, "The potential energy deduced from the phase of scattered waves," *Doklady Akad. Nauk USSR 104,* 695-698.

Marchi, R. P., and C. R. Mueller, 1963, "Validity of the WKB approximation in the interpretation of molecular beam scattering data," *J. Chem. Phys. 38,* 740-744.

Mason, E. A., and L. Monchick, 1964, "Supernumerary rainbows in molecular scattering, *J. Chem. Phys. 41,* 2221-2222.

__________, 1967, "Methods for the determination of intermolecular forces," pp. 329-387 in Hirschfelder, ed., 1967.

McDowell, M. R. C., ed., 1964, *Atomic Collision Processes,* North Holland, Amsterdam.

Miller, S. C., Jr., and R. H. Good, Jr., 1953, "A WKB-type approximation to the Schrödinger equation," *Phys. Rev., 91,* 174-179.

Miller, W. H., 1969, "WKB solution of inversion problems for potential scattering," *J. Chem. Phys. 51,* 3631-3638.

__________, 1971, "Additional WKB inversion relations for bound state and scattering problems, *J. Chem. Phys. 54,* 4174-4177.

__________, 1975, "Semiclassical quantization of nonseparable systems: A new look at periodic orbit theory," *J. Chem. Phys., 63,* 996-999.

Morse, P. M., 1929, "Diatomic molecules according to the wave mechanics II. Vibrational levels," *Phys. Rev., 34,* 57-67.

Mott, N. F., and H. S. W. Massey, 1965, *The Theory of Atomic Collisions,* 3rd edn., Clarendon, Oxford.

Mueller, C. R., and R. P. Marchi, 1963, "WKB analysis of molecular beam scattering and the Lennard-Jones potential," *J. Chem. Phys. 38,* 745-748.

Newman, W. I., and W. R. Thorson, 1972, "New method for rapid numerical solution of the one-dimensional Schrödinger equation," *Phys. Rev. Let., 29,* 1350-1353.

Newton, R. G., 1972, *The Gel'fand Levitan Method of the Inverse Scattering Problem*, Indiana University Press, Bloomington, Indiana.

Oldham, K. B., and J. Spanier, 1974, *The Fractional Calculus*, Academic Press, New York and London.

Pauli, W., 1933, "Die allgemeinen Prinzipien der Wellenmechanik," pp. 83-272 in Geiger and Scheel, eds., 1933; see pp. 170-173 for the cited JWKB connection formulas.

Pauli, W., ed., 1955, *Niels Bohr and the Development of Physics*, McGraw-Hill, New York [cited under Wheeler (1955)].

Pauly, H., and J. P. Toennies, 1965, "The study of intermolecular potentials with molecular beams at thermal energies," *Adv. Atom. Mol. Phys.*, ed. by D. R. Bates and I. Estermann, Academic Press, London *1*, 195-344.

Pechukas, P., 1972, "Semiclassical approximation of multidimensional bound states," *J. Chem. Phys.*, *57*, 5577-5594.

Porter, R. N., L. M. Raff, and W. H. Miller, 1975, "Quasiclassical selection of initial coordinates and momenta for a rotating Morse oscillator," *J. Chem. Phys.*, *63*, 2214-2218.

Pöschl, G., and E. Teller, 1933, "Bemerkungen zur Quantenmechanik des anharmonischen Oszillators," *Zeits. f. Physik*, *83*, 143-151.

Powers, R. T., 1971, "Frequencies of radial oscillation and revolution as affected by features of a central potential," pp. 235-242 in Verde, ed., 1971.

Rees, A. L. G., 1947, "The calculation of potential-energy curves from band-spectroscopic data," *Proc. Phys. Soc.* (London) *59*, 998-1008.

Riblet, H. J., 1948, "Note on the maximum directivity of an antenna," *Proc. I.R.E.*, *36*, 620-623.

Richardson, M. J., 1971, "Solution of the Mathieu equation in the WKB approximation," *Amer. J. Phys.*, *39*, 560-565.

Ross, B., ed., 1975, *Fractional Calculus and Its Applications*, Springer, Berlin.

Ross, J., ed., 1966, *Molecular Beams*, Vol. 10 of *Advances in Chemical Physics*, Wiley-Interscience, New York [see Bernstein (1966)].

Rydberg, R., 1931, "Graphische "Darstellung einiger bandenspektroskopischer Ergebnisse," *Zeits. f. Physik*, *73*, 376-385.

__________, 1933, "Über einige Potentialkurven des Quecksilberhydrids," *Zeits. f. Physik*, *80*, 514-524.

Sabatier, P. C., 1965, "On the asymptotic approximation for the elastic scattering by a potential. I. - Monotonic potential and uncritical range," *Il Nuovo Cimento 37*, 1180-1227.

Sabatier, P. C., 1966, "Le problème inverse à énergie donnée en mécanique quantique," Thèse Faculté des Sciences Orsay

____________, 1972, "Complete solution of the inverse scattering problem at fixed energy," *J. Math. Phys. 13,* 675-699.

____________, 1974a, "Construction of potentials from the cross section at a given energy by approximate methods," lecture at the "Summer Seminar on Inverse Problems," organized by the American Mathematical Society at University of California, Los Angeles, 5-14 August 1974, issued as a preprint from Universite des Sciences et Techniques du Languedoc, Montpellier, July 1974. 32 pp.

____________, 1974b, "Construction of the scattering amplitude from the elastic cross section at a given energy," lecture at the "Summer Seminar on Inverse Problems," organized by the American Mathematical Society at University of California, Los Angeles, 5-14 August 1974, issued as a preprint from Université des Sciences et Techniques du Languedoc, Montpellier, July 1974. 34 pp.

____________, 1974c, "Construction of potentials from the scattering amplitude at a given energy," lecture at the "Summer Seminar on Inverse Problems," organized by the American Mathematical Society at University of California. Los Angeles, 5-14 August 1974, issued as a preprint from Université des Sciences et Techniques du Languedoc, Montpellier, July 1974. 29 pp.

____________, 1974d, "Approche des problemes inverses," issued as a preprint from Université des Sciences et Techniques du Languedoc, Montpellier. 23 pp.

Sabatier, P. E., and F. Quyen Van Phu, 1971, "Numerical computations in the inverse-scattering problem at fixed energy," *Phys. Rev. D4,* 127-132.

Sanada, J., ed., 1968, *Proceedings of the* [Tokyo] *International Conference on Nuclear Structure, Supp. to J. of Phys. Soc. of Japan, 24* [cited under Tombrello (1968)].

Schelkunoff, S. A., 1942, "Directive antenna systems," *U. S. Patent No. 2,286,839.*

____________, 1943, "A mathematical theory of linear arrays," *Bell Sys. Tech. Jour. 22,* 80-107.

Schelkunoff, S. A., and H. T. Friis, 1952, "Superdirective antennas," pp. 195-198 in their *Antennas: Theory and Practice,* Wiley, New York. Appreciation is expressed here to John R. Pierce for reference to the work of Friis and Schelkunoff.

Scott, A. C., F. Y. F. Chu, and D. W. McLaughlin, 1973, "The soliton: a new concept in applied science," Proc. Inst. Electrical and Electronics Eng. *61,* 1443-1483.

Segrè, E., ed., 1962, *Enrico Fermi Collected Papers*, University of Chicago Press, Vol. I [cited under Fermi (1928)].

Slater, J. C., 1963, *Quantum Theory of Molecules and Solids*, McGraw-Hill, New York.

Smith, F. J., 1964, "The numerical evaluation of the classical angle of deflection and of the JWKB phase shift," *Physica 30*, 497-504.

Stearns, C. O., 1961, "Computed performance of moderate size, super-gain antennas," *U. S. Nat. Bur. Stand. Rep. No. 6797*.

Thomas, T. D., 1968, "Compound nuclear reactions induced by heavy ions,' *Ann. Rev. Nuclear Sci., 18*, 343-406.

Tombrello, T. A., 1968, "Few-nucleon systems and reactions," pp. 63-75 (and bibliography on pp. 615-627) in Sanada, ed., 1968.

Van Horn, H. M., and E. E. Salpeter, 1967, "WKB approximation in three dimensions," *Phys. Rev., 157*, 751-758.

Verdi, M., ed., 1971, *Atti del Convegno Mendeleeviano, "Periodicita e Simmetrie nella Struttura Elementare della Materia,"* Accademia delle Scienze di Torino, Torino [cited under Powers (1971) and Wheeler (1971)].

Wald, S. S., and P. Lu, 1974, "Application of the higher order modified WKB method to the Lennard-Jones potential," *J. Chem. Phys., 61*, 4680-4685.

Wheeler, J. A., 1936, "The dependence of nuclear forces on velocity," *Phys. Rev. 50*, 643-649.

____________, 1937, "On the mathematical description of light nuclei by the method of resonating group structure," *Phys. Rev. 52*, 1107-1122.

____________, 1955a, "Nuclear fission and nuclear stability," pp. 163-184 in Pauli, ed., 1955.

____________, 1955b, "Scattering and potential," *Phys. Rev. 99*, 630.

____________, 1971, "From Mendeléev's atom to the collapsing star," pp. 189-233, in Verde, ed., 1971.

Yaru, N., 1951, "A note on super-gain antenna arrays," *Proc. I. R. E., 39*, 1081-1085.

Yost, F. L., J. A. Wheeler, and G. Breit, 1936, "Coulomb wave functions in repulsive fields," *Phys. Rev. 49*, 174-189.

Zener, C., 1932, "Non-adiabatic crossing of energy levels," *Proc. Roy. Soc.* (London) *A 137*, 696-702.

*Casimir, H. B. G., 1968, "On supergain antennae," pp. 73-79 in Puppi, G., ed., 1968, *Old and new problems in Elementary Particles: A volume dedicated to Gilberto Bernardini in his sixtieth birthday*, Academic Press, New York.

INSTABILITY PHENOMENA IN THE EXTERNAL FIELD PROBLEM FOR TWO CLASSES OF RELATIVISTIC WAVE EQUATIONS

A. S. Wightman

1. *Introduction*

A relativistic wave equation is a system of partial differential equations of the form

$$(\beta^\mu \partial_\mu + \rho)\psi(x) = 0 \tag{1}$$

where ψ stands for a function on space-time with N components $\psi_i, i = 1, \cdots, N$, and the $\beta^\mu, \mu = 0, 1, 2, 3$ and ρ are $N \times N$ numerical matrices satisfying

$$\begin{aligned} S_1(A^{-1})\beta^\mu S_2(A) &= \Lambda(A)^\mu{}_\nu \beta^\nu \\ S_1(A^{-1})\rho\, S_2(A) &= \rho\ . \end{aligned} \tag{2}$$

Here $A \to S_1(A)$ and $A \to S_2(A)$ are two representations of the group $SL(2, C)$.

The external field problem for the wave equation (1) is the problem of solving

$$[\beta^\mu \partial_\mu + \rho + B(x)]\psi(x) = 0 \tag{1'}$$

where $B(x)$ is some $N \times N$ matrix valued function of space-time.

In the context of relativistic field theory there are requirements on the admissibility of solutions of (1) and (3) that arise when one insists that the corresponding quantum mechanical theory have a reasonable interpretation in terms of particles. Quite general sufficient conditions of this kind are known for the free equation (1) [1]. The additional requirements arising from the physical interpretation of the external field problem (1′) can be regarded as stability requirements on the free theory: it should continue to make sense when perturbed by the external field, $B(x)$ [2].

There are several classes of perturbations, $B(x)$, significant for applications. First of all, smooth local perturbations for which $B(x)$ is infinitely differentiable and has a compact support are physically interesting because they yield a simple extension of the Schrödinger theory of scattering by a potential in which the amplitude for scattering by the external field has to be supplemented by the probability amplitude for pair creation from the vacuum and by the probability amplitude for the vacuum to remain the vacuum; for fermions, the vacuum-to-vacuum amplitude is the Fredholm determinant of the scattering integral equation with Feynman boundary conditions and so gives a physical interpretation of the determinant, a quantity which normally plays a role only as an auxiliary mathematical tool.

Just as in Schrödinger potential scattering the theory of the solution of (1′) can be extended to admit $B(x)$ which are locally quite rough as well as to $B(x)$ which are quite long-range in space. The extension to $B(x)$ constant in time is delicate as would be expected on physical grounds. A static potential capable of creating pairs has an infinitely long time to create them. Since the produced pairs do not interact with each other, one would anticipate that there might be troubles associated with the accumulation of an infinite charge density and that, in fact, turns out to be the case [3] [4].

A second significant class of perturbations $B(x)$ arises when one considers the solution of certain Lagrangean field theories of interacting quantized fields. When such theories are treated with the methods of Euclidean field theory, the expressions for the Schwinger functions (≡ Euclidean Green's functions) involve functional integrals over Euclidean fields. When the coupling is bilinear in one field and linear in another (as it is for a Yukawa coupling), the functional integral over the field appearing bilinearly can be carried out explicitly and yields an integrand involving the Euclidean propagator of the field that was integrated over. There also appears the vacuum-vacuum amplitude in the external field. The class of admissible external fields is here determined by the probability measure appearing in the functional integral. For the simplest super-renormalizable models in space-times of dimension two ($P(\phi)_2$ and Y_2), the measure is absolutely continuous with respect to the free field measure, and the free field measure dictates that with probability one the fields are very rough, not being signed measures at any point. For $\lambda(\phi^4)_3$, the measure is not equivalent to the free field measure, but the roughness is not mitigated. In renormalizable but not super-renormalizable, not to speak of nonrenormalizable theories, there is as yet little known about the interacting measure. However, it seems unlikely that with probability one the fields will be much smoother. Thus, for the purposes of the theory of interacting quantized fields, one wants to have the solution of the external field problem (1′) for quite rough fields [5].

The present article is concerned with the first class of smooth perturbations. Here, for the Dirac equation for spin one-half particles, there is a completely satisfactory theory, but for general wave equations and general perturbations there are notorious difficulties. One of the most important of these is the Velo-Zwanziger phenomenon [6] [7], the breakdown of causality in the presence of a suitable perturbation. This phenomenon takes place even for spin zero (with symmetric tensor coupling) and for spin one (with electric quadrupole coupling). However its effects are most dramatic for particles with spin $\geq 3/2$. There the Velo-Zwanziger phenomenon joins a list of difficulties which have plagued the theory of higher spin particles nearly since its inception. Since these difficulties are important for what follows, they deserve a preliminary description.

a) If the wave equation (1) has an associated conserved current, conventional field theory uses it to define the scalar product in the single particle state space. In general, the scalar product so defined does not yield a positive norm for positive energy solutions of the wave equation.

b) If one alters the scalar product (by dropping the contribution from negative norm states or changing its sign), one gets a theory in which the basic commutation relations contain extra derivatives in addition to those that would appear in the normal process of quantization. This alteration in the in field commutation relations can give rise to the kind of characteristic instability phenomenon discussed in [2]. For example, if in the presence of a suitable smooth perturbation, B(x), of compact support, the ingoing field, ψ^{in}, is constructed so as to satisfy the altered commutation relations, it can happen that the outgoing field, ψ^{out}, fails to satisfy the same commutation relations. This is a disaster for the physical interpretation of the theory since the introduction of an asymptotic particle interpretation depends on the existence of ingoing and outgoing fields satisfying the same commutation relations and of unique ingoing and outgoing vacuum state vectors, Ψ_0^{in} and Ψ_0^{out}, which describe states in which no particles are present in the distant past or distant future, respectively.

The main result of the present article is a proof that the instability difficulty just described actually occurs in two classes of theories: Iverson's theory of Kursonoglu's equations [8] and Hurley's theory for a class of equations for a single mass and spin [9][10]. As will be seen, the result can be obtained for a considerably more general class which includes these, but does not as yet provide an argument for the current conjecture that every wave equation that does not have the Velo-Zwanziger acausality pathology will founder on the instability phenomenon discussed here if it describes particles of spin $\geq 3/2$.

Since the details that follow involve a good deal of obscuring algebraic machinery, I will outline here the essentials of the argument.

First, let me recall how one proves the commutation relations for the out field in the Capri formalism for the standard case in which

$$\rho = m\,1, \quad S_1(A) = S_2(A) = S(A) \qquad A \in SL(2,\mathbb{C}) \tag{3}$$

and there exist a hermitean matrix, η, and a matrix, C, non-singular, and such that

$$-(\beta^\mu)^* = \eta\,\beta^\mu\,\eta^{-1} \tag{4}$$

$$S(A)^* = \eta S(A^{-1})\eta^{-1} \tag{5}$$

$$\overline{\beta^\mu} = C\beta^\mu C^{-1} \tag{6}$$

$$\overline{S(A)} = CS(A)C^{-1} \tag{7}$$

$$\eta = (-1)^\sigma C^T \eta C^{-1} \tag{8}$$

where $(-1)^\sigma$ is $+1$ on the subspace where $S(A)$ is single-valued as a representation of the restricted Lorentz group and -1 where it is double-valued.

The in field, ψ^{in}, is assumed given as a free field, i.e. a field satisfying (1) and the commutation relations

$$\begin{aligned} [\psi^{in}(x), \psi^{in}(y)]_\pm &= 0 \\ [\psi^{in}(x), \psi^{in+}(y)]_\pm &= \frac{1}{i} S(x-y) . \end{aligned} \tag{9}$$

Here

$$\psi^{in+}(x) = (\psi^{in}(x))^* \eta \tag{10}$$

$$S(x) = S_R(x) - S_A(x) \tag{11}$$

and

$$S_{\substack{R\\A}}(x) = d(i\partial)\Delta_{\substack{R\\A}}(m_1^2, m_2^2, \cdots, m_n^2; x) \tag{12}$$

where

$$\Delta_{\substack{R\\A}}(m_1^2, m_2^2, \cdots, m_n^2; x) = (2\pi)^{-4} \int d^4p \left[\prod_{j=1}^{n} (-p^2+m^2) \right]^{-1}_{\substack{R\\A}} \exp - ip\cdot x , \tag{13}$$

$m_1^2 \cdots m_n^2$ is the mass spectrum of the equation i.e. the set of zeros of $\det(-\not\xi + m)$ (where $\not\xi = i\beta^\mu \xi_\mu$) as a polynomial in ξ^2, and $d(\not\xi)$ is the so-called Klein-Gordon divisor of the wave equation i.e. the uniquely determined matrix, polynomial in $\not\xi$, satisfying

$$d(\xi)(-\not\xi+m) = \prod_{j=1}^{n} (-\xi^2+m_j^2)1 \ . \tag{14}$$

Explicitly,

$$d(\xi) = -\frac{1}{m}\prod_{j=1}^{n}(\xi^2-m_j^2)\sum_{r=0}^{q-1}\left(\frac{\not\xi}{m}\right)^r$$

$$+\frac{1}{m^2}(\not\xi+m)\left(\frac{\not\xi}{m}\right)^q\left\{\sum_{r=0}^{n-1}\left(\frac{\xi^2}{m^2}\right)^r\left[(\xi^2)^n-(\xi^2)^{n-1}\sum_{j=1}^{n}m_j^2\right.\right.$$

$$+(\xi^2)^{n-2}\sum_{j_1<j_2}m_{j_1}^2 m_{j_2}^2-\cdots(-1)^r(\xi^2)^{n-r}\sum_{j_1<j_2<\cdots j_r}m_{j_1}^2\cdots j_{m_r}^2$$

$$\left.\left.-\prod_{j=1}^{n}(\xi^2-m_j^2)\right]\right\} . \tag{15}$$

The solution of the perturbed equation (1′) under the assumption

$$(B(x))^* = \eta B(x)\eta^{-1} \tag{16}$$

is given by

$$\psi(f) = \psi^{in}(T_R^{-1}f) \tag{17}$$

where f is a test function which is infinitely differentiable and of fast decrease i.e. $\epsilon\ \mathcal{S}(R^4)$, and T_R is a mapping of $\mathcal{S}(R^4)$ onto itself defined by

$$(T_R f)(x) = f(x) + \int d^4y\, f(y)\, S_R(y-x)\, B(x) \tag{18}$$

The out field, ψ^{out}, is given by

$$\psi^{out}(f) = \psi^{in}(T_R^{-1}T_A f) \tag{19}$$

where

$$(T_A f)(x) = f(x) + \int d^4y\, f(y)\, S_A(y-x)\, B(x). \tag{20}$$

From the formulae (17) and (19), for ψ and ψ^{out} in terms of ψ^{in}, it is easy to compute formulae for the adjoint fields

$$\psi^{+}(f) = (\psi(\overline{\eta \bar{f}}))^{*} = \psi^{in+}(T_R'^{-1} f) \tag{21}$$

$$\psi^{out+}(f) = (\psi^{out}(\overline{\eta \bar{f}}))^{*} = \psi^{in+}(T_R'^{-1}\, T_A' f) \tag{22}$$

where

$$(T_R' f)(x) = f(x) + B(x) \int S_A(x-y)\, d^4y\, f(y) \tag{23}$$

$$(T_A' f)(x) = f(x) + B(x) \int S_R(x-y)\, d^4y\, f(y) \tag{24}$$

From the explicit formulae (19) and (22) for ψ^{out} and ψ^{out+}, it is clear that

$$[\psi^{out}(f), \psi^{out}(g)]_{\pm} = 0 \tag{25}$$

and

$$[\psi^{out}(f), \psi^{out+}(g)]_{\pm} = [\psi^{in}(T_R^{-1} T_A f), \psi^{in+}(T_R'^{-1} T_A' g)]_{\pm} \tag{26}$$

$$= \iint d^4x\, d^4y (T_R^{-1} T_A f)(x)\, S(x-y)\, (T_R'^{-1} T_A' g)(y)$$

Thus, the problem of establishing that the out field has the same commutation relations as the in field is equivalent to showing that the right-hand side of (26) reduces to

$$\iint d^4x\, d^4y\, f(x)\, S(x-y)\, g(y)\ . \tag{27}$$

The requisite elementary identities were derived in [12], pp. 23-24. The essential point about the derivation is that it makes use of the equality of the S that appears on the right-hand side of the commutation relations and the S defined as $S_R - S_A$. In a theory with altered commutation relations for the in field, we may expect the derivation of the identity to fail.

More explicitly, note that because both $S_R(T_R f - f, g)$ and $S_R(f, T'_A g - g)$ are equal to

$$\iiint f(x)\, S_R(x-y)\, B(y)\, S_R(y-z)\, g(z)\, d^4x\, d^4y\, d^4z \ ,$$

we have

$$S_R(T_R f, g) = S_R(f, T'_A g) \ . \tag{28}$$

Similarly, because both $S_A(T_A f - f, y)$ and $S_A(f, T'_R g - g)$ are equal to

$$\iiint f(x)\, S_A(x-y)\, B(y)\, S_A(y-z)\, g(z)\, d^4x\, d^4y\, d^4z$$

we have

$$S_A(T_A f, g) = S_A(f, T'_R g) \ . \tag{29}$$

On the other hand, because both $S_A(T_R f - f, g)$ and $S_R(f, T'_R g - g)$ are equal to

$$\iiint f(x)\, S_R(x-y)\, B(y)\, S_A(y-z)\, g(z)\, d^4x\, d^4y\, d^4z$$

we have

$$S_R(f, T'_R g) - S_A(T_R f, g) = S_R(f, g) - S_A(f, g) \ . \tag{30}$$

Finally, because both $S_R(T_A f - f, g)$ and $S_A(f, T'_A g - g)$ are equal to

$$\iiint f(x)\, S_A(x-y)\, B(y)\, S_R(y-z)\, g(z)\, d^4x\, d^4y\, d^4z \ ,$$

we have

$$S_R(T_A f, g) = S_A(f, T'_A g) = S_R(f, g) - S_A(f, g) \quad (31)$$

Now what has to be evaluated is

$$S_R(T_R^{-1} T_A f, T_R'^{-1} T'_A g) - S_A(T_R^{-1} T_A f, T_R'^{-1} T'_A g) . \quad (32)$$

To arrive at this expression, start with (30), with f replaced there by $T_R^{-1} T_A f$ and g by $T_R'^{-1} T'_A g$:

$$S_R(T_R^{-1} T_A f, T'_A g) - S_A(T_A f, T_R'^{-1} T'_A g) = (32) .$$

If (28) is applied in the first term on the left-hand side and (29) in the second, we have

$$S_R(T_A f, g) - S_A(f, T'_A g) = (32)$$

and (31) says this is

$$S_R(f, g) - S_A(f, g)$$

which was to be proved.

This argument shows (26) coincides with (27) and therefore that the out field has the same commutation relations as the in field so long as the $S(x-y)$ in (26) and (27) are interpreted as $S_R(x-y) - S_A(x-y)$. It is typical of the theories discussed in the following sections that in (26) instead of $S_R(x-y) - S_A(x-y)$, there appears a differential operator acting on it. The way the identities (28)...(31) are proved makes it plausible that the presence of the differential operator will destroy the proof; the purpose of Sections 2 and 3 is to demonstrate that. The argument is elementary. $[\psi^{out}(x), \psi^{out+}(y)]_{\pm}$ ought to be independent of the perturbation, B. To show that it is not, it suffices to calculate the lowest order B dependent term in perturbation theory and to check that it does not vanish.

2. *Instability of a Class of Multi-Mass Equations*

The class of wave equations considered in this section is restricted by the requirements (3)...(8) and is assumed to have a mass spectrum $m_1 \cdots m_n$. For simplicity, it is assumed that these numbers are distinct and positive and that associated with each one is a set of solutions of the wave equation of a unique spin. Further, it is assumed that the scalar product defined by the current when restricted to solutions of a given mass, is either positive on positive energy solutions or negative there, possibly different for different masses. The generalized Harish-Chandra relation

$$\prod_{j=1}^{n} \left[(\xi)^2 - \xi^2 \left(\frac{m^2}{m_j^2} \right) \right] \xi^q = 0 \tag{33}$$

then follows [1], and from it one can derive a formula for the projection operator, $E_{m_\ell}(p)$, onto the states of definite mass, m_ℓ, and energy-momentum, p, (a specification which includes the sign of energy). Writing $m_\ell = \frac{m}{\lambda_\ell}$ with $\lambda_\ell > 0$, one has

$$E_{m_\ell}(p) = \varepsilon_\ell \sum_j u_j^{\lambda_\ell}(p) \otimes u_j^{\lambda_\ell}(p)^+ = \prod_{j \neq \ell} \frac{\left(\frac{-\not{p}}{\sqrt{p^2}}\right)^2 - \lambda_j^2}{\lambda_\ell^2 - \lambda_j^2} \frac{\left(\frac{\not{p}/}{\sqrt{p^2}} + \lambda_\ell\right)}{2\lambda_\ell} \left(\frac{\not{p}}{\sqrt{p^2}\,\lambda_\ell}\right)^q \tag{34}$$

where

$$\varepsilon_\ell = \pm 1 \ \text{ if } \ u_j^{\lambda_\ell}(p)^+ u_j^{\lambda_\ell}(p) = \pm 1 \tag{35}$$

and the $u_j^{\lambda_\ell}(p)$ are solutions of the Dirac equation

$$\left(-\not{p} + \frac{m}{\lambda_\ell}\right) u_j^{\lambda_\ell}(p) = 0 \ . \tag{36}$$

They are presumed orthogonal for distinct j

$$u_j^{\lambda_\ell}(p)^+ u_k^{\lambda_\ell}(p) = 0 \qquad j \neq k \tag{37}$$

and normalized as in (35).

Using this formula, one can define an $S^{(+)}(m_\ell, x)$ which satisfies the homogeneous equation

$$(\beta^\mu \partial_\mu + m) S^{(+)}(m_\ell, x) = 0 \tag{38}$$

and arises only from the mass m_ℓ positive energy solutions

$$S^{(+)}(m_\ell, x) = \frac{i}{2(2\pi)^3} \int_{H^+_{m_\ell}} d\Omega_{m_\ell}(p)\, \varepsilon_\ell\, E_{m_\ell}(p) \exp[-ip \cdot x] . \tag{39}$$

$H^+_{m_\ell}$ stands for the positive energy hyperboloid $p^2 = m_\ell^2$, $p^0 > 0$. The ε_ℓ is included in this definition to make (39) have the traditional form when expressed in terms of the $u_j^{\lambda_\ell}$. Using $S^{(+)}(m_\ell, x)$, one can define a scalar product

$$(\phi, \chi)_{m_\ell} = \varepsilon_\ell \iint_\Sigma d\sigma^\mu(x) \phi^+(x) \beta_\mu i^{-1} S^{(+)}(m_\ell, x-y) \beta_\nu \chi(y) d\sigma^\nu(y) \tag{40}$$

which receives contributions only from the parts of ϕ and χ satisfying the Dirac equation (36) with mass, m_ℓ. The factor, ε_ℓ, is included to make the scalar product positive on positive energy solutions. The integrations run over any space-like surface Σ and are, for soiutions of (1), independent of Σ. The contributions from the different components of the mass spectrum can be combined with arbitrary weights to give a family of possible scalar products

$$\sum_{\ell=1}^{n} a_\ell (\phi, \chi)_{m_\ell}, \; a_1, \cdots, a_n \geq 0 . \tag{41}$$

By starting with such a scalar product defined on smooth solutions of the wave equation, one gets by completion a Hilbert space of states, $\mathcal{H}_1$, suitable as a space of single-particle states. The passage from the single-particle theory to a corresponding theory of a free quantized field goes by standard procedures. I will use them in the form presented in [1]. The state space for the field theory is the Fock space

$$\mathcal{F}_\varepsilon(\mathcal{H}_1) \otimes \mathcal{F}_\varepsilon(\mathcal{H}_1) \tag{42}$$

where

$$\begin{gathered} \mathcal{F}_\varepsilon(\mathcal{H}_1) = \bigoplus_{n=0}^{\infty} \mathcal{H}^{(n)} \\ \mathcal{H}^{(0)} = \mathbf{C}, \quad \mathcal{H}^{(n)} = (\mathcal{H}_1{}^{\otimes n})_\varepsilon \end{gathered} \tag{43}$$

$\varepsilon = a$ (anti-symmetric) or s (symmetric). The annihilation and creation operators for particles $a(\chi)$, $a^*(\Phi)$ act on the first of the two factors of the tensor product while the annihilation and creation operators $b(\chi)$, $b^*(\Phi)$ for the anti-particles act on the second

$$\psi(f) = a(\Pi_+ f) + b^*(\Pi_- f) \ . \tag{44}$$

(Here, strictly speaking, one should write $a(\Pi_+ f) \otimes 1$ in place of $a(\Pi_+ f)$

$$\left\{ \begin{matrix} 1 & \text{integer spin} \\ (-1)^N & \text{half odd integer spin} \end{matrix} \right\} \otimes a^*(\Pi_- f)$$

in place of $b^*(\Pi_- f)$ where N is the number of particles, but such refinements will be eschewed.) Π_+ is a linear mapping from the test function space to the dual of $\mathcal{H}_1$, while Π_- is a mapping from the test function space to $\mathcal{H}_1$ itself. The relation between $\Pi_\pm$ and the preceding machinery will become visible shortly.

The commutation relations between $a(\chi)$ and $a^*(\Phi)$ are expressed directly in terms of the value $\langle \chi, \Phi \rangle$ of the linear functional χ on the vector Φ:

$$[a(\chi), a^*(\Phi)]_\pm = \langle \chi, \Phi \rangle \, 1 \ . \tag{45}$$

$\langle \chi, \Phi \rangle$ can be expressed in terms of the scalar product between vectors in $\mathcal{H}_1$ because there is an anti-linear bijection, J, of the space of linear functionals onto $\mathcal{H}_1$ given by

$$<\chi,\Phi> = (J\chi,\Phi) \ . \tag{46}$$

Thus, the commutation relations between the field ψ and the field ψ^+ defined as in (21) is

$$\begin{aligned}[\psi(f),\psi^+(g)]_{\pm} &= [a(\Pi_+f), a(\Pi_+\overline{\eta g})^*]_{\pm} + [b^*(\Pi_-f),(b^*(\Pi_-\overline{\eta g}))^*]_{\pm} \\ &= \{<\Pi_+f, J\Pi_+\overline{\eta g}> \pm <J^{-1}\Pi_-\overline{\eta g},\Pi_-f>\}1 \ .\end{aligned} \tag{47}$$

Now there remains the task of showing how the right-hand side of this relation reduces to an expression of the form

$$\iint f(x)\, d^4x\, E(x-y)\, d^4y\, g(y) \ .$$

For this one needs the explicit form of the mappings $\Pi_{\pm}$ and of the scalar product in $\mathcal{H}_1$. Let $\Pi(p)$ be the projection operator onto the linear manifold of solutions of

$$(-\not{p}+m)\Phi(p) = 0 \tag{48}$$

defined by

$$\Pi(p) = \sum_{j=1}^{n} E_{m_\ell}(p) \tag{49}$$

with E_{m_ℓ} defined in (34). The scalar product (41) written in momentum space is

$$(\Phi,\Psi) = \sum_{j=1}^{n} \int d\Omega_{m_j}(p)\Phi(p)^+\mathfrak{N}_j(p)\Psi(p) \tag{50}$$

where $\mathfrak{N}_j(p) = a_j E_{m_j}(p)$ satisfies

$$\Phi(p)^+\mathfrak{N}_j(p)\Phi(p) \geq 0 \tag{51}$$

and

$$S(A^{-1})\mathfrak{N}_j(p)\,S(A) = \mathfrak{N}_j(\Lambda(A^{-1})p) \ . \tag{52}$$

The elements of $\mathcal{H}_1$ are equivalence classes of Φ's satisfying

$$\Pi(p)\Phi(p) = \Phi(p) \tag{53}$$

for all positive energy p on the mass shell, two vectors lying in the same equivalence class if they differ by a vector of length zero in the norm defined by (50). The dual of $\mathcal{H}_1$ consists of all linear functionals of the form

$$<\chi,\Phi> = \sum_{j=1}^{n} \int d\Omega_{m_j}(p)\chi(p)\mathfrak{N}_j(p)\Phi(p) \tag{54}$$

where $\chi(p)$ satisfies the transpose of (53):

$$\Pi(p)^T\chi(p) = \chi(p) . \tag{55}$$

The bijection J is given by

$$(J\chi)(p) = \eta^{-1}\overline{\chi(p)} , \tag{56}$$

the mappings $\Pi_{\pm}$ by

$$(\Pi_+ f)(p) = \sqrt{\pi}\,\Pi(p)^T\hat{f}(p) \tag{57}$$

$$(\Pi_- f)(p) = \sqrt{\pi}\,\Pi(p)(C^T\eta)^{-1}\,\hat{f}(p) \tag{58}$$

where p runs over the positive energy mass shell.

With this machinery in hand, we can evaluate the first term of (47)

$$\begin{aligned} <\Pi_+ f, J\Pi_+\bar{\eta}\bar{g}> &= \sum_{j=1}^{n} \int d\Omega_{m_j}(p)(\Pi_+ f)(p)\mathfrak{N}_j(p)(J\Pi_+\bar{\eta}\bar{g})(p) \\ &= \pi \sum_{j=1}^{n} \int d\Omega_{m_j}(p)\hat{f}(p)\Pi(p)\mathfrak{N}_j(p)\,\eta^{-1}\,\Pi(p)^*\,\eta\hat{g}(-p) . \end{aligned} \tag{59}$$

Using the fact that $\mathfrak{N}_j(p)$ commutes with $\Pi(p)$, which follows from the way the scalar product was constructed, and

$$\eta^{-1}\,\Pi(p)^{*}\,\eta = \Pi(p) \tag{60}$$

we have that (59) is

$$\pi \sum_{j=1}^{n} \int d\Omega_{m_j}(p)\,\hat{f}(p)\,\mathfrak{N}_j(p)\,\Pi(p)\,\hat{g}(-p)$$

$$= \sum_{j=1}^{n} \iint d^4x\, f(x)\,\varepsilon_j\,\mathfrak{N}_j(i\partial)\,\frac{1}{i}\,S^{(+)}(m_j, x-y)\,g(y)\,d^4y \tag{61}$$

$$= \iint d^4x\, f(x)\left\{\left[\sum_{j=1}^{n} \varepsilon_j\,\mathfrak{N}_j(i\partial)\right]\frac{1}{i}\,S^{(+)}(x-y)\right\} g(y)\,d^4y$$

where

$$S^{(+)}(x) = \sum_{j=1}^{n} S^{(+)}(m_j, x)\ . \tag{62}$$

If all the $\varepsilon_j = 1$, $\sum_{j=1}^{n} \varepsilon_j\mathfrak{N}_j(i\partial) = \Pi(i\partial)$ which acts like the identity operator when acting on $S^{(+)}$. In that case, (61) reduces to the standard result. (See [1], p. 467, equation (3.16).) The second term is slightly more complicated

$$< J^{-1}\Pi\_\overline{\eta g}, \Pi\_f > = \pi \sum_{j=1}^{n} \int d\Omega_{m_j}(p)\,\overline{[\eta\Pi(p)(C^T\eta)^{-1}}\,\eta\hat{g}(p)]\,\mathfrak{N}_j(p)\Pi(p)(C^T\eta)^{-1}\hat{f}(-p)$$

$$= \pi \sum_{j=1}^{n} \int d\Omega_{m_j}(p)\,\hat{g}(p)\,\eta^T C\Pi(p)\,\mathfrak{N}_j(p)\,\Pi(p)\,\eta^{-1}(C^T)^{-1}\hat{f}(p)$$

$$= (-1)^{\sigma}\,\pi \sum_{j=1}^{n} \int d\Omega_{m_j}(p)\,\hat{f}(-p)\,C^{-1}\overline{\Pi(p)}C\,C^{-1}[\eta\,\mathfrak{N}_j(p)\eta^{-1}]^T C\hat{g}(p)$$

$$= -(-1)^{\sigma} \iint d^4x\, f(x) \left\{ \left[\sum_{j=1}^{n} \varepsilon_j \mathfrak{N}_j(i\partial) \right] \frac{1}{i} S^{(-)}(m_j, x-y) \right\} g(y)\, d^4y$$

$$= -(-1)^{\sigma} \iint d^4x\, f(x) \left\{ \left[\sum_{j=1}^{n} \varepsilon_j \mathfrak{N}_j(i\partial) \right] \frac{1}{i} S^{(-)}(x-y) \right\} g(y)\, d^4y \qquad (63)$$

where

$$S^{(-)}(m_j, x) = \frac{-i\varepsilon_j}{2(2\pi)^3} \int_{H^+_{m_j}} d\Omega_{m_j}(p)\, \Pi(-p) \exp ip \cdot x$$

$$(64)$$

$$S^{(-)}(x) = \sum_{j=1}^{n} S^{(-)}(m_j, x)$$

and equations (8), (60)

$$C^{-1}\, \overline{\Pi(p)}\, C = \Pi(-p) \qquad (65)$$

and

$$\eta\, \mathfrak{N}_j(p)\, \eta^{-1} = \mathfrak{N}_j(p)^* \qquad (66)$$

have been used. Again, if all the $\varepsilon_j = +1$, the operator $\sum_{j=1}^{n} \varepsilon_j \mathfrak{N}_j(i\partial)$ would be equivalent to 1 when acting on $S^{(-)}(x-y)$ and (63) would be equivalent to the standard result. (See [1], p. 468, equation (3.20).) Thus, we have finally the desired commutation relations

$$[\psi(f), \psi^+(g)]_{\pm} = \iint f(x)\, d^4x \left\{ \left[\sum_{j=1}^{n} \varepsilon_j \mathfrak{N}_j(i\partial) \right] \frac{1}{i} S(x-y) \right\} d^4y\, g(y) \qquad (67)$$

where

$$S(x) = S^{(+)}(x) + S^{(-)}(x) \ .$$

As Iverson justly remarked about his construction (see p. 59 of [7]), what has been done is to construct a generalized free field with the same mass and spin spectrum in its two-point function as is predicted by the given single-particle equation.

The S appearing in (67) is equal to $S_R - S_A$, so the feature of the commutation relations on which the proof of instability must rest is the occurrence of the operator $\sum_{j=1}^{n} \varepsilon_j \mathfrak{N}_j(i\partial)$.

The proof starts from the expression for the commutator that follows from (22)

$$[\psi^{\mathrm{out}}(f), \psi^{\mathrm{out}+}(g)]_{\pm} = \iint d^4x\,(T_R^{-1} T_A f)(x) \left[\sum_{j=1}^{n} \varepsilon_j \mathfrak{N}_j(i\partial)\, S(x-y)\right] (T_R'^{-1} T_A' g)(y)\, d^4y$$

and uses the expressions

$$\begin{aligned}(T_R^{-1} T_A f)(x) = f(x) &- \int f(y)\, d^4y.\, S(y-x)\, B(x) \\ &+ \int f(z)\, S(z-y)\, B(y)\, d^4z\, d^4y\, S_R(y, x; B)\, B(x)\end{aligned} \tag{69}$$

$$\begin{aligned}(T_R'^{-1} T_A' g)(y) = g(y) + B(y) &\int S(y-x)\, g(x)\, d^4x \\ - B(y) &\int S_A(y, z; B)\, B(z)\, S(z-w)\, g(w)\, d^4z\, d^4w\ .\end{aligned} \tag{70}$$

(See [1], p. 420, equation (4.3).) The resulting expansion contains nine terms and is well suited to the evaluation of the perturbation expansion of the commutator in B. When $B = 0$, the only nonvanishing term coincides with the right-hand side of (67). The linear terms in B are

$$\iint f(x)\left\{\left[\sum_{j=1}^{n} \epsilon_j \mathfrak{N}_j(i\partial)\right] \frac{1}{i} S(x-y)\right\} B(y)\, S(y-z)\, g(z)\, d^4x\, d^4y\, d^4z$$

$$- \iint f(y)\, S(y-x)\, B(x)\left\{\left[\sum_{j=1}^{n} \epsilon_j \mathfrak{N}_j(i\partial)\, \frac{1}{i}\right] S(x-z)\right\} g(z)\, d^4y\, d^4z\, d^4x \;. \tag{71}$$

If the operator $\sum_{j=1}^{n} \epsilon_j \mathfrak{N}_j(i\partial)$ were absent, these two terms would cancel. However, when not all of the ϵ_j are not equal, they do not. The essential point is that when the expression is rewritten in Fourier transform, it involves a double sum over the mass spectrum

$$\sum_{j=1}^{n} \sum_{k=1}^{n} \iint d\Omega_{m_j}(p)\, d\Omega_{m_k}(q)$$

$$\left\{\hat{f}(p)\epsilon_j a_j E_{m_j}(p)\hat{B}(p-q)E_{m_k}(q)\hat{g}(-q) - \hat{f}(p)E_{m_j}(p)\hat{B}(p-q)\epsilon_k a_k E_{m_k}(q)\hat{g}(-q)\right\} . \tag{72}$$

If $\hat{f}$ is chosen to vanish except on the plus m_j mass hyperboloid and $\hat{g}$ except on the minus m_k mass hyperboloid where $\epsilon_j = -1$ and $\epsilon_k = +1$, the terms will add instead of cancelling, and for appropriately chosen B will be nonvanishing. The possibility that one of the a_k vanishes (or both) leads in the end to the absurd conclusion that all vanish.

Thus any multimass equation (1) whose β^μ and ρ matrices satisfy (3)... (8) which has retarded and advanced fundamental solutions in the presence of a perturbation B satisfying (16) will be unstable if the scalar product defined by its conserved current attributes both positive and negative norms to some positive energy solutions of the wave equation.

3. *Instability for a Class of Equations Studied by Hurley*

There is a class of wave equations, discovered and rediscovered in the last quarter of a century which uses the representation

$$([s,0] \oplus [s-1/2,1/2]) \oplus ([0,s] \oplus [1/2,s-1/2]) \tag{73}$$

of SL(2, C) where s is any non-negative integer or half-odd integer. It was realized quite early that with the customary definitions, the resulting equation describes parity doublets of particles of mass m and spin s. One of the members of the doublet has all its states of negative norm and the other of positive norm. Various methods have been proposed to eliminate the states of negative norm. Some have the property that the introduction of an external vector potential via a minimal coupling ansatz leads to an equation which does not have the Velo-Zwanziger acausality phenomenon. For details and complete references, see the work of W. J. Hurley [9][10][11]. In this section we rewrite Hurley's proposal in the notation used above and show that it suffers from the same type of instability that afflicts Iversen's theory discussed in the preceding section.

Let us begin with the parity doubled theory in an explicit form, given by Hurley. The representation of SL(2, C) is

$$\underline{S} = \left\{\begin{array}{c|c} S_+ & 0 \\ \hline 0 & S_- \end{array}\right\} \tag{74}$$

where

$$\begin{aligned} S_+ &= (s,0) \oplus (s-1/2,1/2) \\ S_- &= (0,s) \oplus (1/2,s-1/2) \ . \end{aligned} \tag{75}$$

The β matrices are

$$\underline{\beta}^{\mu} = \left(\begin{array}{c|c} \beta^{\mu} & 0 \\ \hline 0 & \beta_{\mu} \end{array}\right) \tag{76}$$

where $\beta^i = -\beta_i$ $i = 1,2,3$, $\beta^0 = \beta_0$ and the β^{μ} are $(6s+1)\times(6s+1)$ matrices given explicitly in [9]. (In this section, to facilitate the reader's comparison with [9], we use $i\underline{\beta}^{\mu}$ in place of our former β^{μ} so the wave equation is $(-i\beta^{\mu}\partial_{\mu}+m)\psi = 0$.) The space inversion operator in this basis is

$$\mathscr{P} = \left(\begin{array}{c|c} 0 & 1 \\ \hline 1 & 0 \end{array}\right) : \tag{77}$$

To see the indefiniteness of the scalar product in detail, look at the wave equation in momentum space for a particle at rest

$$(-\underline{\beta}^0 p_0 + m) u = 0 . \tag{78}$$

Since

$$\underline{\beta}^0 = \left\{\begin{array}{ccc|ccc} 1 & 0 & 0 & & & \\ 0 & -1 & 0 & & 0 & \\ 0 & 0 & 0 & & & \\ \hline & & & 1 & 0 & 0 \\ & 0 & & 0 & -1 & 0 \\ & & & 0 & 0 & 0 \end{array}\right\} \tag{79}$$

where the 1 is a $(2s+1) \times (2s+1)$ matrix and the -1 a $(2s-1) \times (2s-1)$ matrix, it is clear that the positive energy solutions are spanned by

$$\begin{pmatrix} 1 \\ 0 \\ 0 \\ 0 \\ 0 \\ 0 \end{pmatrix} \quad \text{and} \quad \begin{pmatrix} 0 \\ 0 \\ 0 \\ 1 \\ 0 \\ 0 \end{pmatrix} . \tag{80}$$

The particular linear combinations

$$\frac{1}{\sqrt{2}} \begin{pmatrix} 1 \\ 0 \\ 0 \\ \pm 1 \\ 0 \\ 0 \end{pmatrix} \tag{81}$$

have $\begin{smallmatrix}\text{even}\\\text{odd}\end{smallmatrix}$ parity respectively. Furthermore,

$$u^+ \underline{\beta}^0 u = \bar{u} \eta \underline{\beta}^0 u = \pm 1 \tag{82}$$

for the $\begin{smallmatrix}\text{even}\\\text{odd}\end{smallmatrix}$ parity states. Here the fact that $\eta = \mathscr{P}$ has been used.

The analogue of the procedure employed above to obtain a positive scalar product is to insert the operator

$$\frac{a_1}{2}(1+\mathcal{P}) - \frac{a_2}{2}(1-\mathcal{P}) = \frac{a_1-a_2}{2} + \frac{a_1+a_2}{2}\mathcal{P} \tag{83}$$

in the scalar product provided by the current. The extra complication here is that the states whose norm has to be changed in sign are degenerate in mass with states for which the norm is not changed. To get the scalar product for arbitrary momentum, one boosts to velocity $\vec{p}/E$.

$$\begin{aligned} &\bar{u}\eta\left[\frac{a_1}{2}(1+\mathcal{P}) - \frac{a_2}{2}(1-\mathcal{P})\right]u = \\ &\bar{u}\underline{S}(\exp(\chi/2\vec{n}\cdot\vec{\tau}))^*\mathcal{P}\mathcal{P}^{-1}\underline{S}(\exp-\chi/2\vec{n}\cdot\vec{\tau})^*\mathcal{P} \\ &\left[\frac{a_1}{2}(1+\mathcal{P}) - \frac{a_2}{2}(1-\mathcal{P})\right]\underline{S}(\exp-\chi/2\vec{n}\cdot\vec{\tau})\underline{S}(\exp-\chi/2\vec{n}\cdot\vec{\tau})u \\ &= \bar{u}(p)\eta\mathcal{P}^{-1}S(\exp-\chi/2\vec{n}\cdot\vec{\tau})^*\mathcal{P}\left[\frac{a_1}{2}(1+\mathcal{P}) - \frac{a_2}{2}(1-\mathcal{P})\right]\underline{S}(\exp-\chi/2\vec{n}\cdot\vec{\tau})u(p) \end{aligned} \tag{84}$$

where $\vec{n}$ is a unit vector in the direction of $\vec{p}$, $\tanh\chi = v/c$, and the $\vec{\tau}$ are the Pauli matrices. Now

$$\mathcal{P}^{-1}\underline{S}(\exp-\chi/2\vec{n}\cdot\vec{\tau})\mathcal{P} = \underline{S}(\exp\chi/2\vec{n}\cdot\vec{\tau})\ . \tag{85}$$

Thus, what has to be evaluated is

$$\begin{aligned} &\underline{S}(\exp\chi/2\vec{n}\cdot\vec{\tau})\left[\frac{a_1-a_2}{2} + \frac{a_1+a_2}{2}\mathcal{P}\right]\underline{S}(\exp-\chi/2\vec{n}\cdot\vec{\tau}) \\ &= 1/2\begin{Bmatrix} S_+ & 0 \\ 0 & S_- \end{Bmatrix}\begin{Bmatrix} a_1-a_2 & a_1+a_2 \\ a_1+a_2 & a_1-a_2 \end{Bmatrix}\begin{Bmatrix} S_+^{-1} & 0 \\ 0 & S_-^{-1} \end{Bmatrix} \\ &= 1/2\begin{Bmatrix} (a_1-a_2) & (a_1+a_2)S_+S_-^{-1} \\ (a_1+a_2)S_-S_+^{-1} & (a_1-a_2) \end{Bmatrix}. \end{aligned} \tag{86}$$

From the construction of S_+ and S_- we have

$$S_+(A) = S_-(A^*)^{-1} \tag{87}$$

so for the off-diagonal elements of (86)

$$S_+(A)\, S_-(A)^{-1} = S_+(A)\, S_+(A^*) = S_+(AA^*) \tag{88}$$

and

$$S_-(A)\, S_+(A)^{-1} = S_-(A)\, S_-(A^*) = S_-(AA^*)\ . \tag{89}$$

Since in the case at hand $A = \exp \chi/2\ \vec{n}\cdot\vec{\tau}$

$$\begin{aligned} AA^* &= \exp \chi\ \vec{n}\cdot\vec{\tau} = \cosh \chi + \sinh \chi\ \vec{n}\cdot\vec{\tau} \\ &= \frac{1}{m}\,(p^0 + \vec{p}\cdot\vec{\tau}) = \underset{\sim}{p}/m\ . \end{aligned} \tag{90}$$

Thus, the scalar product is

$$(u, v) = \int_{H_m^+} d\Omega_m(p)\, u^+(p)\, \mathfrak{N}(p)\, u(p) \tag{91}$$

where

$$\mathfrak{N}(p) = 1/2 \begin{pmatrix} (a_1 - a_2) & (a_1 + a_2) S_+(\underset{\sim}{p}/m) \\ (a_1 + a_2) S_-(\underset{\sim}{p}/m) & (a_1 - a_2) \end{pmatrix}.$$

The Lorentz transformation properties of $\mathfrak{N}(p)$ are easily deduced:

$$\begin{aligned} \underline{S}(A)^{-1}\mathfrak{N}(p)\underline{S}(p) &= 1/2 \begin{pmatrix} (a_1 - a_2) & (a_1 + a_2) S_+(A)^{-1} S_+(\underset{\sim}{p}/m) S_-(A) \\ (a_1 + a_2) S_-(A)^{-1} S_-(p/m) S_+(A) & (a_1 - a_2) \end{pmatrix} \\ &= \mathfrak{N}(\Lambda(A)^{-1} p) \end{aligned} \tag{92}$$

because

$$S_+(A^{-1}) S_+(\underset{\sim}{p}/m) S_+((A^{-1})^*) = S_+(A^{-1}\underset{\sim}{p}/m(A^{-1})^*) = S_+(\underset{\frown}{\Lambda(A^{-1})p}/m) \tag{93}$$

and analogously

$$S_-(A^{-1})S_-(\underset{\sim}{p}/m)S_-((A^{-1})^*) = S_-(\underset{\frown}{\Lambda(A^{-1})}p/m). \tag{94}$$

Thus, if u and v are functions on the plus hyperboloid $p^2 = m^2$, $p^0 > 0$ with $2(6s+1)$ components, the sesquilinear form (91) is positive on the positive energy solutions of

$$(-\not{p}+m)u(p) = 0 \tag{95}$$

and the continuous representation of the Poincaré group

$$(U(a, A)u)(p) = \exp ip \cdot a\, \underline{S}(A)u(\Lambda(A^{-1})p) \tag{96}$$

leaves the scalar product (91) invariant

$$(U(a, A)u, U(a, A)v) = (u, v) . \tag{97}$$

This information being given, the construction of a free field theory along the lines described in the preceding section goes just as before, with the simplification that here the mass spectrum consists of the single value m. The commutation relations turn out to be

$$[\psi(f), \psi^+(g)]_{\pm} = \iint d^4x\, f(x)\mathfrak{N}(i\partial)\frac{1}{i}\, S(m, x-y)g(y)\,d^4y . \tag{98}$$

The proof of instability can again be carried out by calculation of the corrections linear in B to the commutator of the out field, ψ^{out}, with its adjoint field, ψ^{out+}. The resulting expression is up to a non-vanishing numerical factor

$$\iint d\Omega_m(p)d\Omega_m(q)\{\hat{f}(p)\mathfrak{N}(p)\Pi(p)\hat{B}(p-q)\Pi(q)g(-q) - \hat{f}(p)\mathfrak{N}(p)\hat{B}(p-q)\mathfrak{N}(q)\Pi(q)\hat{g}(-q)\} . \tag{99}$$

It is not difficult to see that this is nonvanishing for suitable $\hat{f}$, $\hat{g}$, and $\hat{B}$. However, I will not spell out the argument in detail because the theory whose demise is thereby assured is not Hurley's!

What Hurley did was to start with states of definite intrinsic helicity. (He was not the first to do so; Joseph Harris' Purdue thesis of 1955 on the same wave equation was entitled *Enantiomorphic Particles of Arbitrary Spin* which indicates he was following a similar idea. For nonchemist readers, it should perhaps be explained that two distinguishable structures are enantimorphs if they are related by the operation of space inversion.) In the rest system, the states are precisely those given in (80); they will be denoted u_R and u_L, respectively. They have zero length in the scalar product provided by the current

$$u_R{}^+ \beta^0 u_R = 0 = u_L{}^+ \beta^0 u_L \ . \tag{100}$$

Note

$$u_R = \frac{1}{\sqrt{2}}(u_E + u_0) \qquad u_L = \frac{1}{\sqrt{2}}(u_E - u_0) \tag{101}$$

where u_E, u_0 are the states of definite parity given in (81). The scalar product in Hurley's theory can be expressed in two equivalent forms. In the first, one uses the $2(6s+1)$ component formalism but chooses a scalar product that annihilates the last $(6s+1)$ components, and thereby leaves out the contribution of left-handed states u_L. This amounts to taking $a_1 = a_2 = 1/2$ in (81), but then multiplying it by

$$\left\{\begin{array}{c|c} 1 & 0 \\ \hline 0 & 0 \end{array}\right\} . \tag{102}$$

The calculations (84)...(90) then yield a scalar product

$$u^+(p)\mathfrak{N}(p)\left\{\begin{array}{c|c} 1 & 0 \\ \hline 0 & 0 \end{array}\right\} u(p) = \overline{u_+(p)}\, S_+(\tilde{p}/m)\, u_+(p) \tag{103}$$

apart from a normalization factor. Here $u_+(p)$ has $(6s+1)$ components, the first $6s+1$ components of $u(p)$, the bar is complex conjugation and $\tilde{p} = p^0 - \vec{p}\cdot\vec{\tau}$. The second formalism employs the $(6s+1)$ component u_+ and a scalar product given by the right-hand side of (103). It is instructive to work it out in detail because there is no η matrix and as a consequence, one has to deal with several phenomena somewhat concealed in the preceding treatment. The most prominent is the fact that the representations of $SL(2, \mathbb{C}): A \to S_\pm(A)$ are not left invariant (up to equivalence) under the automorphism $A \to (A^*)^{-1}$ but rather are interchanged

$$S_+((A^*)^{-1}) = S_-(A) \ . \tag{104}$$

In the $2(6s+1)$ component formalism, the analogous equation is

$$\underline{S}((A^*)^{-1}) = \eta \underline{S}(A) \eta^{-1} \ . \tag{105}$$

The behavior (104) implies that the transformation law of the single particle states is different from that of the field itself. If, for example, the transformation law of the field is

$$U(a, A)\psi(x) U(a, A)^{-1} = S_+(A^{-1})\psi(\Lambda(A)x + a) \ , \tag{106}$$

then the one-particle states will transform like the vectors $\psi(x)^*\Psi_0$ where Ψ_0 is the vacuum state, and therefore will involve the representation $A \to \overline{S_+(A)}$ which is equivalent to $A \to S_-(A)$. This equivalence is closely related to

$$\underline{S}((A^{-1})^T) = \underline{C}^T \eta \underline{S}(A) (\underline{C}^T \eta)^{-1} \tag{107}$$

which follows from (5) and (7). The point is that

$$\underline{C} = \left\{ \begin{array}{c|c} 0 & C \\ \hline (-1)^{2s} C & 0 \end{array} \right\}, \qquad \eta = \left\{ \begin{array}{c|c} 0 & 1 \\ \hline 1 & 0 \end{array} \right\} \tag{108}$$

so

$$\eta^T \underline{C} = \left\{\begin{array}{c|c} (-1)^{2s}C & 0 \\ \hline 0 & C \end{array}\right\} \tag{109}$$

and

$$S_{\pm}((A^{-1})^T) = CS_{\pm}(A)C^{-1} = S_{\pm}(\zeta)S_{\pm}(A)S_{\pm}(\zeta)^{-1} \tag{110}$$

where $\zeta = \begin{pmatrix} 0 & 1 \\ -1 & 0 \end{pmatrix}$. On the other hand, from

$$\underline{\beta}^{\mu *} = \underline{\eta}\underline{\beta}^{\mu}\eta^{-1} - \overline{\underline{\beta}^{\mu}} = \underline{C}\underline{\beta}^{\mu}\underline{C}^{-1} \tag{111}$$

there follows

$$-\underline{\beta}^{\mu T} = (\eta^T \underline{C})\underline{\beta}^{\mu}(\eta^T \underline{C})^{-1} \tag{112}$$

and so, denoting the $(6s+1) \times (6s+1)$ components of β^{μ} by $\beta^{\mu}_{\varepsilon}$, $\varepsilon = \pm 1$,

$$-\beta^{\mu T}_{\varepsilon} = C\beta^{\mu}_{\varepsilon}C^{-1} . \tag{113}$$

The detailed structure of C is given by Hurley as

$$C = \begin{Bmatrix} 0 & \mathfrak{D}^{(s)}(\zeta) & 0 \\ \mathfrak{D}^{(s)}(\zeta) & 0 & 0 \\ 0 & 0 & \mathfrak{D}^{(s-1)}(\zeta) \end{Bmatrix} \tag{114}$$

which implies

$$C^T = (-1)^{2s}C, \ \ \overline{C} = C, \ \ C^* = C^{-1} . \tag{115}$$

To obtain the space of states, we use a one-particle Hilbert space $\mathcal{H}^{(1)}$ consisting of equivalence classes of $(6s+1)$ component functions $\phi(p)$ defined on the positive energy hyperboloid of mass m, satisfying

$$(-\beta^{\mu}_{-} p_{\mu} + m)\phi(p) = 0 \tag{116}$$

and of finite norm with respect to the scalar product

$$(\phi, \psi) = \int d\Omega_m(p)\overline{\phi(p)}\, S_-(\tilde{p}/m)\psi(p) \ . \qquad (117)$$

The representation of ISL(2, C) we take as

$$(U^{(1)}(a, A)\phi)(p) = \exp(ip\cdot a)\, S_-(A)\phi(\Lambda(A^{-1})p) \ . \qquad (118)$$

It should be emphasized that this representation is unitary equivalent to

$$(U^{(\bar{1})}(a, A)\phi)(p) = \exp(ip\cdot a)\, S_+(A)\phi(\Lambda(A^{-1})p) \qquad (119)$$

acting on the solutions of

$$(-\beta_+^\mu p_\mu + m)\phi(p) = 0 \qquad (120)$$

with the scalar product

$$(\phi, \psi) = \int d\Omega_m(p)\overline{\phi(p)}\, S_+(\tilde{p}/m)\psi(p) \ . \qquad (121)$$

The only advantage of (116), (117), and (118) is that they permit a simple construction of a local field. The physical content of the transformation law is the same in the two cases. The unitarity of the representations (118) and (119) is a consequence of the identities

$$S_\varepsilon(A)^* S_\varepsilon(\tilde{p}/m)\, S_\varepsilon(A) = S_\varepsilon(A^*\tilde{p}/mA) = S_\varepsilon(\widetilde{\Lambda(A^{-1})p}/m) \ . \qquad (122)$$

For the one anti-particle subspace, $\mathcal{H}^{(\bar{1})}$, we use the solutions of (120) equipped with the scalar product (121) and with the transformation law (119).

The next step in the construction requires the projection operator onto the solutions of (116), which will be denoted, $\Pi_-(p)$, the analogous projection onto the solutions of (120) being $\Pi_+(p)$. These projections may be obtained from the Harish-Chandra relation

$$[(\beta^0)^2 - 1]\beta^0 = 0 \qquad (123)$$

in the manner described by E. Speer (see [13], pp. 110-112). The result is

$$\Pi_\varepsilon(p) = \frac{1}{2m}\,[\beta^\mu_\varepsilon\, p_\mu + m]\left(\frac{\beta^\mu_\varepsilon\, p_\mu}{m}\right), \quad \varepsilon = \pm 1 . \tag{124}$$

Since

$$S_\varepsilon(A)^{-1}\,\beta^\mu_\varepsilon\, S_\varepsilon(A) = \Lambda(A)^\mu_\nu\,\beta^\nu_\varepsilon , \tag{125}$$

$\Pi_\varepsilon(p)$ satisfies

$$S_\varepsilon(A)^{-1}\Pi_\varepsilon(p)\,S_\varepsilon(A) = \Pi_\varepsilon(\Lambda(A^{-1})\,p) . \tag{126}$$

The state space of the field theory is

$$\mathcal{H} = \mathcal{F}_\varepsilon(\mathcal{H}^{(1)}) \otimes \mathcal{F}_\varepsilon(\mathcal{H}^{(\bar{1})}) \tag{127}$$

with $\varepsilon = s$ or a according to the integer or half-odd integer character of the spin s. The transformation law of states is

$$U(a, A) = \Gamma(U^{(1)}(a, A)) \otimes \Gamma(U^{(\bar{1})}(a, A)) \tag{128}$$

where

$$\Gamma(V) = \bigoplus_{n=0}^{\infty} V^{\otimes n} \tag{129}$$

with $V^{\otimes 0} = 1$, V being a bounded operator in $\mathcal{H}^{(1)}$ or $\mathcal{H}^{(\bar{1})}$, respectively.

The expression for the field is in terms of annihilation and creation operators on the Fock spaces $\mathcal{F}_\varepsilon(\mathcal{H}^{(1)})$ and $\mathcal{F}_\varepsilon(\mathcal{H}^{(\bar{1})})$ and their definition in turn involves mappings, J, of the dual spaces, $\mathcal{H}^{(1)*}$ and $\mathcal{H}^{(\bar{1})*}$, of the single-particle and anti-particle spaces, $\mathcal{H}^{(1)}$ and $\mathcal{H}^{(\bar{1})}$, onto $\mathcal{H}^{(1)}$ and $\mathcal{H}^{(\bar{1})}$, respectively. In the preceding theories the definition of J involved the hermitizing matrix, η, which is not available here. Here J is defined

$$(J\chi)(p) = \overline{\chi(p)} \tag{130}$$

where $\chi \in \mathcal{H}^{(1)*}$ and similarly for $\mathcal{H}^{(\bar{1})*}$. The important point is that the linear functional χ evaluated on a vector, Φ, in $\mathcal{H}^{(1)}$ is given by

$$<\chi,\Phi> = (J\chi,\Phi) = \int d\Omega_m(p)\chi(p)S_-(\tilde{p}/m)\Phi(p) \tag{131}$$

while for $\chi \in \mathcal{H}^{(\bar{1})*}$ and $\Phi \in \mathcal{H}^{(\bar{1})}$

$$<\chi,\Phi> = (J\chi,\Phi) = \int d\Omega_m(p)\chi(p)S_+(\tilde{p}/m)\Phi(p) . \tag{132}$$

The definition of the field is

$$\psi(f) = a(\Pi_+ f) + b^+(\Pi_- f) \tag{133}$$

where as usual a is a shorthand for $a \otimes 1$ and b for $\left\{\begin{matrix} 1 \\ (-1) \end{matrix} N\right\} \otimes a^+$, a and a^+ being annihilation and creation operators, respectively. The mapping Π_+ of the test function space, $\mathcal{S}$, into the dual, $\mathcal{H}^{(1)*}$, of the single-space, $\mathcal{H}^{(1)}$, is given by

$$(\Pi_+ f)(p) = \sqrt{\pi}\,\overline{\Pi_-(p)}\,\hat{f}(p) \tag{134}$$

while the mapping Π_- of the test function space, $\mathcal{S}$, into the one anti-particle space, $\mathcal{H}^{(\bar{1})}$, is given by

$$(\Pi_- f)(p) = \sqrt{\pi}\,\Pi_+(p)C^{-1}\hat{f}(-p) . \tag{135}$$

(The mappings, Π_ε, should not be confused with the finite dimensional matrices, $\Pi_\varepsilon(p)$!)

With these definitions, the transformation law

$$U(a, A)\psi(f)U(a, A)^{-1} = \psi(\{a, A\}f) \tag{136}$$

holds with

$$(\{a, A\}f)(x) = S_+(A^{-1})^T f(\Lambda(A^{-1})(x-a)) . \tag{137}$$

(This is just (106) in smeared form.) The conditions that have to be checked are just those of equation (1.43) of [1].

$$J^{-1}U^{(1)}(a,A)\,J\Pi_+ f = \Pi_+\{a,A\}f \tag{138}$$

$$U^{(\bar{1})}(a,A)\,\Pi_- f = \Pi_-\{a,A\}f\ . \tag{139}$$

The first reads

$$\begin{aligned}\exp(-ip\cdot a)\overline{S_-(A)}(\Pi_+ f)(\Lambda(A^{-1})p) &= \sqrt{\pi}\,\exp(-ip\cdot a)\overline{S_-(A)}\,\overline{\Pi_-(\Lambda(A^{-1})p)}\hat{f}(\Lambda(A^{-1})p)\\ &= \sqrt{\pi}\,\overline{\Pi(p)}\exp(-ip\cdot a)S_+((A^{-1})^T)\hat{f}(\Lambda(A^{-1})p)\\ &= (\Pi_+\{a,A\}f)(p)\end{aligned} \tag{140}$$

where (104) has been used in the form

$$\overline{S_-(A)} = S_+((A^{-1})^T)\ . \tag{141}$$

The second reads

$$\begin{aligned}\exp(ip\cdot a)S_+(A)(\Pi_- f)(\Lambda(A^{-1})p) &= \sqrt{\pi}\,\exp(ip\cdot a)S_+(A)\Pi_+(\Lambda(A^{-1})p)C^{-1}\hat{f}(-\Lambda(A^{-1})p)\\ &= \sqrt{\pi}\,\Pi_+(p)\exp(ip\cdot a)S_+(A)C^{-1}\hat{f}(-\Lambda(A^{-1})p)\\ &= \sqrt{\pi}\,\Pi_+(p)C^{-1}\exp(ip\cdot a)S_+((A^{-1})^T)\hat{f}(-\Lambda(A^{-1})p)\\ &= (\Pi_-\{a,A\}f)(p)\end{aligned} \tag{142}$$

where (110) has been used in the penultimate step.

The commutation relations have next to be determined. The relation

$$[\psi(f),\psi(g)]_\pm = 0 \tag{143}$$

holds for the usual trivial kinematical reasons. On the other hand, since $a(\Pi_+ g)^*) = a^+(J\Pi_+ g)$,

$$\begin{aligned}[\psi(f),\psi(g)^*]_\pm &= [a(\Pi_+ f), a(\Pi_+ g)^*]_\pm + [b^+(\Pi_- f),(b^+(\Pi_- g))^*]_\pm\\ &= (J\Pi_+ f, J\Pi_+ g) \pm (\Pi_- g, \Pi_- f)\ .\end{aligned} \tag{144}$$

The first term is

$$(J\Pi_+ f, J\Pi_+ g) = \int d\Omega_m(p)\,(\Pi_+ f)(p)\, S_-(\tilde{p}/m)\,\overline{(\Pi_+ g)(p)}$$

$$= \pi \int d\Omega_m(p)\, \hat{f}(p)\, \Pi_-(p)^*\, S_-(\tilde{p}/m)\, \Pi_-(p)\, \overline{\hat{g}(p)}\,. \quad (145)$$

The second term is

$$(\Pi_- g, \Pi_- f) = \pi \int d\Omega_m(p)\, \overline{\hat{g}(-p)}\, (C^{-1})^*\, \Pi_+(p)^*\, S_+(\tilde{p}/m)\, \Pi_+(p)\, C^{-1}\, \hat{f}(-p)$$

$$= \pi \int d\Omega_m(p) \hat{f}(-p)\, [C\Pi_+(p)C^{-1}]^T [CS_+(\tilde{p}/m)C^{-1}]^T [(C^{-1})^* \Pi_+(p)^* C^{-1}]^T \hat{g}(-p)$$

$$= (-1)^{2s}\, \pi \int d\Omega_m(p) \hat{f}(-p)\, \Pi_-(-p)^*\, S_-(-\tilde{p}/m)\, \Pi_-(-p)\, \hat{g}(-p) \quad (146)$$

where the following identities have been used

$$[C\Pi_+(p)\, C^{-1}]^T = \Pi_+(-p) \quad (147)$$

$$\Pi_-(-p) = \Pi_+(-p)^* \quad (148)$$

$$[CS_+(\tilde{p}/m)\, C^{-1}]^T = S_+(\underset{\sim}{p}/m) \quad (149)$$

$$S_+(\underset{\sim}{p}/m) = S_-(\tilde{p}/m) = (-1)^{2s}\, S_-(-\tilde{p}/m) \quad (150)$$

and (115).

Notice that

$$\Pi_\varepsilon(p)^*\, S_\varepsilon(\tilde{p}/m) = S_\varepsilon(\tilde{p}/m)\, \Pi_\varepsilon(p) \quad (151)$$

so that in (145) and (146) all the projection operators can be written either on the right or on the left. Thus,

$$(J\Pi_+ f, J\Pi_+ g) = \iint d^4x\, d^4y\, f(x)\ \ S_-(i\tilde{\partial})\,\frac{1}{i}\, S^{(+)}(x-y)\ \overline{g(y)} \qquad (152)$$

where

$$S^{(+)}(x) = \frac{i}{2(2\pi)^3} \int_{H_m^+} d\Omega_m(p) \exp(-ip \cdot x)\, \Pi_-(p) \qquad (153)$$

and

$$(-1)^{2S}(\Pi_- g, \Pi_- f) = \iint d^4x\, d^4y\, f(x)\, S_-(i\tilde{\partial}/m)\,\frac{1}{i}\, S^{(-)}(x-y)\, \overline{g(y)} \qquad (154)$$

where

$$S^{(-)}(x) = \frac{-i}{2(2\pi)^3} \int_{H_m^+} d\Omega_m(p) \exp(ip \cdot x)\, \Pi_-(-p) \qquad (155)$$

and therefore

$$[\psi(f), \psi(g)^*] = \iint d^4x\, d^4y\, f(x)\, S_-(i\tilde{\partial}/m)\,\frac{1}{i}\, S(x-y)\, \overline{g(y)} \qquad (156)$$

where

$$S(x) = S^{(+)}(x) + S^{(-)}(x) \ . \qquad (157)$$

This completes the construction except for the verification that ψ does indeed satisfy the wave equation. To that end, let

$$f(x) = (-i\beta_+^{\mu T}\, \partial_\mu + m)\, h(x) \qquad (158)$$

then

$$
\begin{aligned}
(\Pi_+ f)(p) &= \sqrt{\pi}\ \Pi_-(p)\,(-\beta_+^{\mu T}\, p_\mu + m)\, \hat{h}(p) \\
&= \sqrt{\pi}\ \Pi_+(p)^T(-\beta_+^{\mu T}\, p_\mu + m)\, \hat{h}(p) \\
&= \sqrt{\pi}\ [(\beta_+^\mu\, p_\mu + m)\,\Pi_+(p)]^T\, \hat{h}(p) \\
&= 0
\end{aligned}
\qquad (159)
$$

and

$$
\begin{aligned}
(\Pi_- f)(p) &= \sqrt{\pi}\, \Pi_+(p) C^{-1} (\beta_+^{\mu T} p_\mu + m) \hat{h}(-p) \\
&= \sqrt{\pi}\, \Pi_+(p) (-\beta_+^{\mu} p_\mu + m) C^{-1} \hat{h}(-p) \\
&= \sqrt{\pi}\, (-\beta_+^{\mu} p_\mu + m) \Pi_+(p) C^{-1} \hat{h}(-p) \\
&= 0 \ . \qquad (160)
\end{aligned}
$$

The field, ψ, is local in the sense that the commutators or anti-commutators

$$[\psi(x), \psi(y)]_\pm , \quad [\psi(x), \psi^*(y)]_\pm$$

vanish for space-like separations $(x-y)$, but it is not canonical; the latter commutator or anti-commutator with $x^0 - y^0 = 0$ is not a constant (possibly matrix) multiple of $\delta(\vec{x}-\vec{y})$, but in general, contains spatial derivatives of the delta function.

In [11], Hurley puts strong emphasis on a field

$$\tilde{\phi} = S_-(-i\underset{\sim}{\partial}/m)^T \phi^* \ . \qquad (161)$$

The commutation relations of ϕ and $\tilde{\phi}$ can be deduced directly from (156)

$$[\psi(x), S_-(-i\partial/m)\phi^*(y)]_\pm = \frac{1}{2m}(i\beta^\mu \partial_\mu + m)\left(\frac{i\beta^\mu p_\mu}{m}\right)\frac{1}{i}\Delta(x-y) \qquad (162)$$

in agreement with (9.5) of [11].

Now, at last, we turn to the external field problem for Hurley's theory. The differential equation is

$$[\beta_+^\mu \partial_\mu + m + B(x)]\psi(x) = 0 \qquad (163)$$

or in the customary Yang-Feldman form

$$\psi(x) = \psi^{in}(x) - \int S_R(x-y) B(y) \psi(y) d^4y \qquad (164)$$

which is reduced as usual to

$$\psi(f) = \psi^{in}(T_R^{-1} f) \qquad \psi^{out}(f) = \psi^{in}(T_R^{-1} T_A f) \tag{165}$$

with T_R and T_A defined as usual by (18) and (20). For the adjoint field,

$$\psi^*(f) = \psi(\bar{f})^* = \psi^{in*}(T_R'^{-1} f) \tag{166}$$

where

$$(T_R' f)(x) = f(x) + B(x)^* \int S_R(y-x)^* f(y) \, d^4y \, . \tag{167}$$

Clearly,

$$[\psi(f), \psi(g)]_{\pm} = 0 \tag{168}$$

and

$$[\psi(f), \psi^*(g)]_{\pm} = [\psi^{in}(T_R^{-1} f), \psi^{in*}(T_R'^{-1} g)]_{\pm}$$
$$= \iint (T_R^{-1} f)(x) \, d^4x \, S\_(i\bar{\partial}/m) \frac{1}{i} S(x-y) (T_R'^{-1} g)(y) \, d^4y \, . \tag{169}$$

If the theory were to go according to the standard pattern, the right-hand side would be written

$$\iint d^4x \, f(x) \frac{1}{i} S(x, y; B) \, g(y) \, d^4y \tag{170}$$

where

$$S(x, y; B) = \text{“}S_R\text{”}(x, y; B) - \text{“}S_A\text{”}(x, y; B) \tag{171}$$

and "S_R"$(x, y; B)$ and "S_A"$(x, y; B)$ are respectively retarded and advanced:

$$\text{“}S_R\text{”}(x, y; B) = 0 \text{ unless } (x-y)^2 \geq 0 \text{ and } x^0 - y^0 \geq 0 \tag{172}$$

$$\text{“}S_A\text{”}(x, y; B) = 0 \text{ unless } (x-y)^2 \geq 0 \text{ and } x^0 - y^0 \leq 0 \, . \tag{173}$$

The quotation marks around S_R and S_A are there because these retarded and advanced functions cannot be the standard retarded and advanced fundamental solutions of the wave equation which occur in the usual treatment; the presence of the operator $S_-(i\tilde{\partial}/m)$ in (156) prevents that. It is not clear that any decomposition (171) exists and more generally, although $S(x, y; B)$ surely exists as a tempered distribution and satisfies the homogeneous wave equation in each variable separately, it is not obvious that it vanishes when $x-y$ is space-like.

Rather than pursue this matter further, since my objective is to show the instability of Hurley's theory, I will compute the commutator or anti-commutator of the out fields, ψ^{out} and ψ^{out*}, and show that it differs from the commutator or anti-commutator of the in fields, ψ^{in} and ψ^{in*}.

The calculation runs parallel to that for Iversen's theory given in Section 2. The formula for $T_R^{-1}T_A$ is (69) as before. However, that for $T_R'^{-1}T_A'$ is different since the definition of T_R' and T_A' has been altered.

$$(T_R'^{-1}T_A'f)(x) = f(x) - B(x)^* \int S(y-x)^* f(y)\, d^4y$$
$$+ B(x)^* \iint S_R(y, x; B)^* B(y)^* S(z-y)^* f(z)\, d^4z\, d^4y \ . \tag{174}$$

Thus,

$$[\psi^{out}(f), \psi^{out*}(g)]_\pm = [\psi^{in}(T_R^{-1}T_A f), \psi^{in*}(T_R'^{-1}T_A' g)]_\pm$$
$$= \iint d^4x\, d^4y\, (T_R^{-1}T_A f)(x)\, S_-(i\tilde{\partial}/m)\, \frac{1}{i}\, S(x-y)\, (T_R'^{-1}T_A' g)(y) \ . \tag{175}$$

As before, there are nine terms of which those linear in B are

$$-\iiint f(z)\, d^4z\, S(z-x)\, B(x)\, S_-(i\tilde{\partial}/m)\, \frac{1}{i}\, S(x-y)\, g(y)\, d^4x\, d^4y$$
$$-\iiint f(x)\, S_-(i\tilde{\partial}/m)\, \frac{1}{i}\, S(x-y)\, B(y)^* \int S(z-y)^* g(z)\, d^4x\, d^4y\, d^4z \ . \tag{176}$$

In momentum space, this expression becomes

$$-i \iint \operatorname{sgn} p^0 \delta(p^2 - m^2) \operatorname{sgn} r^0 \delta(r^2 - m^2) d^4p\, d^4r \tag{177}$$

$$\hat{f}(p)[\Pi_-(p)\hat{B}(p-r)S_-(\tilde{r}/m)\Pi_-(r) - S_-(\tilde{p}/m)\Pi_-(p)\hat{B}(-(p-r))^*\Pi_-(r)^*]\hat{g}(-r)\,.$$

There is no reason at all why this expression should vanish for all $\hat{f}$ and $\hat{g} \in \mathcal{S}$, even if B is chosen to be a minimum coupling perturbation, $\hat{B}(q) = e\hat{A}_\mu(q)\beta^\mu$, and direct calculation shows it does not.

Thus, Hurley's proposal, like Iversen's does not give rise to a reasonable particle interpretation.

4. *Remarks*

Since the first papers on relativistic wave equations for particles of spin $\geq 3/2$, there has been a general belief that there are some essential difficulties in the theory in the presence of external fields. These difficulties were initially displayed under the assumption that the fields of the theory yield representations of the canonical commutation relations or anti-commutation relations. That left open the question whether the difficulties would persist if the canonical formalism was not assumed valid. A decisive step forward was taken by Velo and Zwanziger when they showed the existence of acausal propagation in certain higher spin theories without making any assumption about the validity of the canonical formalism. The arguments of the present paper make possible a further step toward an understanding of the problem since they isolate a source of difficulty which can occur even for theories whose causality properties are impeccable.

Since there is no wave equation known to escape both sets of difficulties, it is natural to conjecture that none, in fact, exists. It would appear that with some improvement in techniques for handling the β^μ and ρ matrices and the possible scalar products, the proof or disproof of this conjecture should be within reach. A positive answer would give a precise sense to the primacy of particles of spin 0, 1/2, and 1 in quantum field theory. While, in the opinion of the present author, the argument against higher spin fundamental fields would not be physically very profound, it would at least be better than the naked prejudice which theories with such fields faced for many years. The possibility that all these difficulties are mere artifacts produced by the inadequacy of the approximations inherent in describing interacting fields by external fields has to be borne in mind.

Acknowledgement

The essentials of this paper were developed in response to the work of W. J. Hurley and G. Iversen, and were circulated in the form of a letter to F. Rohrlich dated July 26, 1972, and one to W. Hurley dated March 21, 1974. I thank Hurley and Iversen for discussions. Part of the argument was elaborated in collaboration with A. Z. Capri while I enjoyed the hospitality of the Physics Department of the University of Alberta.

JOSEPH HENRY LABORATORIES OF PHYSICS
PRINCETON UNIVERSITY

REFERENCES

[1] There is a vast literature on the subject. Part of it is summarized in convenient mathematical form in A. S. Wightman *Relativistic Wave Equations as Singular Hyperbolic Systems*, pp. 441-477 in *Partial Differential Equations*, Proc. of Symposia in Pure Math. Vol. XXIII Amer. Math. Soc. 1973.

[2] A. S. Wightman, *The Stability of Representations of the Poincaré Group*, Proc. Fifth Coral Gables Conference 1968, pp. 298-312.

[3] H. Snyder and J. Weinberg, *Stationary States of Scalar and Vector Fields*, Phys. Rev. *57* (1940), 307-314, L; I. Schiff, H. Snyder, and J. Weinberg, *On the Existence of Stationary States of the Mesotron Field*, Phys. Rev. *57* (1940), 315-318.

[4] B. Schroer, R. Seiler, and A. Swieca, *Indefinite Metric and Stationary External Interactions of Quantized Fields*, Phys. Rev. *D2* (1970), 2938-2944.

[5] See, for example, the discussion on pp. 2-25 of *Renormalization Theory*, Reidel, to appear 1976, and the references quoted there.

[6] G. Velo and D. Zwanziger, *Propagation and Quantization of Rarita-Schwinger Waves in an External Electromagnetic Potential*, Phys. Rev. *186* (1969), 1337-1341; *Non-causality and Other Defects of Interaction Lagrangians for Particles with Spin One and Higher*, Phys. Rev. *188* (1969), 2218-2222.

[7] Further relevant literature is quoted in *Troubles in the External Field Problem for Invariant Wave Equations, Tracts in Mathematical and Natural Sciences*, Vol. 4, Gordon and Breach 1971.

[8] G. Iverson, *Some Remarks on the Supermultiplet Theory*, pp. 44-64 in [7].

[9] W. J. Hurley, *Relativistic Wave Equations for Particles with Arbitrary Spin*, Phys. Rev. *D4* (1971), 3605-3616.

[10] W. J. Hurley, *Consistent Description of Higher-Spin Fields*, Phys. Rev. Letts. *29* (1972), 1475-1477.

[11] W. J. Hurley, *Invariant Bilinear Forms and the Discrete Symmetries for Relativistic Arbitrary-Spin Fields*, Phys. Rev. *D10* (1974), 1185-1200.

[12] A. S. Wightman, *Partial Differential Equations and Relativistic Quantum Field Theory*, pp. 1-52 in *Lecture Series in Differential Equations*, Vol. II, A. K. Aziz Ed., Van Nostrand Math. Studies #19 (1969).

[13] E. Speer, *Generalized Feynman Amplitudes*, Annals of Math. Study No. 62, Princeton Press 1969.

Library of Congress Cataloging in Publication Data
Main entry under title:

Studies in mathematical physics.

(Princeton series in physics)
CONTENTS: Abarbanel, H. D. I. The inverse r-squared force: an introduction to its symmetries.--Babbitt, D. Certain Hilbert spaces of analytic functions associated with the Heisenberg group.--Barnes, J. F., Brascamp, H. J., and Lieb, E. H. Lower bound for the ground state energy of the Schrödinger equation using the sharp form of Young's inequality. [etc.]
1. Mathematical physics--Addresses, essays, lectures. 2. Bargmann, V. I. Bargmann, V. II. Lieb, Elliott H. III. Simon, Barry. IV. Wightman, A. S.

Library of Congress Cataloging in Publication Data

QC20.5.S78 530.1'5 76-4057
ISBN 0-691-08180-8

GPSR Authorized Representative: Easy Access System Europe - Mustamäe tee 50, 10621 Tallinn, Estonia, gpsr.requests@easproject.com

www.ingramcontent.com/pod-product-compliance
Lightning Source LLC
Chambersburg PA
CBHW060425220326
41598CB00021BA/2294
* 9 7 8 0 6 9 1 6 1 7 0 9 1 *